Molecular Pharmacognosy

Lu-qi Huang
Editor

Molecular Pharmacognosy

Second Edition

Editor
Lu-qi Huang
National Resource Center for Chinese
Materia Medica
China Academy of Chinese Medical
Sciences
Beijing, China

ISBN 978-981-32-9036-5   ISBN 978-981-32-9034-1 (eBook)
https://doi.org/10.1007/978-981-32-9034-1

1st edition: © Shanghai Scientific and Technical Publishers and Springer Science+Business Media Dordrecht 2013

This Springer imprint is published by the registered company Springer Nature Singapore Pte Ltd.
The registered company address is: 152 Beach Road, #21-01/04 Gateway East, Singapore 189721, Singapore

Preface

With almost 200 years of development, pharmacognosy exists as an applied science with comparably impeccable theory and technology. As times advance, it faces various problems that cannot be solved by current technology and methodology, for example, the exact identification of the "species" level of medicinal plants, sustainable use of rare and endangered medicinal resources, and directional control of medical material quality. All of these issues greatly need to be settled by introducing new technologies and methodologies.

Ever since the discovery of DNA's double-helix structure and semi-conservative replication, molecular biology technology has permeated into almost all fields of life science, thus generating quite a number of interdisciplinary subjects. It can be clearly anticipated that molecular biology will lead the development of life science throughout the twenty-first century. Pharmacognosy will combine with molecular biology and bring about new areas of study.

During this collision between pharmacognosy and molecular biology, how can we grasp and choose the binding point of the two? What is the theoretical basis for their binding? This is the problem I often pondered upon during my graduate studies. After continuous study and thought, as well as discussions with my teachers and classmates, I put forward the concept of "molecular pharmacognosy" in an article titled "Anticipation on the Application of Molecular Biological Technology in Pharmacognosy", published in 1995. I had never thought that its publication would arouse such strong resonance among so many scholars. They raised their own ideas which encouraged me to do further research. Along with a 5-year research quest and practice, the rudiments of a new field, "molecular pharmacognosy", came into being with its own theories and technologies.

In the past 15 years, owing to the research work of the author and other researchers at home and abroad, the technology of molecular biology has been widely applied and practised in the relevant fields of pharmacognosy. Molecular pharmacognosy has been further developed theoretically and systematically. Up to now, the Chinese edition of *Molecular Pharmacognosy* has been copied three times, with new research contents added to each edition, thus making the system of molecular pharmacognosy more abundant and gradually more complete. Besides, the course of molecular pharmacognosy has been opened up at China Academy of Chinese Medical Sciences, Beijing University of Chinese Medicine, China Pharmaceutical University, Shenyang Pharmaceutical University, Tongji Medical College of Huazhong University of Science & Technology, Shanghai University of Traditional Chinese Medicine and other institutions of higher education. The innovative teaching material of molecular pharmacognosy was published in 2008. All of these have sped up the advancement of the cultivation of research talents in molecular pharmacognosy, as well as the subject's development as a whole.

Along with deeper studies of molecular pharmacognosy and further perfection of the subject's theoretical system, there are an increasing number of studies in the field worldwide. The internationalization of molecular pharmacognosy, a new subject, is scheduled. Under this circumstance, this internationalized edition of *Molecular Pharmacognosy* was first published in 2013. This time, we have cooperated with experts and scholars in this field from multiple countries in the hope of furthering the internationalized development of molecular pharmacognosy and gradually making the field of molecular pharmacognosy more abundant and complete.

This second edition is divided into ten chapters. The first chapter, "Emerging Molecular Pharmacognosy", mainly introduces the historical background, concepts, and research contents of molecular pharmacognosy, as well as its relations to other subjects. The second chapter is "Molecular Identification of Traditional Medicinal Materials". It mainly discusses the common methods to identify Chinese medical materials and also explores the molecular identification of families, genera, species, and varieties of original Chinese medical plants and animals. The third chapter, "Seeking New Resource Materials for TCM", is centred on the introduction of the development of plant systematics, research methods of plant systematics, theoretical bases of the molecular systematics of medicinal plants, and some case studies on medicinal plant resources. The fourth chapter, "Phylogeography of Medicinal Plants", mainly introduces the background, basic theories, methods, application of molecular phylogeography in biological evolution, application of phylogeography in the evolution of medicinal plants, and a case studies on medicinal plant. The fifth chapter, "Salvation of Rare and Endangered Medicinal Plants", mainly discusses the theory, concept, method, and practice of protecting traditional Chinese medicine resources, and introduces reasons for the destruction of Chinese medicine resources, classification of endangered medicinal plants and animals, and specific examples of Chinese medicine resource protection. The sixth chapter, "Gene Modification of Medical Plant Germplasm Resources", mainly describes the concept, investigation, collection, and evaluation of medical plant germplasm resources. The seventh chapter, "Functional Genome of Medicinal Plants", introduces genomics, transcriptomics, proteomics, metabolomics, epigenomics, etc. The eighth chapter, "Molecular Mechanisms and Gene Regulation for Biosynthesis of Medicinal Plant Active Ingredients", mainly talks about the biosynthetic pathway, related functional genes, and regulation for the biosynthesis of active ingredients in medicinal plants. The ninth chapter, "Synthetic Biology of Active Compounds", contains an overview of synthetic biology in traditional Chinese medicine (TCM).

The chapter describes the biological basis of synthesizing active ingredients of TCM, analyses key links in the study of synthetic biology in TCM, and discusses the application of synthetic biology for the sustainable utilization of TCM resources. The tenth chapter, "The Mechanisms of Dao-di Herb Formation", mainly focuses on China's special Dao-di herbs and the theoretical hypotheses of their formation mechanisms.

The publication of this book is intended to explicate the concepts, theories, and methodologies of molecular pharmacognosy and encourage more people in similar fields to solve more new problems of pharmacognosy by fully utilizing the developing technologies and methodologies of biology within the framework of molecular pharmacognosy.

Considering that molecular pharmacognosy is a new interdisciplinary field that needs to be further improved, we sincerely hope that more professionals will raise more attention and give more support to increasing the development of this emerging discipline.

Beijing, China Lu-qi Huang

Contents

Contributors

Yu-Chung Chiang Department of Biological Sciences, National Sun Yat-sen University, Kaohsiung, Taiwan

Wei Gao School of Pharmaceutical Sciences, Capital Medical University, Beijing, China

Wenyuan Gao Tianjin Key Laboratory for Modern Drug Delivery and High Efficiency, School of Pharmaceutical Science and Technology and Key Laboratory of Systems Bioengineering, Ministry of Education, Tianjin University, Tianjin, People's Republic of China

Thomas Avery Garran National Resource Center for Chinese Materia Medica, China Academy of Chinese Medical Sciences, Beijing, China

Juan Guo National Resource Center for Chinese Materia Medica, China Academy of Chinese Medical Sciences, Beijing, China

Lan-ping Guo National Resource Center for Chinese Materia Medica, China Academy of Chinese Medical Sciences, Beijing, China

Lu-qi Huang National Resource Center for Chinese Materia Medica, China Academy of Chinese Medical Sciences, Beijing, China

Meirong Jia Department of Plant Biology, University of California, Davis, California, USA

Chao Jiang National Resource Center for Chinese Materia Medica, China Academy of Chinese Medical Sciences, Beijing, China

Dan Jiang School of Chinese Materia Medica, Beijing University of Chinese Medicine, Beijing, China

Xiao-Lei Jin Department of Biological Sciences, National Sun Yat-sen University, Kaohsiung, Taiwan

Chuan-zhi Kang National Resource Center for Chinese Materia Medica, China Academy of Chinese Medical Sciences, Beijing, China

Chun-lei Li Biotechnological Institute of Chinese Materia Medica, Jinan University, Guangzhou, China

Ming Li Centre for Protein Science and Crystallography, School of Life Sciences, The Chinese University of Hong Kong, Shatin, NT, Hong Kong, China

Linglong Luo International Center for New Resources of Chinese Materia Medica, Beijing University of Chinese Medicine, Beijing, China

Kee-Yoeup Paek Department of Horticultural Science, Chungbuk National University, Cheongju, Republic of Korea

Paul Pui-Hay Centre for Protein Science and Crystallography, School of Life Sciences, The Chinese University of Hong Kong, Shatin, NT, Hong Kong, China

Pang-Chui Shaw Centre for Protein Science and Crystallography, School of Life Sciences, The Chinese University of Hong Kong, Shatin, NT, Hong Kong, China

Jia-Hui Sun National Resource Center for Chinese Materia Medica, China Academy of Chinese Medical Sciences, Beijing, China

Hexin Tan School of Pharmacy, Second Military Medical University, Shanghai, China

Khabriev Ramil Usmanovich N.A. Semashko National Research Institute of Public Health, Moscow, Russia

Juan Wang Tianjin Key Laboratory for Modern Drug Delivery and High Efficiency, School of Pharmaceutical Science and Technology and Key Laboratory of Systems Bioengineering, Ministry of Education, Tianjin University, Tianjin, People's Republic of China

Sheng Wang National Resource Center for Chinese Materia Medica, China Academy of Chinese Medical Sciences, Beijing, China

Xueyong Wang International Center for New Resources of Chinese Materia Medica, Beijing University of Chinese Medicine, Beijing, China

Jia Hui Wu International Center for New Resources of Chinese Materia Medica, Beijing University of Chinese Medicine, Beijing, China

Wen Juan Xu International Center for New Resources of Chinese Materia Medica, Beijing University of Chinese Medicine, Beijing, China

Jian Yang National Resource Center for Chinese Materia Medica, China Academy of Chinese Medical Sciences, Beijing, China

Qing-Jun Yuan National Resource Center for Chinese Materia Medica, China Academy of Chinese Medical Sciences, Beijing, China

Yuan Yuan National Resource Center for Chinese Materia Medica, China Academy of Chinese Medical Sciences, Beijing, China

Rong-min Yu Biotechnological Institute of Chinese Materia Medica, Jinan University, Guangzhou, China

Philipp Zerbe Plant Biology Department, College of Biological Sciences, University of California, Davis, Davis, CA, USA

Lei Zhang School of Pharmacy, Second Military Medical University, Shanghai, China

Yifeng Zhang School of Traditional Chinese Medicine, Capital Medical University, Beijing, China

Zhi-Yong Zhang School of Agricultural Sciences, Jiangxi Agricultural University, Nanchang, Jiangxi, China

Yaqiu Zhao Nanjing University of Chinese Medicine, Nanjing, China

Jian-hua Zhu Institute of Molecular Plant Sciences, The University of Edinburgh, Edinburgh, Scotland, UK

Chapter 1
Emerging Molecular Pharmacognosy

Lu-qi Huang, Yaqiu Zhao, and Yuan Yuan

Abstract Molecular pharmacognosy is a science that uses molecular biology technology to study crude drugs, which refers to either fresh or simply processed products derived from plants, animals, and minerals, to be used directly as a natural medicine in medical care or as a raw material for the production of medicine. Molecular pharmacognosy has been given the following main research areas: systematic assortment of varieties of Chinese herbs and study of quality standardization, conservation of medicinal plant and animal biodiversity and research of sustainable utilization of crude drugs resources, medicinal plant marker breeding and new variety cultivation, gene regulation of metabolic pathway and directional control of the quality of Chinese herbal medicines, the use of genetic engineering and tissue culture technique to achieve high-level expression and production of natural active ingredients or genetically modified ingredients, genetic engineering and green pollution-free medicinal plant.

1.1 Historical Development of Pharmacognosy

As a subject, pharmacognosy has gone through four stages of development: pharmacognosy in ancient times, pharmacognosy in early modern times, pharmacognosy in modern times, and the period of natural pharmacognosy.

L.-q. Huang (✉) · Y. Yuan
National Resource Center for Chinese Materia Medica, China Academy of Chinese Medical Sciences, Beijing, China
e-mail: huangluqi01@126.com; y_yuan0732@163.com

Y. Zhao
Nanjing University of Chinese Medicine, Nanjing, China
e-mail: Zhaoyq_1@163.com

L.-q. Huang (ed.), *Molecular Pharmacognosy*,
https://doi.org/10.1007/978-981-32-9034-1_1

1.1.1 Pharmacognosy in Ancient Times (Before the Nineteenth Century)

Foreign medicine originated in Egypt and India. Around 1500 B.C., the use of medicines was recorded in "Papytus" of Egypt and later in "Ajur Veda" of India. In "Papytus," crocus, dried ox bile, castor oil, and so on were mentioned. Around 77 A.D., Dioscorides, a Greek doctor, kept a record of about 600 kinds of crude drugs in his compiled book "De Materia Medica," a book that played an important role in pharmacology and botany until the fifteenth century. Ancient Rome also promoted the development of medicine. "Naturalis Historia," written by Pliny the Elder (23–79 A.D.), gave a brief account of nearly 1000 species of plants, most of which could be used as medicines. Galen (131–200 A.D.) also recorded many prescriptions and preparations for crude drugs. Wild herbs were used to treat illnesses in the Soviet Union before the eleventh century.

In ancient China, there were numerous herbal books with different contents in each period. The "Compendium of Materia Medica," written by Li Shizhen (1518–1593 A.D.), represents the peak developments of herbalism of the time.

From ancient times to the middle of the nineteenth century, pharmacology was in its traditional stage for all countries of the world. At that time, knowledge about medicines came mainly from the senses and practical experiences; the major contents of any medicinal book mostly comprised medical effectiveness, while the names, origins, morphologies, and sensory characteristics to identify, etc. were touched upon. Due to the underdevelopment of science, people could not know all of the details of medical plants. Besides, it was hard for people to conclusively agree because of differences in location and personal experience.

1.1.2 Pharmacognosy in Early Modern Times (1815–1930)

At the beginning of the nineteenth century, pharmacognosy began to emerge. In 1815, C.A. Seydler, a German who proposed the term "pharmakognosie" in his book "Analecta Pharmacognostica," was referred to as the Father of Pharmacognosy. Etymologically, pharmakognosie means the knowledge of crude drugs. In 1825, Martius, a German scholar, established Pharmakognosie as a subject in college, and then a new discipline named Pharmakognosie emerged in the field of natural science. According to Martius, pharmacognosy, as a part of merchandising, was a study to research drug bases taken from nature to test their purity and to check their impurities or adulterants. After this, German scholars published works, named "Pharmacognosie," on plant and animal drugs successively.

Japanese early studies of pharmacy were based on quoting and researching Chinese herbs. After the Meiji Restoration, Japan tried to absorb the new scientific achievements of other countries and developed herbalism into pharmakognosie. In 1880, Gendō Oi, a Japanese scholar, translated pharmakognosie as pharmacognosy.

In 1803, Derosne, a French chemist, found that alkaloids are a component of crude drugs. In 1806, Sertürner, a German pharmacist, clarified that the cell was the basic unit of plant structure; after this, microscopes were used to research the internal structure of crude drugs. In 1857, Schleiden published a book named "Grundniss der Pharmakognosie des Pflanzenreiches" (foundation of Pharmacognosy of plants), in which he gave a detailed description of the microstructure of many plant crude drugs. Later, Berg in 1865 and Vogl in 1887 successively published anatomical atlases of crude drugs, which furthered the development of identifying crude drugs using microscopes; this became one of the most important methods of identifying crude drugs. Meanwhile, qualitative and quantitative methods of chemistry were also used in crude drug identification. Fluorescence analysis and chromatography were used in sequence in the latter half of the nineteenth century and in the beginning of the twentieth century, both of which enriched the research field of pharmacognosy and promoted a greater development than did the earlier method of identifying crude drugs by shape and smell.

All in all, the reason why pharmacognosy has become an independent subject is closely related to the historic development of international traffic and trade. In the first half of the nineteenth century, the rapid progress in international trade gave rise to an increase in the variety of medicines and resulted in enlarging the scope of raw materials and products. Crude drugs were often sold after being broken or ground into powder. In order to increase profits, some merchants took advantage of the difficulty in identifying powder and mixed drugs with low prices into those with high prices; some even presented fake goods as high-quality products. Therefore, the problem of how to identify the authenticity and quality of crude drugs arose. With the advancements in bioscience and increasingly widespread applications of microscopy, pharmacognosy made great progress in the middle of the nineteenth century and finally became an independent subject. The early works of pharmacognosy were based on establishing quality standards for crude drugs in business.

1.1.3 Pharmacognosy in Modern Times (1930 to the Late 1990s)

Since the 1930s, developments in biology and chemistry have enriched the means of studying crude drugs. The development of the bioassay to determine the intensity of drug action (biological potency) advanced the study of crude drug active ingredients and strengthened evaluations of quality. Chemical and physical methods, such as colorimetry, spectrophotometry, and fluorescence analysis, were all gradually applied to the identification of crude drugs. When pharmacognosy developed in the lines of morphology and chemistry, many new disciplines emerged. For example, with the accumulation of the chemical compositions of plants in type and number, a new subdiscipline—plant chemotaxonomy—came into being through the exploration of the chemical composition of plants and their genetic relationships.

This new discipline not only had taxonomic significance but also promoted the emergence of new sources of crude medicines. At the same time, chemical research on the components of marine life developed by leaps and bounds. From marine algae, sponges, coelenterates, annelids, molluscs, mosses, echinoderms, etc., new biologically active substances were discovered, which promoted the development of marine natural resources, which in turn produced a new subdiscipline, marine pharmacognosy.

1.1.4 Period of Natural Pharmacognosy (End of the Twentieth Century to Early Twenty-First Century)

Through its first three stages of development, pharmacognosy became an established applied discipline by perfecting its technologies and theories. In the 1970s and 1980s, many universities cancelled their "pharmacognosy" courses. However, at the end of the twentieth century, with humans' "returning to nature" and the uprising of modern life science, pharmacognosy returned with a renewed vitality and broad prospects. Progresses in the separation of the chemical composition, structure determination, and quantitative technology made it possible for proton nuclear magnetic resonance (^{1}H NMR), carbon-13 (^{13}C) NMR, DNA fingerprint identification, etc. to be used in the identification of crude drugs, in promoting their standardization and in normalization.

Through constant exploration, a batch of new developing points were ensured. In 1995, Huang Luqi [1] first mentioned the concept of molecular pharmacognosy in his paper "Prospect of Application of Molecular Biology Technology to Pharmacognosy." This paper aroused a strong resonance in scientific circles. People with similar interests were encouraged to put forward their ideas and encouraged him to further his exploration and systematize the theory. Based on this resonance and encouragement, he worked with great effort for years and published "Molecular Pharmacognosy" in the Peking University Medical Press in June of 2000 [2]. This book allowed original pharmacognosy to enter a new era, giving rise to the birth of a new interdiscipline, molecular pharmacognosy. This book released its second edition in 2006 [3]. In the same year, "Molecular Pharmacognosy" entered the undergraduate textbook series [4]. In 2015, "Molecular Pharmacognosy" issued its third edition [5] and won the fourth Chinese Government Publishing Award.

1.2 Concepts of Molecular Pharmacognosy and Its Development

1.2.1 Generation of Molecular Pharmacognosy

Watson and Crick's discovery of the structure of DNA in 1953 marked a new era for life science, significantly changing life medicine and the thinking of scholars in relevant fields; since then, people have begun to reunderstand life's nature and laws from the biological macromolecular level. Although this discovery had no direct significance on the development of pharmacognosy, it still had an immeasurable impact on the whole life science. Molecular biology developed rapidly and penetrated into the field of applied biomedicine, which brought into existence a large number of interdisciplinary fields and frontier disciplines. Genetic engineering technologies based on molecular cloning and reorganization rose, and related tissue culture technologies, especially molecular marker technology based on polymerase chain reaction (PCR), began to spring up like mushrooms; this contributed toward the rapid development of pharmacognosy and a full extension and enrichment of its research areas and methods. Conflictions and mutual integrations between pharmacognosy and molecular biology brought forth a new interdisciplinary, termed molecular pharmacognosy.

Molecular pharmacognosy had the following three theoretical bases: first, the development of molecular biology brought all branches related to biology to the molecular level [1]. Pharmacognosy, with a major focus on plant and animal crude drugs, touched upon many biological theories and methods, so was no exception. Second, crude drugs had a major origin in animals and plants; their cells contained DNA, the material basis to store, duplicate, and transmit genetic information. DNA was also the material base of molecular biology, so pharmacognosy had the combined material bases of DNA and molecular biology. This made molecular biology theoretically and methodologically apply to pharmacognosy. The study of crude drugs of animals and plants in pharmacognosy progressed from the population, organism, tissue, organ, and cell levels to the genetic level. Third, the formation and development of molecular pharmacognosy is inspired by many disciplines, such as genomics, molecular systematics, molecular ecology, conservation biology, biochemistry, medicinal plant breeding, etc., which are mainly related to postgraduate pharmacology at the molecular level of nucleic acids and proteins [6]. Therefore, the advancement and modernization of pharmacognosy had a close relationship with molecular biology; this inevitably promoted the study of pharmacognosy to reach the molecular level.

1.2.2 Concept of Molecular Pharmacognosy

Molecular pharmacognosy studies the classification, identification, cultivation, and protection of crude drugs, as well as the production of effective elements at the molecular level. Based on the theories and methods of pharmacognosy and molecular biology, molecular pharmacognosy is a promising and prospective interdisciplinary in pharmacognosy [4]. It can be said that molecular pharmacognosy carries on the traditional contents and missions of pharmacognosy while endowing it with new applications and challenges.

The sources of crude drugs are mainly plants and animals, though there are a few minerals that are considered crude drugs. The study objects of molecular pharmacognosy are limited to the floral and faunal sources of crude drugs. According to the different organism levels and gradual combination relation, molecular pharmacognosy can be divided into six major biological levels: gene, cell, organ, organism, population, and community. This biological mode, based on multiple levels, is called the biological spectrum. Each unique level in the biological spectrum was discovered by a historical process. Generally speaking, developing the levels of organisms in the micro- and macrodirections brought about the gradual discovery of all levels, as was the case for the discovery of cells and genes. Studies of the unique scientific questions at each level have given rise to independent branches in the field of life science. Branches, such as molecular biology and cytobiology, could be mutually promotive but could never be substitutes for each other. Currently, pharmacognosy is mainly focused on research at the tissue, organ, organism, and population levels, upon which relatively mature and independent theories and methods, such as pharmacognosy histology and morphology, have been brought forth. Molecular pharmacognosy deals with crude drugs at the genetic level, with its theories and methodologies based on molecular biology.

1.2.3 Major Concerns and Main Applications for Molecular Pharmacognosy

With pharmacognosy as its theoretical base, molecular pharmacognosy presents problems to the field of pharmacognosy whose major contents can be summarized by authenticity and excellence, as follows:

(a) Discerning the false from the genuine so as to settle the problem of variety confusion: due to the rise in scope of use and dosage of medicines, there are increasing numbers of plant and animal homonyms and materials with similar appearances; these can be taken as the same drugs in different regions, thus leading to variety confusion. As such, it is necessary to discern the false from the genuine in terms of their origins and distribution areas. Only in this way can quality be guaranteed.

(b) Quality assessment: a systematic study should be conducted on crude drugs with multiple origins and genuine quality, including the place of origin, harvesting, processing, storage, and the influence of transportation upon active ingredients, to confirm high-quality varieties and the factors that may have effects on them. More than that, excellent varieties should be researched and cultured to achieve fast growth, high-quality, and high yield in order to meet the increasing demand for medication.

The scientific connotations on authenticity and excellence mentioned above are related to differences in their DNA (except for mineral medicines). The fake and the genuine may have different DNA compositions due to differences in their origins; thus, DNA should be used to separate the genuine from the fake. Therefore, one scientific intention of studying molecular pharmacognosy is to research DNA and its ability to determine the authenticity and excellence of crude drugs.

Molecular pharmacognosy has been given the following main applications:

(i) *Systematic assortment of varieties of Chinese herbs and the study of quality standardization*

Assortment of varieties of Chinese herbs, based on classical taxonomy, can be applied to systematic assortment, classification, and identification, but too many human factors are involved, especially for planted groups of Chinese herbal medicines, such as the identification of authentic raw materials, which still remains a problem. The development of species biology and molecular systematics provides an effective method and basis for the study of systems, evolution, classification, and identification. The modern species concept has been widely accepted by taxonomists, and pharmacognosists cannot ignore theories or fruits in species biology and molecular systematics. Breakthroughs in molecular systematics and the application of biological engineering technology have provided crude drug classification and identification with an effective method and basis to test molecules. The most commonly used techniques for the systematic assortment of varieties of Chinese herbs and the study of quality standardization are based on the combination of PCR and electrophoresis techniques, such as random amplified polymorphic DNA (RAPD), simple sequence repeat (SSR), allele-specific PCR, multiplex amplification-refractory mutation system (MARMS), anchored primer amplification polymorphism DNA (APAPD), and PCR- restriction fragment length polymorphism (RFLP) [7], based on DNA sequencing such as single nucleotide polymorphism (SNP) and DNA barcoding technologies. In recent years, these techniques have been applied to a variety of herbal and animal drugs, such as *Panax ginseng* [8], *Bupleurum chinese* [9], *Dendrobium officinale* [10], *Solegnathus hardwickii* [11], and *Saiga tatarica* [12]. To establish a method and system to identify crude drugs based on species biology, molecular systematics, Chinese medicinal resources, and herbalism, and to further the development of the systematic assortment and quality standardization of Chinese herbal medicines at the population, individual, and even genetic level, is one of the major concerns of molecular pharmacognosy.

(ii) *Conservation of medicinal plant and animal biodiversity and research into the sustainable utilization of crude drug resources*

DNA diversity is the essence of biodiversity. Molecular makers based on DNA polymorphism analysis and molecular systematics based on genome sequence analysis can directly test DNA variation patterns and determine the key units to protect. The study of the molecular systematics of medicinal plants and animals may presume the development status and the degree of endangerment of populations, thus making new operative methods for the measurement of biodiversity and countermeasures to protect rare medicinal plant and animal resources necessary.

According to Wang et al. [13], the molecular phylogeographic analysis of the single relictual species *Cathaya argyrophylla* was carried out to elucidate its endangerment mechanism. Gao et al. [14] studied the relationship between the origin, geographical pattern of *Taxus wallichiana* and that of the flora of East Asia. Yuan et al. [15] revealed a highly differentiated phylogeography pattern of the vulnerable plant *Dipentodon sinicus*, which was the result of the uplift of the Qinghai-Tibet Plateau. Tian et al. [16] analyzed the migration route and refuge of *Pinus kwangtungensis* in the glacial period.

Moreover, the application of research findings in molecular systematics based on DNA polymorphism makes finding and enlarging the scope of medicinal plant and animal resources more effective and efficient. Expounding the relevance among the genetic relationships of DNA molecules, active ingredients and efficacy in the combination of chemical taxonomy, and obtaining the molecular genetic background of important chemical elements so as to identify whether unknown plants have the genes to produce specific chemical compositions are shortcuts to find and enlarge the scope of crude drug materials using molecular systematics.

(iii) *Medicinal plant marker breeding and new variety cultivation*

In the process of exploring molecular theory and practice, it can be said that the molecular detection of genetic diversity and molecular systematics lays the basis for understanding and transforming nature. Exploring and harnessing molecular markers with important traits is the reason for us to understand and transform nature. With the support of cell and genetic engineering, molecular pharmacognosy research becomes more practical. The quantitative trait loci (QTL) method, which combines breeding technology with the key medicinal plant and animal genetic linkage map constructed using molecular genetic markers, makes it possible to provide information on the mapping of the quantitative trait loci, such as the quantity of genes with the target trait, genetic effects, ways of interaction between genes, and decomposition of the quantitative traits, which could never have been provided by traditional quantitative genetics.

With the rapid development of high-throughput sequencing technology and the reduction of sequencing costs, a new generation of high-throughput sequencing technology is used to sequence plants across the genome and generate rich transcriptome data, which contain a large number of EST sequences [17]. For example, Gai et al. [18] used the Roche 454 GS FLX platform to sequence the *de novo* transcriptome of

Paeonia suffruticosa. In total, 625,342 expressed sequence tags (ESTs) with an average length of 358.1 bp were obtained, and 23,652 sequence overlap groups (contigs) and single sequences (singletons) were obtained after splicing, of which 15,284 were longer than 300 bp. In total, 2253 SSR loci were obtained from 454 ESTs. Then 149 SSR loci were selected to design the primers. A total of 121 pairs of primers could successfully extend the bands. Among them, 73 pairs of primers showed polymorphism in the PCR products. These sequences provided genetic resources for studying the physiological function of *P. suffruticosa*.

(iv) *Gene regulation of the metabolic pathway and directional control of the quality of Chinese herbal medicines*

Increasing attention should be paid to the basic research into secondary metabolite biosynthesis; especially, research into the gene regulation of key enzymes will be particularly noticeable and could become one of the most challenging and promising directions in molecular pharmacognosy; the major sources of active ingredients in Chinese herbs are secondary metabolites. The presence or absence of secondary metabolites and their quantity decide the quality of Chinese herbal medicine; thus, carrying out genetic engineering and improving the active ingredient contents of Chinese herbal medicines will help ease the pressure on their resources.

Through antisense technology, the amount of a target compound can be increased, and the synthetic approaches of other compounds are inhibited. For example, Mol et al. [19] successfully regulated the activities of cinnamyl alcohol dehydrogenase in the hairy roots of *Linum flavum* and inhibited the synthesis of the lignin molecule in the branched metabolism to increase the amount of the anticancer compound 5-methoxypodophyllotoxin. Besides, RNAi technology has been widely used in the identification of biological gene function, as well as the screening and breeding of new genes. Schweizer et al. [20] produced effective interference in the function of the dihydroxyflavonol 4-reductase gene encoding maize and barley at the single cell level using the RNAi method, and the accumulation of red anthocyanin relative to cells was decreased.

(v) *Genetic engineering and tissue culture techniques to achieve high-level expression and the production of natural active ingredients or genetically modified ingredients*

The use of genetically modified organisms as bioreactors to produce exogenous gene-encoding goods is among the most attractive prospects in genetic engineering, and thus these bioreactors are called “new-generation pharmaceutical factories.” They have many advantages: they can express complex natural proteins in a natural state; they can be obtained continuously from animal milk and blood and extracted from the bodies of plants; they can also pass through the digestive tract. In addition, for plant or animal proteins with strong active ingredients that also possess side effects, such as scorpion venom and trichosanthin, the sequence that causes toxicity can be removed or its expression can be inhibited, thus strengthening the expression of the active compounds.

Hairy root and crown gall cultures—new technologies in combination of plant genetic engineering and cell engineering in recent years—have opened up a new road for the research and development of active pharmaceutical ingredient production. With the improvements in expression efficiency and the expansion of the scope of receptor plants, biotechnology will certainly bring a new impetus to research into adding new genetic characteristics to traditional crude drugs; meanwhile, with the development of new bioreactor technologies and the establishment and improvement of efficient cell cultures, the commercialization and industrialization of natural medicine biotechnology will be sped up.

Gao used transcriptome sequencing combined with metabolome analysis to analyze the hairy root cultures of *Salvia miltiorrhiza* at different induction time points [21] and obtained 20,972 genes, of which 6358 showed different expressions at different induction time points. Focusing on the analysis of genes related to the biosynthesis and regulation of tanshinone, a total of 70 transcription factors related to the accumulation and regulation of tanshinone were detected.

(vi) *Genetic engineering and pollution-free medicinal plants*

The problem of pesticide pollution in medicinal plants has aroused public concern; it damages the environment, endangers public health, and limits the export of Chinese herbal medicines. Thus, advocating pollution-free medicinal plants and controlling pests without industrial chemicals when growing medicinal plants have become goals for people to achieve. Improving the ability of medicinal plants against pests via genetic engineering is also one of the applications for molecular pharmacognosy.

1.3 Relation of Molecular Pharmacognosy to Other Disciplines

As a new and pioneering interdisciplinary in pharmacognosy, still being developed, molecular pharmacognosy combines pharmacognosy and molecular biology. With its birth, pharmacognosy has inevitably developed into an intensive microstudy. Pharmacognosy itself is a basic multidisciplinary science for applications, and molecular biology is based on modern sciences, such as life medicine; thus, molecular pharmacognosy is compatible for cross-disciplinary, interdisciplinary, and multidisciplinary studies with rich connotations and denotation.

Molecular pharmacognosy is closely related to the following fields of study.

Identification of TCMs: with the purpose of identifying the genuine and the fake and the good and the bad and guaranteeing safety and efficacy, molecular pharmacognosy differs from traditional identification methods, such as original plant identification, character identification, microscopic identification, and physical and chemical identification, in using DNA molecular genetic technology to directly analyze the polymorphism of genetic materials so as to determine differences in the intrinsic and external performance between different varieties of TCMs, thus providing a new convenient and accurate method.

Resource science of traditional Chinese medicine: this field of study aims to determine the varieties, quantities, geographical location, temporal and spatial variation, rational exploitation, and scientific management of TCM resources. Studying TCM resource distribution, this field aims to develop cost-effective optimization techniques to make a reasonable arrangement for TCM resources to be harvested, processed, and utilized so that society, the economy, and ecology can all achieve a coordinated and balanced program of development and provide sufficient high-quality raw materials to the health care and pharmaceutical industries. Therefore, resource science of TCM is a comprehensive natural science and an emerging multidisciplinary, cross-disciplinary, interdisciplinary science with a management nature. The concept originated in the early 1980s and was established as an independent discipline in the late 1980s. "Resource Science of Traditional Chinese Medicine," the first book used for teaching, was compiled and published in May 1993 by Zhou Ronghan, a professor at the College of Traditional Chinese Medicine of China Pharmaceutical University in Nanjing. Resource science and molecular pharmacognosy have a close relationship; the latter lays a new theoretical basis for resource identification, germplasm diversity detection, and searching for and expanding new varieties and resources of TCM, thus promoting the development of the resource science of TCM.

Pharmaceutical botany: as a science that deals with the study of pharmaceutical plants' morphology and structure, as well as taxonomic knowledge and methods in botany, pharmaceutical botany shows a major concern for the systematic study of botanical knowledge to research the identification and classification of pharmaceutical plants and to investigate their resources, organize the types of Chinese herbal medicines, and ensure the accurate and effective application of medication. Molecular pharmacognosy will provide genetic evidence for pharmaceutical plants in their systemic evolution and aid in the search for new drug sources and the cultivation of new varieties.

References

1. Huang LQ. Prospects for application of molecular biotechnique to pharmacognosy. China J Chin Mater Med. 1995;11:643–645+702.
2. Huang LQ. Molecular pharmacognosy. The first edition. Beijing: Peking University Medical Press; 2000. p. 1–375.
3. Huang LQ. Molecular pharmacognosy. Beijing: Peking University Medical Press; 2006. p. 1–678.
4. Huang LQ, Xiao PG. Molecular pharmacognosy. Beijing: China Press of Traditional Chinese Medicine; 2008. p. 1–189.
5. Huang LQ, Liu CX. Molecular pharmacognosy. Beijing: China Science Publishing; 2015. p. 1–678.
6. Huang LQ, Xiao PG, Guo LP, et al. Molecular pharmacognosy: An emerging edge discipline. Sci China. 2009;39(12):1101–10.

7. Hong X, Wang ZT, Zhi bH. Development and application of technology for DNA molecular identification in traditional Chinese medicine. Mod Tradit Chin Med Mater Med-World Sci Technol. 2003;5(2):24–30.
8. Chen ML. Identification of ginseng from the analytical level by RAPD and PCR-RFLP methods. International Journal of Traditional Chinese Medicine|Int J Trad Chin Med. 2002;24 (5):304–5.
9. Ying W, Liu CS, Liu YF, et al. ITS sequence identification of *Radix Bupleuri*. China J Chin Mater Med. 2005;30(10):732.
10. Dong XM, Yuan Y, et al. Molecular ID for populations of *Dendrobium officinale* of Yunnan and Anhui province based on SSR marker. Mod Chin Med. 2017;19(5):617–24.
11. Liu FY, Jin Y, Yuan Y, et al. Survey on origin of medicinal commodities of Syngnathus based on morphology and DNA sequencing identification. World Chin Med. 2018;13(02):241–7.
12. Jiang C, Jin Y, Yuan Y, et al. Molecular authentication of Saigae Tataricae Cornu Tampon by specific PCR. World Chin Med. 2018;13(02):252–5.
13. Wang HW, Ge S. Phylogeography of the endangered Cathaya argyrophylla (Pinaceae) inferred from sequence variation of mitochondrial and nuclear DNA. Mol Ecol. 2006;15(13):4109–23.
14. Gao LM, Möller M, Zhang XM, et al. High variation and strong phylogeographic pattern among cpDNA haplotypes in Taxus wallichiana (Taxaceae) in China and North Vietnam. Mol Ecol. 2007;16(22):4684–98.
15. Yuan QJ, Zhang ZY, Peng H, et al. Chloroplast phylogeography of Dipentodon (Dipentodontaceae) in southwest China and northern Vietnam. Mol Ecol. 2008;17(4):1054–65.
16. Tian S, Li DR, Wang HW, et al. Clear genetic structure of Pinus kwangtungensis (Pinaceae) revealed by a plastid DNA fragment with a novel minisatellite. Ann Bot. 2008;102(1):69–78.
17. Simon SA, et al. Short-read sequencing technologies for transcriptional analyses. Annu Rev Plant Biol. 2009;60(1):305.
18. Gai S, et al. Transcriptome analysis of tree peony during chilling requirement fulfillment: assembling, annotation and markers discovering. Gene. 2012;497(2):256–62.
19. Oostdam A, Mol JNM, van der Plas LHW. Establishment of hairy root cultures of Linum flavum producing the lignan 5- methoxypodophyllotoxin. Plant Cell Rep. 1993;12(7–8):474–7.
20. Schweizer P, Pokorny J, Schulze P, et al. Double-stranded RNA interference with gene function at the single-cell level in cereals. Plant J. 2000;24(6):895–903.
21. Gao W, et al. Combining metabolomics and transcriptomics to characterize tanshinone biosynthesis in Salvia miltiorrhiza. BMC Genomics. 2014;15(1):73.

Chapter 2
Molecular Identification of Traditional Medicinal Materials

Ming Li, Chao Jiang, Paul Pui-Hay, Pang-Chui Shaw, and Yuan Yuan

Traditional medicines are consumed by 80% of the population in the world for health maintenance and disease treatment. The adulteration and substitution of source materials are life-threatening problems that have grown along with its popularity. Consequently, a reliable identification method is important for the safety and quality assurance of traditional Chinese medicine (TCM) materials. Molecular techniques provide an alternative means to conventional organoleptic and chemical identification methods and are superior in terms of their accuracy, sensitivity, resolution and reproducibility. Since the early 1990s, a number of molecular techniques have been developed to identify traditional medicinal materials based on DNA fingerprinting, specific amplification, DNA sequencing, DNA microarrays and fluorescence detection techniques. Molecular techniques are capable of differentiating traditional medicinal materials from their adulterants and substitutes in closely related species, subspecies, variants, cultivars and species from different localities, and, in some cases, they can distinguish the growth year and herb quality. This chapter introduces the major molecular identification techniques and reviews their applications in the identification of animal and botanical medicinal materials.

Ming Li and Chao Jiang are equally contributed for this chapter.

M. Li · P. Pui-Hay · P.-C. Shaw (✉)
Centre for Protein Science and Crystallography, School of Life Sciences, The Chinese University of Hong Kong, Shatin, NT, Hong Kong, China
e-mail: liming@cuhk.edu.hk; paulbut@hotmail.com; pcshaw@cuhk.edu.hk

C. Jiang · Y. Yuan (✉)
National Resource Center for Chinese Materia Medica, China Academy of Chinese Medical Sciences, Beijing, China
e-mail: jiangchao0411@126.com; y_yuan0732@163.com

L.-q. Huang (ed.), *Molecular Pharmacognosy*,
https://doi.org/10.1007/978-981-32-9034-1_2

2.1 Introduction

Traditional medicines have a long and well-documented history, having been used for thousands of years in traditional Oriental, Ayurvedic and Latin American medicine to prevent diseases and improve health. Over 80% of the world's population use traditional medicines to maintain health and cure diseases [1]. A fundamental prerequisite for the proper delivery of healthcare with traditional medicines is the use of authentic herbal materials. For historical and geographical reasons, TCM materials share similar morphological appearances, textures and microscopic characteristics, and these species are exposed to contaminants, adulterants and counterfeits in traditional medicine markets. The inaccurate identification of TCMs may compromise their therapeutic efficacy and could pose a threat to the medicine's safety. In the early 1990s in Belgium, rapidly progressive interstitial fibrosis and terminal renal failure were observed in some 80 women taking a regimen of herbal medicinal slimming products made from various herbs, including Stephaniae Tetrandrae Radix (Fangji) and Magnoliae Officinalis Cortex (Houpo) [2]. It was later revealed that the herb Fangji was adulterated by another herb, Guangfangji, derived from *Aristolochia fangchi*, which contains carcinogenic aristolochic acids [3]. Many more cases of aristolochic acid nephropathy have subsequently been reported in many Western and Asian countries [4, 5]. In 2000, the United States Food and Drug Administration alerted consumers against the use of herbs containing aristolochic acids [6]. Despite this warning, the marketing of such herbs continues [7]. In 2004 in Hong Kong, aristolochic acid nephropathy was diagnosed in three patients after the prolonged consumption of herbal regimens prescribed with the non-harmful anti-cancer herb Solani Lyrati Herba (Baiying), derived from *Solanum lyratum* [8]; further investigation revealed that the ingested material was the aristolochic acid-containing herb Aristolochiae Mollissimae Herba (Xungufeng) derived from *Aristolochia mollissima* [9]. Safety issues regarding the use of Periplocae Cortex (Xiangjiapi) derived from *Periploca sepium* have also been reported. Most Xiangjiapi-induced poisoning issues are found to have resulted from the replacement of Periplocae Cortex with Acanthopanacis Cortex (Wujiapi) derived from *Acanthopanax gracilistylus* [10, 11]. Investigations revealed that the most lethal cases were caused by the mislabelling of non-harmful herbal materials that were replaced with morphologically similar poisonous species. Adulteration could be due to (i) erroneous adulteration caused by the two herbs having similar features or the absence of distinguishable characters, (ii) the intentional substitution of high-value material with inexpensive substances, (iii) misuse caused by the two herbs having similar common names and (iv) the historical use of local substitutes. In order to ensure the safety, efficacy and quality of traditional medicines and their products, the accurate identification of TCM materials is essential.

There are a number of effective identification methods that have evolved with the improvement of technologies. Morphological, microscopic and chemical identification methods have been widely used to determine herbal materials from ancient times to the present. In the past, the identification of medicinal material was based on the description of morphological features, as stated in 'Shengnong Bencaojing' (~200 A.D.). In a later record, 'Bencao Gangmu' (1593 A.D.), the morphological features of materials were graphically illustrated. In 1857, German botanist Schleiden introduced microscopy to identify herbal drugs in 'Grundniss der Pharmakognosie des Pflanzenreiches'; this then became the main method of identification in modern pharmacognosy worldwide. Nowadays, chemical profiles obtained from thin-layer chromatography (TLC), high-pressure liquid chromatography (HPLC), gas chromatography (GC) or liquid chromatography/mass spectrometry (LC/MS) are applied to increase the accuracy of identification. In the 1990s, the introduction of molecular techniques was a major breakthrough in the identification of traditional medicine [12, 13]. In recent years, DNA-based molecular approaches have become a popular species identification tool for their high specificity, robustness and reliability, from the original plant to commercial products. Molecular techniques provide international-standard references for organism identification. 'Pharmacopoeia of the People's Republic of China (2015 edition)' and its supplement officially recorded specific polymerase chain reaction (PCR) as the standard method for identifying the three animal medicinal materials Zaocys (Wushaoshe), Agkistrodon (Qishe) and Bungarus Parvus (Jinqianbaihuashe); PCR-restriction fragment length polymorphism (PCR-RFLP) as the method for identifying Fritillariae Cirrhosae Bulbus (Chuanbeimu); and DNA barcoding as the principal method for the molecular identification of TCM materials. 'The Japanese Pharmacopoeia' (JP XVI) accepts the amplification-refractory mutation system (ARMS, also known as allele-specific PCR) and PCR-RFLP as the purity test methods for Atractylodes Lancea Rhizome (Cangzhu) and Atractylodes Rhizome (Baizhu). It is foreseeable that molecular protocols will be included for more medicinal materials in future editions. This chapter reviews and comments on the most commonly used molecular identification techniques. An account of the strategies and examples of identifying plant and animal medicinal materials at different taxonomic levels are also included.

2.2 Principles for Identifying Medicinal Material

The molecular identification of Chinese traditional medicine has stemmed from laboratory research into application, but establishing an appropriate DNA molecular identifying method still poses intractable problems. The molecular identification of TCM has a scientific and objective basis, follows a certain systematic research background and adopts practical principles to establish a case-by-case multi-class identification system.

The essence of DNA identification is delimiting species; how to delimit the inter- and intra-species of a particular medicinal material is the bottleneck of TCM identification. Molecular systematics is the fundamental basis of TCM molecular identification. Biasness, one-sidedness and superficiality are inevitable when identifying methods with the absence of systematics analysis. Therefore, a two-step identify method is proposed: (i) establishing a molecular systematic database that includes all species (medicinal and non-medicinal) of the particular genus that needs to be identified and (ii) comparing the identified TCM with the database to determine its identity. Number of species in the database and the comprehensiveness of the database determine the reliability of the identification system. On the other hand, because of the variety of TCMs and their complex origins, artificial changes in gene flow between populations and species can further strengthen hybridisation and introgression in medicinal plants and animals. The establishment of molecular identification methods for some medicinal materials cannot prove that other medicinal materials also have the same conditions for establishing molecular identification methods. A case-by-case analysis principle should be implemented for each specific TCM [14].

2.3 Methodologies for Identifying Medicinal Materials

Accurate species identification is critical for controlling the efficacy and safety of medicinal material products. A reliable method is required to validate the authenticity of herbal products for production quality, particularly for herbal markets. The traditional identification methods based on organoleptic and microscopic features, such as shape, colour, texture, odour, tissue arrangement and cell components, are simple and inexpensive. However, insufficient informative characteristics in the medicinal materials may lead to low accuracy and limited resolution. Alternatively, chemical profiling has become a standard practice for species identification and quality control. However, chemical components can vary because of a number of factors, including the growth stage of the material, harvest time, locality, storage condition, processing method and manufacturing procedure. The presence of large amounts of proteins, polysaccharides, resins, tannins and secondary metabolites make the reproduction of chemical analysis difficult [15].

Molecular identification based on the variation in DNA sequences of different organisms provides an alternative approach. In principle, the genetic makeup is unique to a species independent of body parts, growth stage and environment. Therefore, DNA-based identification methods are less sensitive to biological, physiological, physical and environmental factors. In addition, benefiting from the development of PCR, a small amount of a sample is sufficient for carrying out the identification process. These advantages are particularly important in identifying shredded material or powder, not to mention expensive materials with limited supplies [16]. Furthermore, DNA is relatively stable and may be extractable from herbarium specimens, processed food and commercial products. There are several main molecular techniques being used, namely DNA fingerprinting, specific amplification, DNA sequencing, DNA microarray and fluorescence detection.

2.3.1 DNA Fingerprinting

DNA fingerprinting explores the DNA polymorphism in the whole genome or in a specific region of the sample. The polymorphic patterns are usually visualised by agarose gel, polyacrylamide gel or capillary electrophoresis. Because the quantity and quality of DNA in a medicinal material may be poorly preserved due to post-harvest processing and storage, most DNA fingerprinting methods undergo an amplification phase. Nowadays, DNA fingerprinting identification methods in medicinal materials include random amplified polymorphism, anchored primer amplification polymorphism and amplified fragment length polymorphism (AFLP). Most DNA fingerprinting methods suffer from poor reproducibility between laboratories because of the requirement of consistent PCR amplification conditions. Nowadays, DNA fingerprinting has departed from the previous molecular identification methods for medicinal materials.

2.3.1.1 Random Amplified Polymorphism

Random amplified polymorphism uses whole genome fingerprints and contains a series of methods (RAPD [17], arbitrarily primed PCR [AP-PCR] [18] and direct amplification of length polymorphisms [DALP]) for marker analysis. This method utilises short PCR primers consisting of random sequences usually in the size range of 8–20 nucleotides in length. Single primer anneals to the genomic DNA template at a number of sites and acts as both the forward and reverse primers. Random amplified polymorphism does not require prior knowledge of the genome and can be used to examine multiple loci simultaneously. Its markers are dominant and are usually unable to distinguish homozygous loci from heterozygous loci. There have been many publications using this approach to identify medicinal material. Cheung et al. [12] found that all AP-PCR fingerprints generated using three primers (M13 forward, M13 reverse and Gal-K primers) successfully differentiated the dried roots of oriental ginseng (*Panax ginseng*) from those of American ginseng (*P. quinquefolius*). Cheng et al. [13] employed RAPD to determine the components in a Chinese herbal prescription, wherein primer OPP-10 generated three specific markers (200 bp, 440 bp and 500 bp markers specific to *Astragalus membranaceus*, *Atractylodes macrocephala* and *Saposhnikovia divaricata*, respectively) for simultaneously identifying three species in prescription Yu-Ping-Feng San. A similar approach was subsequently applied to the identification of other medicinal species, including *Angelica sinensis* (Danggui), *Cordyceps sinensis* (Dongchongxiacao) and *Akebia quinata* (Mutong) [19–21].

2.3.1.2 Anchored Primer Amplification Polymorphism

Anchored primer amplification polymorphism is also a whole-genome scanning fingerprinting technique, which uses PCR primers anchored in SSRs (inter-simple sequence repeat [ISSR] polymorphism [22]), open reading frames (target region amplification polymorphism [23] and sequence-related amplified polymorphism [24]) or start codon flanking sequences (start codon targeted polymorphism [25]). Anchored primer amplification polymorphism also does not require prior knowledge of the genome and is mostly used in population genetic diversity analysis or for differentiating TCMs and their adulterants in variants, cultivars and species from different localities. ISSR polymorphism is the major anchor amplified polymorphic method of TCM authentication. The ISSR polymorphism technique uses microsatellites, usually 16–25 bp long, as primers in a single primer PCR reaction targeting multiple genomic loci to mainly amplify the ISSR sequences of different sizes. The PCR primers are based on a repeat sequence, such as $(CA)_n$, or with a degenerate 3′-anchor, such as $(CA)_8RG$ or $(AGC)_6TY$. Most anchored primer amplification polymorphism methods are user-friendly, and well-chosen primers can provide reasonably accurate fingerprinting results visualised by agarose gel or capillary electrophoresis. ISSR polymorphism has been employed to differentiate Huajuhong (Citri Grandis Exocarpium) derived from *Citrus grandis* 'Tomentosa' from other *Citrus* variants [26]. A total of six ISSR primers ($[CA]_8G$, $[GT]_8A$, $[AC]_8G$, $[CA]_8RG$, $[AC]_8YT$ and $BHB[GA])_7$) were used to reveal the relationship between 23 *Citrus* samples. The six primers generated 57 bands, of which 52 (91.2%) were polymorphic across all 23 *Citrus* samples. Cladistic analysis based on the band polymorphism of the ISSR fingerprints showed that the cultivar *C. grandis* 'Tomentosa' was clearly distinguished from *C. grandis* and other *Citrus* species. ISSR fingerprinting was also applied to identify the *Dendrobium*, *Rhubarb* and *Coptis* species [27–29]. It was also used to study the genetic diversity of *Descurainia sophia*, *Astragalus mongholicus* and *Cassia tora* [30–32].

2.3.1.3 Amplified Fragment Length Polymorphism

The principle of AFLP is to amplify a subset of DNA restriction fragments from the genomic DNA using restriction enzymes [33]. The genomic DNA is first digested with restriction enzymes (e.g. *Eco*RI and *Mse*I) at various restriction sites in multiple loci to generate restriction fragments with sticky ends. Synthetic adaptors are then ligated to these ends, which act as the annealing sites of specific primers for subsequent amplification by PCR under stringent conditions. The separation of amplified fragments is performed using high-resolution polyacrylamide gel and visualised using autoradiography, fluorescence or silver-staining techniques. Similar to AP-PCR and RAPD, AFLP screens multiple loci of the whole genome randomly and simultaneously and does not require prior knowledge of the sequence information. AFLP can detect more loci and generate more polymorphic fragments than

RAPD and can be used to differentiate closely related species [34]. However, DNA degradation in medicinal material may affect the reproducibility of the polymorphic patterns. Our group used AFLP to differentiate closely related medicinal species, such as oriental ginseng (*P. ginseng*) and American ginseng (*P. quinquefolius*) from various localities [35, 36]. Other examples of using AFLP include the identification of *Panax japonicus*, medicinal *Plectranthus* sp. and *Cannabis sativa* [37–39].

2.3.2 Specific Amplification

Specific amplification explores DNA polymorphism in a specific region of the sample. Polymorphic patterns are usually visualised by agarose, polyacrylamide gel or capillary electrophoresis. Specific amplification is the most commonly used method of molecular identifying TCMs. Specific amplification identification methods have been listed in 'Pharmacopoeia of the People's Republic of China (2015 edition)' for the identification of Zaocys (Wushaoshe), Agkistrodon (Qishe), Bungarus Parvus (Jinqianbaihuashe) and Fritillariae Cirrhosae Bulbus (Chuanbeimu) and 'The Japanese Pharmacopoeia' (JP XVI) for testing the purity of Atractylodes Lancea Rhizome (Cangzhu) and Atractylodes Rhizome (Baizhu).

2.3.2.1 Specific PCR

Specific PCR is a DNA identification method based on the difference in DNA sequences between the specific regions of an authentic TCM and its adulterants. Its premise is the design of authentic TCM-specific primers and the establishment of PCR amplification and reaction product-detection methods to distinguish authentic TCMs from their adulterants by the absence/presence or the number of migration bands in a gel electrophoresis pattern. Because the identified primer is designed to be entirely identical with the authentic TCM DNA sequence and different from that of the adulterants, this method could specifically amplify template DNA from authentic TCMs and ultimately generate authentic TCM-specific DNA bonds in a gel electrophoresis pattern. Primers could be located in different regions with several (specific PCR) mismatches or a single nucleotide polymorphism mismatch (allele-specific PCR or ARMS-PCR). This method could also simultaneously determine more than one TCM by using several pairs of specific primers in the same PCR reaction system (multiple PCRs). Specific PCRs are capable of differentiating TCM materials and their adulterants at the genus, species, subspecies, variant and cultivar levels with their corresponding specific primers. Specific regions often come from sequencing and are used to compare the nucleotide differences between authentic TCMs and their adulterants or obtain polymorphic fragments from a whole genome fingerprint, such as RAPD or ISSR. For example, *Dendrobium* species are difficult to identify because of their extremely similar morphologies to that of *Dendrobii caulis* (Shihu). However, comparison of the chloroplast *trnL-F* and internal transcribed spacer (ITS) sequences of eight mainstream commercial *Dendrobii caulis* species found several SNPs to be common among the species, and species-specific primers were designed by these different gene loci. The established species-specific multiplex PCR yielded products of 148 bp, 210 bp, 265 bp, 340 bp, 397 bp, 448 bp, 491 bp and 584 bp amplicons in the presence of *D. fimbriatum*, *D. huoshanense*, *D. chrysotoxum*, *D. nobile*, *D. officinale*, *D. strongylanthum*, *D. aphyllum* and *D. devonianum*, respectively [40]. Other similar specific works included the differentiation of *Aristolochia* species, *Ligularia fischeri* and *Angelica dahurica* [41–43].

The polymorphic fragment is cloned and sequenced for designing a pair of specific PCR primers to amplify the concerned polymorphic fragment. The amplification of the polymorphic fragment or the size difference of the fragments in different samples provides a means for differentiating the samples. This technique focuses on a single locus and is usually reproducible under high stringent PCR conditions. To increase the accuracy of differentiation, several sequence-characterised amplified regions (SCARs) of a sample are analysed. SCAR analysis requires prior information of the sequence of the polymorphic fragment for specific primer design. Degradation within the DNA fragment and the presence of PCR inhibitors may lead to false negative results. A 25-bp insertion was found in a RAPD fragment of *P. ginseng,* which was converted to a SCAR marker and used to differentiate between *P. ginseng* and *P. quinquefolius* [44].

2.3.2.2 PCR Restriction-Fragment Length Polymorphism

PCR-RFLP amplifies a specific region of the genome, which is followed by restriction digestion to produce a restriction polymorphic profile. The specific region should be readily amplified using universal or specific primers. Standard DNA barcodes with high sequence variation, such as the ITS region, are good candidate regions to start with. Restriction digestion of the amplified fragments (e.g. SmaI, HinfI and Sau3A1) generates restriction fragments of different sizes. Mutations creating or disrupting a restriction site is the key to producing polymorphic fingerprints for sample discrimination. Although interpretation of PCR-RFLP data is simple, the discriminating ability of DNA polymorphism is less than that of ISSR and AFLP. PCR-RFLP was used to discriminate Lonicerae Japonicae Flos (Jinyinhua) species from their adulterants by amplifying the psbA-trnH and trnL-trnF regions and therefore performing restriction digestion using HinfI and NlaVI. *Lonicera japonica* possesses an endonuclease digestion site and was cleaved into two fragments, while other *Lonicera* species did not have this sequence at the same site; the digestion products were inactive and remained a single band in gel electrophoresis patterns [45]. PCR-RFLP was also use to differentiate Codonopsis Radix (Dangshen) derived from *Codonopsis pilosula*, *C. tangshen*, *C. modesta* and *C. nervosa* var. *macrantha* from their adulterants by digesting the ITS region using Hinf1 and HhaI [46]. A similar approach was applied to identify *Alisma orientale*, Bulbus Fritillariae Cirrhosae (Chuanbeimu) and *Fritillaria pallidiflora* species [47–49].

2.3.2.3 Simple Sequence Repeats

SSRs, also known as micro-satellites, are DNA tracts in which a short base-pair motif is repeated several to many times in tandem (e.g. CAGCAGCAG). These sequences experience frequent mutations that alter the number of repeats [50]. Repeats alter between species, variety or population and cause different polymorphic patterns that can be visualised by polyacrylamide gel or fluorescent capillary electrophoresis; this could be used for the molecular identification of TCMs, particularly in identifying varieties or populations. SSR genotyping needs to amplify the specific region, including the SSRs, using designed specific primers. These SSR markers often originate from RNA-seq, genome sequences, cDNA libraries or enrichment by magnetic beads and sequencing. Because of the high number of variations, the discriminating ability of SSR is greater than that of ISSR and AFLP. SSR polyacrylamide gel electrophoresis patterns were used to discriminate Lonicerae Japonicae Flos (Jinyinhua) species from their adulterants using three markers (jp.ssr4, jp.ssr64 and jp.ssr65) [51]. This method is also used to assess the purity of *Dendrobium officinale* varieties and investigate their genetic diversity [52]. A similar approach was applied to identify *Epimedium sagittatum*, *Centella asiatica* and *Panax ginseng* varieties [53–55].

2.3.2.4 Isothermal Amplification

Conventional PCR amplifies DNA fragments through thermocycles for the denaturing of double-strand DNA, annealing of primers and synthesising of new strands. Isothermal amplification is a technique allowing DNA amplification without thermocycling, and thus DNA amplification can be achieved without PCR machines. These techniques are mostly applied for the on-site detection of viral and bacterial infection in undeveloped regions where laboratory equipment is limited. There are several ways to perform isothermal amplification. The most well-established methods are exemplified by nucleic acid sequence-based amplification (also known as transcription mediated amplification), helicase-dependent amplification, recombinase polymerase amplification (RPA), rolling circle amplification, multiple displacement amplification, loop-mediated isothermal amplification (LAMP) and strand displacement amplification (SDA).

LAMP is a major isothermal amplification technique with impressive specificity, efficiency and rapidity in TCMs [56]. Four special primers designed from six alleles (one allele for both the forward and reverse outer primers, two alleles for the forward inner primer and two alleles for the reverse inner primer) are used to create 'loops' at the end of DNA strands, which significantly speed up the process of SDA; the whole process can be finished in 1 h. Amplification progress can be accelerated by additional LAMP primers to achieve amplification in 30 min. Recently, LAMP has been applied to identify herbal medicinal material, such as differentiating *Curcuma longa* from *C. aromatica* based on the trnK gene

[57]. LAMP was also used to discriminate *Panax ginseng* from *P. japonicus* based on the 18S rRNA gene [58]. LAMP is efficient and sensitive when all the primers match the target DNA. However, primer design is difficult because many combinations of primers are needed. The primer sites should be conserved regions with minimum intra-specific variations. DNA degradation in dried or processed materials may give false negative results. Integrity control of the amplification region may be necessary to prove that any negative amplification is solely because of DNA degradation.

RPA is a recently developed isothermal amplification method that offers high sensitivity and specificity for DNA detection. RPA can complete an entire reaction at a constant low temperature, lower than 40 °C and even at room temperature (15–25 °C). This technique includes a nucleoprotein complex formed by the recombinase enzyme, an oligonucleotide primer and DNA polymerase. The nucleoprotein complex and oligonucleotide primer facilitate strand exchange, and DNA polymerase I, a *Staphylococcus aureus* homologue, elongates the primer. RPA is more rapid, convenient and efficient than traditional PCR amplification. For non-specialists and detection conditions with time and resource constraints, less complicated devices, such as lateral flow strips, can be preferentially selected to detect amplification results. The RPA result could also be determined by fluorescent probes, and gel electrophoresis is more suitable in RPA assays for the rapid identification of TCMs in less-equipped laboratories. When use alkaline lysis method or filter paper method to speedily extract DNA with a portable equipment, RPA could also perform rapid and sensitive on-site identification of TCMs in the field or in herbal markets [59, 60]. A similar approach was applied to identify *Ficus hirta* (Wuzhimaotao), *Ginkgo biloba* (Yinxin) and *Bubalus bubalis* (Shuiniujiao) [61, 62].

2.3.3 DNA Microarray

DNA microarray is a hybridisation-based technology using labelled nucleotide probes to hybridise single or multiple loci in a target genome. The probes are short nucleotide fragments obtained from either restriction digestion or synthetic oligonucleotides. They are fixed on a supporting matrix where the hybridisation of probes and tested DNA samples takes place. Our group amplified the internal transcribed spacer (ITS) of 16 *Dendrobium* species and used them as probes to identify medicinal *Dendrobium* species in a prescription with multiple herbs [63]. The ITS2 regions of the tested samples were labelled with Cy3 fluorescent dye and allowed to hybridise to the ITS probes. A species-specific fluorescent signal was obtained to clearly identify the five medicinal *Dendrobium* species. A similar approach using 5S rDNA intergenic spacers as probes were applied to differentiate *D. officinale* from other closely related *Dendrobium* species [64].

2.3.4 DNA Sequencing

DNA sequencing is one of the most definitive means for identification as this technique can directly assess sequence variations on a defined locus. It also provides informative characters to reveal phylogenetic relationships. With the decreases in sequencing costs, the identification of medicinal materials using DNA sequencing has become a routine practice. The commonly used DNA regions for medicinal material identification include the nuclear ITS and 5S rDNA intergenic spacer (5S), chloroplast *psbA-trnH* intergenic spacer (*psbA-trnH*), large subunit of the ribulose-bisphosphate carboxylase (*rbc*L), maturase K gene (*mat*K), *trnL* intron (*trnL*), *trnL-trnF* intergenic spacer (*trnL-F*), mitochondrial control region (CR), cytochrome c oxidase subunit 1 (COI) and cytochrome b gene (*Cyt* b). These DNA regions have different evolutionary rates and therefore possess different variabilities. For example, the mitochondrial COI region has slowly evolved, and only a few variations were observed in the 1.4 kb COI sequences in flowering plants [65]. However, this region evolves rapidly and is varied enough to discriminate most animal species. To differentiate medicinal material from adulterants derived from closely related species, it is essential to search for DNA regions with high discriminative powers.

In 2003, the concept of barcoding global species by selected DNA region was first proposed [66], and a substantial effort has been put on the screening of appropriate DNA barcodes. DNA barcoding is a DNA sequencing-based taxonomic method used to identify species based on species-specific differences in short regions (proposed to be analogous to a barcode) of their DNA. DNA barcoding is not restricted by morphological characteristics and physiological conditions and allows species authentication without specialist taxonomic knowledge. This method is also standardised to specific DNA barcodes and universal primers, which is favourable for building databases and establishing a universal standard for identification [67, 68].. Cytochrome c oxidase 1 (COI), developed as a universal barcode in animals, sequences divergences in more than 13,320 congeneric pairs, including representatives from 11 phyla, and shows species-level diagnoses in most taxa [69]. In plants, the Consortium for the Barcode of Life (CBOL) Plant Working Group recently recommended the two-locus combination of *mat*K + *rbc*L as the best plant barcode, with a discriminatory efficiency of only 72% [70]. Taxonomists have suggested that a multi-locus method may be necessary to discriminate between plant species. After that, the ITSs of nuclear ribosomal DNA were also incorporated as core barcodes for seed plants [71]. For bacteria and fungi, the 16S and nuclear ITS regions are the most appropriate DNA barcodes, respectively. In herbal medicines, ITS2 was proposed as a core DNA barcode for medicinal plants. In addition, the *psbA-trnH* region was suggested as a complementary barcode. These methods have been approved for incorporation into Supplement 3 of the 'Chinese Pharmacopoeia' (2010 edition). These DNA barcodes have been proven to be useful not only in biodiversity and conservation studies but also in the identification of medicinal materials. For example, the ITS region can differentiate six *Panax* species from their adulterants derived from *Mirabilis jalapa* and *Phytolacca acinosa* [72]. The

ITS region was also used to identify medicinal *Dendrobium* species [73], Muxiang (Aucklandiae Radix, Vladimiriae Radix and Inulae Radix) [74], Baihuasheshecao (Hedyotii Herba) [75] and Leigonteng derived from *Tripterygium wilfordii* [76]. The chloroplast *trnH-psbA* region is another highly varied DNA barcode for identifying Madouling (Aristolochiae Fructus) [77] and Wutou (Aconiti Radix and Aconiti Kusnezoffii Radix) [78]. Apart from the standard DNA barcodes, a few regions are also useful for identifying medicinal materials. For example, the nuclear 5S region was used to identify Danshen derived from *Codonopsis* species [79] and medicinal *Swertia* species [80]. Furthermore, the chloroplast *trnL* region was used to identify Baibu (Stemonae Radix) [81], and the *trnL-F* region was used to identify Madouling (Aristolochiae Fructus) [77].

2.4 Molecular Identification of Botanical Medicinal Materials

Approximately 90% of medicinal materials recorded in the 'Pharmacopoeia of the People's Republic of China' (2015 edition) are derived from botanical sources. The huge international market for herbal medicinal materials suggests the importance of their correct identification. The pharmacological effects of herbal medicinal materials may vary among closely related species, subspecies, variants, cultivars and localities, not to mention the adulterants derived from distantly related species. Apart from conventional organoleptic and chemical methods, molecular approaches provide an alternative and definite method to identify these samples.

2.4.1 Discrimination at the Inter-Family and Inter-Genus Levels

Adulteration of herbal material by distantly related species from different families or genera is common. The molecular identification of these adulterants is relatively easy as their genetic makeups are quite different from those of the genuine species. DNA fingerprinting techniques usually show clear-cut results. For examples, AP-PCR, RAPD and RFLP fingerprints of medicinal *Panax* species in the family Araliaceae showed different patterns from those of adulterants in the families Nyctaginaceae, Phytolaccaceae, Campanulaceae and Talinaceae [72]. DNA sequencing is also useful to discriminate between distantly related species. For example, the *trnL* region is able to distinguish medicinal *Stemona* species in the family Stemonaceae from adulterants in the family Asparagaceae [81]. Similarly, the *trnL-F* and *psbA-trnH* regions were used to distinguish Madouling derived from *Aristolochia* species (Aristolochiaceae) from a substitute derived from *Cardiocrinum* species (Lilicaeae) [77]. The identification of material from different genera can also be achieved using DNA techniques. PCR-RFLP may be applied to differentiate four *Codonopsis* species (Campanulaceae) from two adulterants derived from *Campanumoea* and *Platycodon* species in the family Campanulaceae [46]. DNA sequencing of the ITS region was applied to distinguish 16 medicinal *Dendrobium* species from *Pholidota* species in the same family of Orchidaceae [73]. Although DNA sequencing is useful to differentiate among samples derived from distantly related species, such as at the family and genus levels, choosing a suitable DNA region is crucial. Some DNA regions, such as ITS and 5S, evolve rapidly, and their sequence similarities at the species level in some families are low. For examples, the sequence similarities of the ITS and 5S regions among Muxiang species (Asteraceae) and their toxic adulterants (Aristolochiaceae) were only 56–58% and 20–30%, respectively [74]. Although such low similarity does not affect the differentiation between samples in different families, it may make sequence alignment and phylogenetic tree construction difficult.

2.4.2 Discrimination at the Inter- and Intra-Species Levels

One of the major advantages of molecular identification is its high resolution, which allows differentiation between samples at the inter- or intra-species level. SCAR markers from DNA fingerprinting have been used to readily differentiate closely related species of *P. ginseng* from *P. notoginseng* [44]. DNA microarrays with hybridisation probes, designed based on ITS and 5S sequences, successfully detected several medicinal *Dendrobium* species [63, 64]. Choosing an appropriate DNA region with high variability and discrimination power is crucial for differentiation between closely related species by DNA sequencing. For example, *trnL* is a relatively conserved region, which can differentiate medicinal *Stemona* species (Stemonaceae) from adulterants derived from *Asparagus* species (Asparagaceae), but fails to discriminate the medicinal species of *S. japonica*, *S. sessilifolia* and *S. tuberosa* from another closely related species, *S. parviflora* [81]. On the contrary, the ITS, 5S and *psbA-trnH* regions are highly varied regions that are commonly used for identification at the species level. The ITS region have enough variable sites to discriminate among all 16 medicinal *Dendrobium* species, with inter-specific divergences ranging from 2% to 17% [73]. This region was also used to identify Baihuasheshecao derived from *Hedyotis diffusa* (Rubiaceae) and discriminated among all 14 *Hedyotis* species studied [75]. In fact, the ITS2 region is highly varied and useful for discriminating most medicinal species and therefore has recently been proposed as a DNA barcode for medicinal plants [82]. Although the ITS region shows high sequence variability among species and is the most frequently used region for species identification of herbal medicinal materials, the presence of multiple copies, which may be non-homogeneous, and the problem of secondary structures resulting in poor-quality sequence data are major drawbacks [83, 84]. Molecular cloning prior to DNA sequencing is necessary to solve these problems. Besides, fungal contamination is common in herbal medicinal material and would interfere with the proper amplification of target ITS sequences by universal primers. Specially designed plant-specific primers should be used in such conditions. The 5S region is a highly varied region and is frequently used for species and subspecies differentiation. It readily discriminated *Swertia mussotii* from *S. chirayita*, *S. franchetiana* and *S. wolfangiana*, with inter-specific divergences ranging from 31% to 65% [80]. It also differentiated between Dangshen derived from *Codonopsis pilosula* and *C. pilosula* var. *modesta* with intra- and inter-specific similarities of 95–98% and 70–73%, respectively [79]. In our experience, however, the 5S region sequence is sometimes too varied, making it difficult for sequence alignment. Moreover, this region has multiple copies, and molecular cloning prior to sequencing is essential. The *psbA-trnH* region is a complementary DNA barcoding region showing the highest amplification success and discrimination rates among nine tested loci [85]. It was used to identify 19 *Aconitum* species with an average inter-specific similarity of 85% [78]. The two closely related medicinal species, *A. carmichaeli* and *A. kusnezoffii*, were clearly distinguished by a 56-bp sequence inversion in their trnH-psbA sequences. A disadvantage of the *psbA-trnH* region is the presence of a poly(A) structure, which reduces the successful rate of DNA sequencing. Besides, sequence alignment may be difficult due to the frequent presence of nucleotide insertion and deletion. In spite of the highly discriminative ability at the species level, *psbA-trnH* could not resolve the relationship between *Cardiocrinum giganteum* and its variant *C. giganteum* var. *yunnanense*; however, the *trnL-F* region could [77]. This example demonstrated that there is no single universal locus suitable for differentiating all taxa at different levels. Finding a suitable region that suits the purpose is of the utmost importance.

2.4.3 Discrimination among Germplasm and Geographical Culture Origins

Herbal medicinal materials derived from various germplasm sources or collected from different geographical origins may be traced using molecular techniques. For example, the herb Dendrobii Officinalis Caulis (Tiepishihu) is derived from *Dendrobium officinale* Kimura et Migo. This species includes many different cultivated populations (germplasm sources) in Anhui, Yunnan, Guangxi and Zhejiang, China. SSR fingerprinting using 32 pairs of primers generated 117 polymorphic alleles and 11 populations of *D. officinale* that could be clustered into three branches, which is in accordance with the geographical distribution of its germplasm. A minimum of four pairs of SSR marker primers could be able to differentiate all 11 populations (germplasm) from Yunnan and Anhui [86]. A similar approach was also applied to identify the germplasm of *Lonicera japonica* and cultivars of *Panax ginseng* [55, 87].

PCR-RFLP analysis of the ITS region of *Lonicera japonica* collected from different geographical origins in China showed that the mutation site found in the

ITS region from geo-authentic *L. japonica* can be recognised by the restriction endonuclease *Eco*NI. PCR products from geo-authentic *L. japonica* cannot be digested completely; the cleavage rate of PCR products by *Eco*NI was determined to be more than 70% in all geo-authentic *L. japonica* and less than 20% in non-geo-authentic *L. japonica* and other species from the genus *Lonicera* [88].

2.5 Molecular Identification of Animal Medicinal Materials

Animals account for 7% of all Chinese medicinal materials in 'Pharmacopoeia of the People's Republic of China' (2015 edition). Some of the materials are expensive. These include Ejiao (Asini Cornii Colla) derived from *Equus asinus* and Lurong (Cervi Cornu Pantotrichum) derived from *Cervus nippon* and *C. elaphus*. They are frequently adulterated by less expensive products with similar organoleptic features. Correct identification also avoids the misuse of endangered species.

Both DNA fingerprinting and sequencing have been shown to be useful for differentiating animal medicine. For example, in 'Pharmacopoeia of the People's Republic of China' (2015 edition), several snake species are listed. These entries include Jinqianbaihuashe (Bungarus parvus) derived from *Bungarus multicinctus* (Elapidae), Qishe (Agkistrodon) derived from *Deinagkistrodon acutus* (Viperidae) and Wushaoshe (Zaocys) derived from *Zaocys dhumnades* (Colubridae). The peeled skin of *Elaphe carinata* (Elapidae), *E. taeniura* (Elapidae) and *Z. dhumnades* are used as Shetui (Serpentis Periostracum), and the meat of *A. strauchii* (Viperidae), *B. multicinctus* and *Opheodrys major* (Colubridae) are used as Sherou. Species-specific RAPD fragments for *A. acutus*, *B. multicinctus* and *Z. dhumnades* were used to design SCAR primers to generate specific markers for these species. DNA sequencing of the *Cyt* b region clearly distinguished *Z. dhumnades* and *B. multicinctus* from *N. naja*, *Ophiophagus hannah*, *Ptyas mucosus* and *Python molurus*. It was also found that the *Cyt* b sequence could resolve the phylogenetic relationships of 90 snake species in the families Boidae, Colubridae, Elapidae and Viperidae with sequence similarities ranging from 71% to 93%. A forensically informative nucleotide sequencing approach based on the *Cyt* b sequence revealed that two of the retailed snake meat samples were derived from *Python reticulates*, and four of them were derived from *Python molurus* [89].

2.6 Identifying the Age of Medicinal Material

Growth year is an important index for assessing the quality of TCMs. Most herbal plants are perennial; their bioactive components accumulate over growth time. For example, bioactive secondary compounds accumulate in some medicinal plants, such as *P. ginseng*, *Salvia miltiorrhiza* and *Coptis chinensis*; older plants usually serve as better medicinal herbs. However, in the pursuit of economic efficiency, a number of inappropriate strategies, including the use of growth hormones and swelling agents, as well as continual transplantation, have been employed to simulate age. Therefore, the quality of Chinese herbal medicines is difficult to determine. Molecular markers are expected to be a reliable and effective method for identifying the age of TCMs that complement the traditional methods of age determination; markers of particular note are telomere length and DNA methylation level.

Telomeres, which are specialised structures at the physical ends of eukaryotic chromosomes comprising highly conserved repeated DNA sequences, shorten with each round of DNA replication because DNA polymerases cannot completely replicate the linear DNA molecules. In gymnosperms, telomere length can be used to predict the future replicative capacity of cells [90, 91]. Highly significant correlations between telomere length and age have been observed in humans, Australian sea lions and different stages of barley [92, 93]. Therefore, telomere shortening can be used as a marker of cell replication and aging. Liang et al. [94] combined traditional identification methods and the measurement of telomere length in ginseng plants of known age and found that telomere length in the main root (approximately 1 cm below the rhizome) was the best indicator of age. By using a terminal restriction fragment (TRF) method to determine the telomere lengths in *P. ginseng* specimens of different ages, telomeres in the main roots showed a significant increase in TRF length with plant age, which could be used for age estimation from 2 to 8 years old and be fulfilled with the following mathematical model: $y = 0.827x + 8.231$, where x is age and y is TRF length. As well as *P. ginseng*, TRF length was also related to the age of *G. biloba*, *Silene latifolia*, *Arabidopsis thaliana*, *Hordeum vulgare* and *Pinus sylvestris* [93, 95–97].

2.7 Conclusion and Prospects

With the exponential growth of the international market and the increasing demand for high-quality TCM material, correct identification is a key factor to ensure the safety, efficacy and fair trade of material. Molecular technology provides a reliable and powerful tool for the definite identification of TCM material at various taxonomic levels, including family, genus, species, variant and cultivar. The molecular identification of TCM material may face difficulties with poor quality or a low quantity of DNA content, the presence of secondary metabolites and fungal contamination. Similar to other identification methods, substantial effort is needed to collect authentic reference species for DNA marker development. With the DNA barcode initiative and the reduction of cost in DNA manipulation and next-generation sequencing, we expect that more molecular markers will be developed and increasing numbers of laboratories will be able to carry out the tests. In addition, the field will benefit from the development of rapid DNA extraction methods, pocket-sized PCR machines and speedy isothermal amplification techniques; the instant on-site molecular identification of medicinal material could be a reality in the near future.

2.8 Case Study

2.8.1 Rapid and Robust Authentication of Deer Antler Velvet Product by Fast PCR-RFLP Analysis

Deer antler velvet originates from sika deer (*Cervus nippon* Temminck) and red deer (*Cervus elaphus* L.) of the genus *Cervus* in the family Cervidae. Only unossified antlers are defined as antler velvet in the 'Chinese Pharmacopoeia' and 'Korean Herbal Pharmacopoeia'. Deer antler velvet possesses high market value as a TCM. The demanding production and premium price promote widespread adulteration of antler velvet. Some antlers from other species of Cervidae, such as sambar (*C. unicolor*), reindeer (*Rangifer tarandus*), white-lipped deer (*Przewalskium albirostris*), Père David's deer (*Elaphurus davidianus*) and fallow deer (*Dama dama*), have commonly been incorporated into authentic antler velvet by merchants to obtain economic profits. The prevention of such fraudulent practices constitutes an important part of quality control systems. The incorrect labelling of Chinese *materia medica* represents not only a commercial fraud but also a potential health problem in the clinical practice of Chinese medicine. However, antler velvet from different Cervidae species have similar components; it is difficult to distinguish among species using pharmacognosy identification or common analytical methods, especially when samples are processed into decoction pieces or powders.

Various methods, including DNA-based molecular marker techniques, have been applied to detecting adulteration in antler velvet. Previously, studies report several PCR-based methods, such as RAPD, specific PCR, real-time PCR, multiplex PCR, sequencing analysis and DNA barcoding [98–103]. However, many disadvantages exist, such as poor sensitivity and specificity (RAPD), long consumption time (sequencing and DNA barcoding), the inability to detect contaminants (specific PCR) and the need for stringent conditions (multiplex real-time PCR), within the molecular authentication methods, which thereby limit the applications in routine inspection.

PCR-RFLP is a stable, rapid and inexpensive technique that combines PCR and restriction endonuclease digestion and has received much attention in identifying genetic diversity and developing biomarkers for the authentication of adulterants at the species level [48, 104]. Additionally, this technique can semi-quantitatively detect contamination in authentic samples and can be performed quickly and inexpensively. Herein we describe a fast RFLP analysis method that is coupled with PCR for the rapid and robust detection of adulteration of antler velvet and commercial velvet products. The method is also used to detect contaminated and adulterated antlers in Chinese patent drugs and dietary supplements.

2.8.2 Material and Methods

Antler samples of sika deer (*Cervus nippon*), red deer (*C. elaphus*), sambar (*C. unicolor*), reindeer (*Rangifer tarandus*), white-lipped deer (*P. albirostris*), Père David's deer (*E. davidianus*), fallow deer (*D. dama*), mule deer (*Odocoileus hemionus*) and roe deer (*Capreolus capreolus*) were obtained from the Department of Pharmacognosy (Beijing University of Chinese Medicine, China) and KM Traditional Chinese Herbal City (Bozhou, China). All specimens were morphologically identified before the samples were obtained, and voucher specimens were deposited in the National Resource Centre for Chinese Materia Medica, China Academy of Chinese Medical Sciences, Beijing, China. The total genomic DNA was isolated from dried material using Column Bone DNAout kits (Tiandz, Beijing, China) following the manufacturer's instructions.

Fifty commercial antler products that were claimed to be from sika deer or red deer according to their labels were purchased from the traditional herbal markets and drug stores in different provinces of China.

Antler materials were frozen in liquid nitrogen and were ground to a fine powder with a Retsch MM 400 Mixer Mill (Retsch Technology GmbH, Haan, Germany). Genomic DNA was extracted from all samples using Column Bone DNAout kits following the manufacturer's instructions, with minor modifications. The DNA was eluted with double-distilled, sterile water, and DNA concentration was brought to 10 ng/μL to perform the PCR reaction. To evaluate the test sensitivity, DNA samples were diluted to 40, 8, 1.6, 0.32 and 0.06 ng. In order to remove potential contamination from other samples, all samples were treated with 75% alcohol and sterilised water prior to DNA extraction.

The cytochrome b region sequences of sika deer (10 subspecies, 51 sequences), red deer (11 subspecies, 184 sequences), sambar (five sequences), reindeer (three subspecies, 149 sequences), white-lipped deer (six sequences), mule deer (11 subspecies, 187 sequences), Père David's deer (three sequences), fallow deer (three sequences) and roe deer (75 sequences) were obtained from the GenBank database to eliminate the potential risk of intra-specific variation. The alignment of sequences was performed using BioEdit 7.0.9.0. Specific restriction recognition sites were exploited using Primer Premier Version 5.0 software, and restriction mapping analysis was performed using the WatCut online restriction analysis program (http://watcut.uwaterloo.ca/watcut/watcut/template.php). Each candidate fragment shorter than 200 bp or containing other identity restriction recognition sites was eliminated. Based on the restriction maps, *Dde* I was selected as the candidate restriction endonuclease for discrimination between sika deer, red deer and other deer.

The primers for the amplification of *Cervus* and its sibling species were designed using Primer Premier Version 5.0 software. The amplicon size amplified by the primers should be larger than 200 bp, and the restriction site should be located in the middle of the amplicon. The melting temperature of the primer was 50–55 °C, and self-complementarity and false priming were avoided. Primer BLAST was used to test primers for sequence homology with other animal species. The forward primer LR-F: 5′-AATATTACTAGTATTATTCGCACCAGA-3′ and reverse primer LR-R: 5′-TTCAGAATAGGCATTGGCTG-3′ were synthesised by Shanghai Sangon Biological Engineering Technology & Services Co. Ltd., Shanghai, China.

PCR amplification was performed with a Veriti PCR thermal cycler (Applied Biosystems,CA, USA) containing 2 U SpeedStar DNA polymerase (Takara Biotech Co. Ltd., Dalian, China), 2.5 μL of 10× Fast PCR buffer I (Takara, China), 2 μL of dNTP (2.5 mM each), 0.5 μL of forward and reverse primers (10 mM) and 2 μL of DNA template (approximately 10 ng) for a total volume of 25 μL. The reaction began with an initial denaturation at 95 °C for 1 min, followed by 35 cycles of denaturation at 95 °C for 5 s and annealing at 56 °C for 5 s, and a final extension for 1 min at 72 °C. The PCR products were then analysed by 1.5% agarose gel (Invitrogen, California, USA) electrophoresis with GelGreen dye (TIANGEN, Beijing, China) in 1× TAE buffer for 15 min at 180 V and imaging on a G:BOX gel documentation system (Syngene, UK).

Digestions with *Dde* I (New England Biolabs, Massachusetts, USA) were performed in a Veriti PCR thermal cycler. Then 2.8 μL of 10× Cutsmart buffer (New England Biolabs, USA) and 1 μL of enzymes were added individually to the appropriate wells containing PCR product, followed by incubation at 37 °C for 15 min, as recommended by the manufacturer. The DNA was fractionated by 2.0% agarose gel electrophoresis and visualised by GelGreen dye staining under UV light.

Negative control reactions (using ddH_2O instead of genomic DNA) were performed for every DNA extraction, PCR amplification and restriction enzyme digestion simultaneously to ensure that the authentications were not the result of contamination of the reagents with DNA.

2.8.3 Results

2.8.3.1 Restriction Analysis

Analysis of the restriction maps revealed a restriction site of endonuclease *Dde* I (5′-CTNAG-3′) in a candidate fragment of the *Cyt* b sequence of sika deer and red deer, while other species of deer did not have this sequence at the same site. Thus, the *Cyt* b sequence of sika deer and red deer could be cleaved by *Dde* I into two fragments of 102 bp and 161 bp, respectively. Because of the lack of the *Dde* I restriction site in the *Cyt* b region, the PCR product of other deer was inactive and remained at 263 bp. When using the primers LR-F and LR-R to amplify different species' antler velvet, a single PCR product of approximately 250 bp was observed. After treatment with *Dde* I, sika deer and red deer cleaved into two fragments of approximately 100 and 250 bp, while there were no restriction sites for this enzyme in the sequences of other species, and the PCR products of other deer could not be digested.

2.8.3.2 Optimisation of the PCR-RFLP Conditions

Several factors, including ingredient concentrations, PCR cycling conditions and reaction time, which can affect PCR specificity and efficiency, were optimised. The results were evaluated by 2% agarose gel electrophoresis. Gradient PCR was carried out on an Applied Biosystems Veriti thermal cycler to determine the best annealing temperature (50, 52, 54, 56, 58 and 60 °C) for the deer antler primers. Strong reaction products were observed from 50 to 60 °C with no significant differences. Therefore, 56 °C was selected as the optimised reaction temperature for the PCR amplification of sika deer and its related species. Because cycle number directly influences the abundance of PCR products, in the next step, the cycle number was varied between 25 and 40. Amplification products were observed with all cycle numbers, but a significant improvement was achieved with cycle numbers up to 35, whereas between 35 and 40 cycles, no improvement was observed. Thus, 35 cycles was selected as the optimised condition for PCR amplification. In the same manner, the concentrations of template DNA, dNTP and primers were optimised in the range of 1–200 ng, 0.4–0.8 mM and 0.05–0.2 μM, respectively. The optimum conditions were 8 ng of genomic DNA, 0.8 mM dNTP and 0.2 μM primers. Five commercially available PCR polymerases and four thermal cyclers were used to evaluate the robustness of the PCR-RFLP. The results indicated that the use of different PCR polymerases and thermal cyclers did not influence the sensitivity and accuracy of the PCR approaches.

To distinguish authentic antler velvet and its adulterants by PCR-RFLP, the PCR products were subsequently digested with *Dde* I restriction enzyme. The RFLP method was optimised with regard to digestion time and enzyme units to simplify and standardise the protocol across samples. The *Dde* I restriction activity varied between 2.5 and 20 U. RFLP genotyping patterns were observed with all enzyme activities, but star activity (non-specific digestion outside a restriction enzyme's recognition site) appeared with a high concentration of *Dde* I enzyme. Consequently, 5 U per reaction was defined as the optimal activity of polymerase for the RFLP genotyping. Different digestion times were also tested, and the results showed that complete digestion could be achieved in 15 min to 2 h, whereas at longer than 4 h or overnight, star activity appeared. For rapid genotyping, 15 min was defined as the optimal digestion time. To evaluate test sensitivity, DNA samples were serially diluted to 40, 8, 1.6, 0.32 and 0.06 ng. Digest products were observed for the 0.32 ng DNA template, and there was no significant difference between 1.6 and 40 ng.

2.8.3.3 Commercial Antler Velvet Sample Identification

Using the PCR-RFLP method, 45 commercial products labelled as containing antler velvet were purchased in drug stores and investigated by PCR-RFLP analysis. In all samples, DNA was extracted using column-based nucleic acid extraction kits. After

PCR amplification by primers LR-F and LR-R, all specimens were successfully amplified by the 263 bp fragments. Using *Dde* I to analyse the 263 bp fragments of antler samples, 21 samples had no cutting site; four samples were incompletely digested and had three bands of 263, 102 and 161 bp; and 25 samples were completely cleaved into 102 and 161 bp. The 21 samples with no cutting sites in the 263 bp fragments were determined to be completely fraudulent, and the four incomplete digestions were determined to be admixtures of authentic antler velvet and adulterants. The fraud ratio was calculated as [total commercial samples – total real products/total commercial samples]. According to the results, up to 50.0% of the commercial antler velvet products were fraudulent. The PCR-RFLP results were also confirmed by pharmacognosy inspection and *Cyt* b sequencing. The sequences were searched against the NCBI nucleotide database using the BLASTn program, and the best hits were recorded as sequencing authentication results. The identities determined by PCR-RFLP showed good agreement with the DNA sequencing results.

2.8.4 Discussion

The advantages of PCR techniques have prompted considerable studies about PCR methods for deer-derived identification. PCR-RFLP is an accurate and authentic PCR-based authentication method based on (1) using universal primers to amplify a conserved region, (2) digesting the PCR product with specific restriction endonucleases and (3) separating the digest fragments by agarose gel electrophoresis [105]. 'Chinese Pharmacopoeia' (2015 edition) recommends using PCR-RFLP analysis for the molecular authentication of Bulbus Fritillariae cirrhosae (Chuanbeimu), a commonly used antitussive medicinal herb, which suggests that PCR-RFLP could be used as a standard method to authenticate some herbal medicines. Compared with species-specific PCR and DNA barcoding methods, PCR-RFLP identifies species according to fragment length, using co-dominant markers to simultaneously and semi-quantitatively discriminate between authentic, adulterant and admixture samples. The major disadvantages of PCR-RFLP are that the recognition site is limited and the process is time-consuming. In general, a common PCR-RFLP analysis requires 4–5 h, limiting its application in routine species identification studies. In this study, we developed a fast PCR-RFLP analysis method by combining rapid PCR and restriction endonuclease digestion. Deer antler velvet could be discriminated from its adulterants or admixtures within 1–1.5 h.

PCR-RFLP has been applied to the authentication of deer products from red deer, fallow deer and roe deer [105, 106]. However, in recent years, various types of velvet from other closely related deer, including *R. unicolor*, *R. tarandus*, *P. albirostris*, *E. davidianus*, *D. dama*, *O. hemionus* and *C. capreolus*, have been used as deer velvet in herbal markets. Previous PCR-RFLP methods cannot cover so many emerging adulterants. In this study, we aligned the *Cyt* b sequences of *C. nippon*, *C. elaphus* and seven types of closely related deer, and a *Dde* I endonuclease site was found only in *C. nippon* and *C. elaphus*. A PCR-RFLP method based on the *Dde* I endonuclease sites was developed to authenticate deer velvet and was validated using morphologically identified control samples. We used the established PCR-RFLP method to analyse commercial deer antler velvet and found that up to 50.0% of commercial antler velvet products were fraudulent in the collected samples. The sequencing results indicated that up to eight types of closely related deer velvet were used as authentic deer velvet, including *R. tarandus*, *P. albirostris*, *E. davidianus*, *Rucervus eldi*, *R. unicolor*, *D. dama*, *O. hemionus*, *Alces alces* and *C. capreolus*. *R. tarandus* is the predominant adulterant; up to 28% of deer velvet was *R. tarandus* in the herbal markets and drug stores in this study. This result is in agreement with previous studies of deer products and antler powder, which also found that a large portion of deer velvet was from *R. tarandus*. The fraud ratio varies greatly in different studies from different years (95.8% vs. 62% vs. 50.0%) due to different specifications of the samples [100, 107].

References

1. World Health Organization. Traditional Medicine. http://www.who.int/mediacentre/factsheets/fs134/en/. 2008 Dec.
2. Vanherweghem J, Tielemans C, Abramowicz D, et al. Rapidly progressive interstitial renal fibrosis in young women: association with slimming regimen including Chinese herbs. Lancet. 1993;341:387–91.
3. Vanhaelen M, Vanhaelen-Fastre R, But P, et al. Identification of aristolochic acid in Chinese herbs. Lancet. 1994;343:174.
4. Debelle FD, Vanherweghem JL, Nortier JL. Aristolochic acid nephropathy: a worldwide problem. Kidney Int. 2008;74:158–69.
5. Hoang ML, Chen CH, Sidorenko VS, et al. Mutational signature of Aristolochic acid exposure as revealed by whole-exome sequencing. Sci Transl Med. 2013;5:197ra102.
6. Lewis CJ, Alpert S. Letter to health care professionals on FDA concerned about botanical products, including dietary supplements, containing aristolochic acid. US food and drug administration, centre for food safety and applied nutrition, office of nutritional product, labelling and dietary supplements. 2000
7. Gold LS, Slone TH. Aristolochic acid, an herbal carcinogen, sold on the web after FDA alert. New Engl J Med. 2003;349:1576–7.
8. Lo SH, Wong KS, Arlt VM, et al. Detection of Herba Aristolochia Mollissemae in a patient with unexplained nephropathy. Am J Kidney Dis. 2005;45:407–10.
9. Li M, Au KY, Lam H, et al. Identification of Baiying (Herba Solani Lyrati) commodity and its toxic substitute Xungufeng (Herba Aristolochiae Mollissimae) using DNA barcoding and chemical profiling techniques. Food Chem. 2012;135:1653–8.
10. Guo H, Mao H, Pan G, Zhang H, et al. Antagonism of Cortex Periplocae extract induced catecholamines secretion by *Panax notoginseng* saponins in cultured bovine adrenal medullary cells by drug combinations. J Ethnopharmacol. 2013;147:447–55.
11. Chen S, Pang X, Song J, et al. A renaissance in herbal medicine identification: from morphology to DNA. Biotechnol Adv. 2014;32:1237–44.
12. Cheung KS, Kwan HS, But PPH, et al. Pharmacognostical identification of American and oriental ginseng roots by genomic fingerprinting using arbitrarily primed polymerase chain reaction (AP-PCR). J Ethnopharmacol. 1994;42:67–9.
13. Cheng KT, Tsay HS, Chen CF, et al. Determination of the components in a Chinese prescription, Yu-Ping-Feng San, by RAPD analysis. Planta Med. 1998;64:563–5.

14. Huang LQ, YuanY YQJ, et al. Key problems in development of molecular identification in traditional Chinese medicine. China J Chin Mater Med. 2014;39:3663–7.
15. Shaw PC, Jiang RW, Wong KL. Health food and medicine: combined chemical and molecular technologies for authentication and quality control. In: Ebeler SE, Takeoka GR, Winterhalter P, editors. Authentication of food and wine. Washington DC: American Chemical Society; 2007.
16. Shaw PC, Ngan FN, But PPH. Molecular markers in Chinese medicinal materials. In: Shaw PC, Wang J, But PPH, editors. Authentication of Chinese medicinal materials by DNA technology. Singapore: World Scientific; 2002.
17. Williams JG, Kubelik AR, Livak KJ, et al. DNA polymorphisms amplified by arbitrary primers are useful as genetic markers. Nucleic Acids Res. 1990;18:6531–5.
18. Welsh J, McClelland M. Fingerprinting genomes using PCR with arbitrary primers. Nucleic Acids Res. 1990;18:7213–8.
19. Zhang C, Mei Z, Cheng J, et al. Development of SCAR markers based on improved RAPD amplification fragments and molecular cloning for authentication of herbal medicines *Angelica sinensis*, *Angelica acutiloba* and *Levisticum officinale*. Nat Prod Commun. 2015;10:1743–7.
20. Lam K, Chan G, Xin GZ, Xu H, Ku CF, Chen JP, et al. Authentication of Cordyceps sinensis by DNA analyses: comparison of ITS sequence analysis and RAPD-derived molecular markers. Molecules. 2015;20:22454–62.
21. Moon BC, Ji Y, Lee YM, Kang YM, et al. Authentication of *Akebia quinata* D ECNE. From its common adulterant medicinal plant species based on the RAPD-derived SCAR markers and multiplex-PCR. Genes Genom. 2015;37:23–32.
22. Zietkiewicz E, Rafalski A, Labuda D. Genome fingerprinting by simple sequence repeat (SSR)-anchored polymerase chain reaction amplification. Genomics. 1994;20:176–83.
23. Hu J, Vick BA. Target region amplification polymorphism: a novel marker technique for plant genotyping. Plant Mol Biol Rep. 2003;21:289–94.
24. Li G, Quiros CF. Sequence-related amplified polymorphism (SRAP), a new marker system based on a simple PCR reaction: its application to mapping and gene tagging in brassica. Theor Appl Genet. 2001;103:455–61.
25. Collard BC, Mackill DJ. Start codon targeted (SCoT) polymorphism: a simple, novel DNA marker technique for generating gene-targeted markers in plants. Plant Mol Biol Rep. 2009;27:86.
26. Su C, Wong KL, But PPH, et al. Molecular authentication of the Chinese herb Huajuhong and related medicinal material by DNA sequencing and ISSR marker. J Food Drug Anal. 2010;18:161–70.
27. Yang LC, Deng H, Yi Y, et al. Identification of medical Dendrobium herbs by ISSR marker. J Chin Med Mater. 2010;33:1841–4.
28. Wang XM. Inter-simple sequence repeats (ISSR) molecular fingerprinting markers for authenticating the genuine species of *rhubarb*. J Med Plant Res. 2011;5:758–64.
29. Han K, Wang M, Zhang L, et al. Application of molecular methods in the identification of ingredients in Chinese herbal medicines. Molecules. 2018;23:2728.
30. Saki S, Bagheri H, Deljou A, et al. Evaluation of genetic diversity amongst *Descurainia sophia* L. genotypes by inter-simple sequence repeat (ISSR) marker. Physiol Mol Biol Plants. 2016;22:97–105.
31. Liu Y, Zhang P, Zhang R, et al. Analysis on genetic diversity of Radix Astragali by ISSR markers. Adv Biosci Biotechnol. 2016;7:381.
32. Kumar V, Roy BK. Population authentication of the traditional medicinal plant *Cassia tora* L. based on ISSR markers and FTIR analysis. Sci Rep. 2018;8:10714.
33. Vos P, Hogers R, Bleeker M, et al. AFLP: a new technique for DNA fingerprinting. Nucleic Acids Res. 1995;23:4407–14.
34. Ruselll JR, Fuller JD, Macaulay M, et al. Direct comparison of levels of genetic variation among barley accessions detected by RFLPs, AFLPs, SSRs and RAPDs. Theor Appl Genet. 1997;95:714–22.

35. Ha WY, Yau FCF, Shaw PC, et al. Differentiation of Panax ginseng from P. quinquefolius by amplified fragment length polymorphism. In: Shaw PC, Wang J, But PPH, editors. Authentication of Chinese medicinal materials by DNA technology. Singapore: World Scientific; 2002.
36. Ha WY, Shaw PC, Liu J, et al. Authentication of Panax ginseng and Panax quinquefolius using amplified fragment length polymorphism (AFLP) and directed amplification of minisatellite region DNA (DAMD). J Agric Food Chem. 2002;50:1871–5.
37. Choi YE, Ahn CH, Kim BB, et al. Development of species specific AFLP-derived SCAR marker for authentication of *Panax japonicus* C. A. Meyer. Biol Pharm Bull. 2008;31:135–8.
38. Datwyler SL, Weiblen GD. Genetic variation in hemp and marijuana (Cannabis sativa L.) according to amplified fragment length polymorphisms. J Forensic Sci. 2006;51:371–5.
39. Passinho-Soares H, Felix D, Kaplan MA, et al. Authentication of medicinal plant botanical identity by amplified fragmented length polymorphism dominant DNA marker: inferences from the *Plectranthus* genus. Planta Med. 2006;72:929–31.
40. Jiang C, Luo Y, Yuan Y, et al. Conventional octaplex PCR for the simultaneous identification of eight mainstream closely related Dendrobium species. Ind Crop Prod. 2018;112:569–76.
41. Dechbumroong P, Aumnouypol S, Denduangboripant J, et al. DNA barcoding of *Aristolochia* plants and development of species-specific multiplex PCR to aid HPTLC in ascertainment of *Aristolochia* herbal materials. PLoS One. 2018;13:e0202625.
42. Choi SJ, Ramekar RV, Kim YB, et al. Molecular authentication of two medicinal plants *Ligularia fischeri* and *Ligularia stenocephala* using allele-specific PCR (AS-PCR) strategy. Genes Genom. 2017;39:913–20.
43. Noh P, Kim W, Yang S, et al. Authentication of the herbal medicine Angelicae Dahuricae Radix using an ITS sequence-based multiplex SCAR assay. Molecules. 2018;23:2134.
44. Wang J, Ha WY, Ngan FN, et al. Application of sequence characterized amplified region (SCAR) analysis to authenticate Panax species and their adulterants. Planta Med. 2001;67:781–3.
45. Jiang C, Yuan Y, Chen M, et al. Molecular authentication of multi-species honeysuckle tablets. Genet Mol Res. 2013;12:4827–35.
46. Fu RZ, Wang J, Zhang YB, et al. Differentiation of medicinal *Codonopsis* species from adulterants by polymerase chain reaction-restriction fragment length polymorphism. Planta Med. 1999;65:648–50.
47. Li X, Ding X, Chu B, et al. Molecular authentication of *Alisma orientale* by PCR-RFLP and ARMS. Planta Med. 2007;73:67–70.
48. Wang CZ, Li P, Ding J, et al. Simultaneous identification of Bulbus Fritillariae cirrhosae using PCR-RFLP analysis. Phytomedicine. 2007;14:628–32.
49. Wang CZ, Li P, Ding J, et al. Identification of *Fritillaria pallidiflora* using diagnostic PCR and PCR-RFLP based on nuclear ribosomal DNA internal transcribed spacer sequences. Planta Me. 2005;71:384–6.
50. Fondon JW III, Hammock EA, Hannan AJ, et al. Simple sequence repeats: genetic modulators of brain function and behavior. Trends Neurosci. 2008;31:328–34.
51. Jiang C, Yuan Y, Liu GM, et al. EST-SSR identification of *Lonicera japonica* Thunb. Acta Pharm Sin. 2012;47:803–10.
52. Xie M, Hou B, Han L, et al. Development of microsatellites of *Dendrobium officinale* and its application in purity identification of germplasm. Acta Pharm Sin. 2010;45:667–72.
53. Zeng S, Xiao G, Guo J, et al. Development of a EST dataset and characterization of EST-SSRs in a traditional Chinese medicinal plant, Epimedium sagittatum (Sieb. Et Zucc.) Maxim. BMC Genomics. 2010;11:94.
54. Sakthipriya M, Vishnu SS, Sujith S, et al. Analysis of genetic diversity of Centella asiatica using SSR markers. Int J Appl Sci Biotechnol. 2018;6:103–9.
55. Choi HI, Kim NH, Kim JH, et al. Development of reproducible EST-derived SSR markers and assessment of genetic diversity in *Panax ginseng* cultivars and related species. J Ginseng Res. 2011;35:399.

56. Notomi T, Okayama H, Masubuchi H, et al. Loop-mediated isothermal amplification of DNA. Nucleic Acids Res. 2000;28:63e.
57. Sasaki Y, Nagumo S. Rapid identification of Curcuma longa and C. aromatica by LAMP. Biol Pharm Bull. 2007;30:2229–30.
58. Sasaki Y, Komatsu K, Nagumo S. Rapid detection of *Panax ginseng* by loop-mediated isothermal amplification and its application to authentication of ginseng. Biol Pharm Bull. 2008;31:1806–8.
59. Jiang C, Huang LQ, Yuan Y, et al. Rapid extraction of DNA from Chinese medicinal materials by alkaline lysis. Chin J Pharm Anal. 2013;33:1081–90.
60. Zou Y, Mason MG, Wang Y, et al. Nucleic acid purification from plants, animals and microbes in under 30 seconds. PLoS Biol. 2017;15:e2003916.
61. Tian E, Liu Q, Ye H, et al. A DNA barcode-based RPA assay (BAR-RPA) for rapid identification of the dry root of Ficus hirta (Wuzhimaotao). Molecules. 2017;22:2261.
62. Liu Y, Wang XY, Wei XM, et al. Rapid authentication of *Ginkgo biloba* herbal products using the recombinase polymerase amplification assay. Sci Rep. 2018;8:8002.
63. Zhang YB, Wang J, Wang ZT, et al. DNA microarray for identification of the herb of *dendrobium* species from Chinese medicinal formulations. Planta Med. 2003;69:1172–4.
64. Sze SCW, Zhang YBK, Shaw PC, et al. A DNA microarray for differentiation of the Chinese medicinal herb *Dendrobium officinale* (Fengdou Shihu) by its 5S ribosomal DNA intergenic spacer region. Biotechnol Appl Biochem. 2008;49:149–54.
65. Cho Y, Mower JP, Qiu YL, et al. Mitochondrial substitution rates are extraordinarily elevated and variable in a genus of flowering plants. Proc Natl Acad Sci U S A. 2004;101:17741–6.
66. Hebert PDN, Cywinska A, Ball SL, et al. Biological identifications through DNA barcodes. Proc R Soc B Biol Sci. 2003;270:313–21.
67. Li X, Yang Y, Henry RJ, et al. Plant DNA barcoding: from gene to genome. Biol Rev. 2015;90:157–66.
68. Chen S, Pang XH, Song JY, et al. A renaissance in herbal medicine identification: from morphology to DNA. Biotechnol Adv. 2014;32:1237–44.
69. Hebert PDN, Ratnasingham S, de Waard JR. Barcoding animal life: cytochrome c oxidase subunit 1 divergences among closely related species. Proceedings of the Royal Society B. 2003;270:S96–9.
70. Group CPW, Hollingsworth PM, Forrest LL, et al. A DNA barcode for land plants. Proc Natl Acad Sci U S A. 2009;106:12794–7.
71. Group CPB, Li DZ, Gao LM, et al. Comparative analysis of a large dataset indicates that internal transcribed spacer (ITS) should be incorporated into the core barcode for seed plants. Proc Natl Acad Sci U S A. 2011;108:19641–6.
72. Ngan FG, Shaw PC, But PPH, et al. Molecular authentication of *Panax* species. Phytochemistry. 1999;50:787–91.
73. Lau DT, Shaw PC, Wang J, et al. Authentication of medicinal *Dendrobium* species by the internal transcribed spacer of ribosomal DNA. Planta Med. 2001;67:456–60.
74. Chen F, Chan HY, Wong KL, et al. Authentication of *Saussurea lappa*, an endangered medicinal material, by ITS DNA and 5S rRNA sequencing. Planta Med. 2008;74:889–92.
75. Li M, Jiang RW, Hon PM, et al. Authentication of the anti-tumor herb Baihuasheshecao with bioactive marker compounds and molecular sequences. Food Chem. 2010;119:1239–45.
76. Law SKY, Simmons MP, Techen N, et al. Molecular analyses of the Chinese herb Leigongteng (Tripterygium wilfordii Hook.F.). Phytochemistry. 2011;72:21–6.
77. Li M, Ling KH, Lam H, et al. *Cardiocrinum* seeds as a replacement for *Aristolochia* fruits in treating cough. J Ethnopharmacol. 2010;130:429–32.
78. He J, Wong KL, Shaw PC, et al. Identification of the medicinal plants in Aconitum L. by DNA barcoding technique. Planta Med. 2010;76:1622–8.
79. Zhang YB, Jiang RW, Li SL, et al. Chemical and molecular characterization of Hong Dangshen, a unique medicinal material for diarrhea in Hong Kong. J Chin Pharm Sci. 2007;16:202–7.

80. Yu MT, Wong KL, Zong YY, et al. Identification of Swertia mussotii and its adulterant *Swertia* species by 5S rRNA gene spacer. China J Chin Mater Med. 2008;33:502–4.
81. Jiang RW, Hon PM, Xu YT, et al. Isolation and chemotaxonomic significance of tuberostemospironine-type alkaloids from Stemona tuberosa. Phytochemistry. 2006;67:52–7.
82. Chen S, Yao H, Han J, et al. Validation of the ITS2 region as a novel DNA barcode for identifying medicinal plant species. PLoS One. 2010;5:e8613.
83. Alvarez I, Wendel JF. Ribosomal ITS sequences and plant phylogenetic inference. Mol Phylogenet Evol. 2003;29:417–34.
84. Baldwin BG, Sanderson MJ, Porter JM, et al. The ITS region of nuclear ribosomal DNA: a valuable source of evidence on angiosperm phylogeny. Ann Mo Bot Gard. 1995;82:247–77.
85. Kress WJ, Erickson DL. A two-locus global DNA barcode for land plants: the coding rbcL gene complements the non-coding *trnH-psbA* spacer region. PLoS One. 2007;2:e508.
86. Dong XM, Yuan Y, Zha LP, et al. Molecular ID for populations of *Dendrobium officinale* of Yunnan and Anhui Province based on SSR marker. Modern Chin Med. 2017;19:247–77.
87. Zhu FJ, Zhang SS, Yuan Y, et al. Establishment of DNA identity card and analysis of genetic similarity among 58 varieties in *Lonicera japonica*. China J Chin Mater Med. 2018;43:1825–31.
88. Wang CZ, Li P, Ding JY, et al. Discrimination of *Lonicera japonica* T HUNB. from different geographical origins using restriction fragment length polymorphism analysis. Biol Pharm Bull. 30:779–82.
89. Wong KL, Wang J, But PPH, et al. Application of cytochrome b DNA sequences for the authentication of endangered snake species. Forensic Sci Int. 2004;139:49–55.
90. Murray BG, Friesen N, Heslop-Harrison JS. Molecular cytogenetic analysis of Podocarpus and comparison with other gymnosperm species. Ann Bot. 2002;89:483–9.
91. Allsopp RC, Vaziri H, Patterson C, et al. Telomere length predicts replicative capacity of human fibroblasts. Proc Natl Acad Sci U S A. 1992;89:10114–8.
92. Izzo C, Hamer DJ, Bertozzi T, et al. Telomere length and age in pinnipeds: the endangered Australian sea lion as a case study. Mar Mamm Sci. 2011;27:841–51.
93. Kilian A, Stiff C, Kleinhofs A. Barley telomeres shorten during differentiation but grow in callus culture. Proc Natl Acad Sci U S A. 1995;92:9555–9.
94. Liang J, Jiang C, Peng H, et al. Analysis of the age of *Panax ginseng* based on telomere length and telomerase activity[J]. Sci Rep. 2015;5:7985.
95. Riha K, Fajkus J, Siroky J, et al. Developmental control of telomere lengths and telomerase activity in plants. Plant Cell. 1998;10:1691–8.
96. Aronen T, Ryynänen L. Variation in telomeric repeats of scots pine (*Pinus sylvestris* L.). Tree Genet Genomes. 2012;8:267–75.
97. Zentgraf U, Hinderhofer K, Kolb D. Specific association of a small protein with the telomeric DNA-protein complex during the onset of leaf senescence in *Arabidopsis thaliana*. Plant Mol Biol. 2000;42:429–38.
98. Yuan G, Sun J, Li H, et al. Identification of velvet antler by random amplified polymorphism DNA combined with non-gel sieving capillary electrophoresis. Mitochondrial DNA. 2014;27:1–7.
99. Shim YH, Seong RS, Kim DS, et al. Utilization of real-time PCR to detect Rangifer Cornu contamination in Cervi Parvum Cornu. Arch Pharm Res. 2011;34:237–44.
100. Zha D, Xing X, Yang F. Rapid identification of deer products by multiplex PCR assay. Food Chem. 2011;129:1904–8.
101. Lu K, Lo C, Lin J. Identification of Testudinis Carapax and Cervi Cornu in Kuei-Lu-Erh-Hsien-Chiao by nested PCR and DNA sequencing methods. J Food Drug Anal. 2009;17:151–5.
102. Luo J, Yan D, Song JY, et al. A strategy for trade monitoring and substitution of the organs of threatened animals. Sci Rep. 2013;3:3108.
103. Zhang R, Liu C, Huang LQ. Study on the identification of Cornu Cervi Pantotrichum with DNA barcoding. Chin Pharm J. 2011;4:263–6.

104. Fajardo V, González I, López-Calleja I, et al. PCR-RFLP authentication of meats from red deer (*Cervus elaphus*), fallow deer (*Dama dama*), roe deer (*Capreolus capreolus*), cattle (*Bos taurus*), sheep (*Ovis aries*), and goat (*Capra hircus*). J Agric Food Chem. 2006;54:1144–50.
105. Druml B, Grandits S, Mayer W, et al. Authenticity control of game meat products--a single method to detect and quantify adulteration of fallow deer (*Dama dama*), red deer (*Cervus elaphus*) and sika deer (*Cervus nippon*) by real-time PCR. Food Chem. 2015;170:508–17.
106. Bielikova M, Pangallo D, Turna J. Polymerase chain reaction-restriction fragment length polymorphism (PCR-RFLP) as a molecular discrimination tool for raw and heat-treated game and domestic animal meats. J Food Nutr Res. 2010;49:134–9.
107. Jia J, Shi LC, Xu ZC, et al. Identification of antler powder components based on DNA barcoding technology. Acta Pharm Sin. 2015;10:1356–61.

Chapter 3
Seeking New Resource Materials for TCM

Wenyuan Gao, Juan Wang, and Kee-Yoeup Paek

Systematics plays an important role in identifying medicinal plants and searching for new medicinal plant sources. This chapter introduces a theoretical basis and development and research methods for medicinal plant systematics. Research methods of medicinal plant systematics contain morphological taxonomy, anatomy taxonomy, embryo taxonomy, cell taxonomy, chemical taxonomy, numerical taxonomy and branch taxonomy. We also introduce the methods and application of molecular systematics, which can be used to identify and protect medicinal plants, as well as search for new medicinal plants.

3.1 Introduction

Medicinal plant systematics is the science of studying medicinal plant biodiversity and interrelationships. It explores, describes and interprets biodiversity through the integration of different disciplines and finally obtains a predictive natural classification system. Whether it is to ensure the accurate and reliable molecular identification of Chinese medicine or to find and expand sources of medicine, it is inseparable from the in-depth study of the evolutionary history of its source plants, which is a very important but relatively weakly studied topic for molecular pharmacology research [1].

W. Gao (✉) · J. Wang
Tianjin Key Laboratory for Modern Drug Delivery and High Efficiency, School of Pharmaceutical Science and Technology and Key Laboratory of Systems Bioengineering, Ministry of Education, Tianjin University, Tianjin, People's Republic of China
e-mail: pharmgao@tju.edu.cn; drwangjuan@163.com

K.-Y. Paek
Department of Horticultural Science, Chungbuk National University, Cheongju, Republic of Korea
e-mail: cbnbio@hotmail.com

L.-q. Huang (ed.), *Molecular Pharmacognosy*,
https://doi.org/10.1007/978-981-32-9034-1_3

The development and utilisation of medicinal plant resources have attracted the attention of various countries. In China, a country with rich medicinal plant resources and a long history of applying Chinese herbal medicines, the subject has naturally received more attention. Through long-term practice, 'medicinal plant kinship', based on medicinal plant systematics, has also emerged. Its main task is to explore the inherent relationships among the genetics, chemical compositions and efficacies of medical plants and use this knowledge in practice.

3.2 Theoretical Basis of Searching for New Medicinal Resources

3.2.1 Concept of New Medicinal Resources

The total number of crude medicinal resources (plants, animals and minerals) has not been calculated, though crude medicinal materials have been identified as having biological activity with modern technology. Medicinal plants occupy the most important role in the development of TCM. New medicinal plant resources could be found in the wild, as cultivated species, artificially mutated individuals or cultivars of tissue culture or through rapid propagation.

3.2.2 The Concept of Pharmaphylogeny

During early evolution on Earth, plants gradually formed relative relationships. The species with close relationship show not only similar morphologies but also similar physiological and biochemical characteristics. Therefore, the chemical contents, such as those of secondary metabolites, are often similar. Closely related species also have similar biological activities and potencies. Similarly, species that have the same effects possess close systematic relationships. Propositions of pharmaphylogeny provide theoretical guidance and direction for medicinal plant systematics.

Pharmaphylogeny as a discipline is interdisciplinary and marginal and studies the relationships, chemical constituents and efficacy of medicinal plants. Its topics range from plant taxonomy, plant phylogenetics, phytochemistry, pharmacology to numerical taxonomy, genomics and informatics [2].

3.2.3 The Background of Pharmaphylogeny

Traditional taxonomy takes the morphology and shape of plants as the basis, which has limitations and artificiality. There are some complex issues in the field of taxonomy, such as phylogeny. Traditional taxonomy could not analyse such a phenomenon. The development of science affords the opportunity for the development of chromosome classification, DNA molecular hybridisation and microscopic classification technologies for the improvement of morphological classification [3]. All of these classifications supplement the traditional method. The evolution of pharmaphylogeny can be described as having two stages:

1. In the 1970s, plant chemotaxonomy played an important role in exploring the distribution of the chemical constituents of medicinal plants, which provided a basis for molecular classification. Plant chemotaxonomy also revealed the laws of plant phylogeny at the molecular level. Plant chemotaxonomy, which studies the evolution of medicinal plants, is based on their chemical composition and traditional morphological taxonomy. Distributions of the chemical constituents of many species of plants have been studied worldwide. Previous studies only focused on the distribution of certain chemical compounds. Therefore, there were no systematic studies, and no theories were formed. On the other hand, the combination of pharmacological effects and efficacies were systematically implemented.
2. From the late 1970s, P.G. Xiao successively conducted a comprehensive study on belladonna, *Berberis*, rhubarb, *Aconitum*, *Thalictrum*, *Lithospermum*, puhuan and azaleas using phytochemistry, pharmacology, plant phylogenetics, numerical taxonomy and computer technology. From these studies, the relationship among pharmaphylogeny, chemical composition and efficacy (biological activity and traditional effects) was revealed, and this theory could be put into practice. This theory made great strides in expanding the resources of medicinal plants, discovering new medicinal resources, finding substitutes for imported drugs and providing guidance for the basic research of Chinese herbs.

3.3 Overview of the Phylogeny of Medicinal Plants

The context of evolutionary biology is phylogeny; it provides connections between all groups of organisms as understood by ancestor/descendant relationships. Not only is phylogeny important for understanding palaeontology, but palaeontology in turn contributes to phylogeny. Many groups of organisms are now extinct, and without their fossils we would not have as clear a picture of how modern life is interrelated. Systematics is the study of the diversification of life on Earth, both in the past and at present, and the relationships among living organisms through time. Taxonomy is the science of naming and classifying the diversity of organisms, and it is a major part of systematics that includes four components: description, identification, nomenclature and classification. Plant systematics is an introduction to the morphology, evolution and classification of land plants. It contains two parts: studies on the classification system and studies on the methods of constructing the classification system. The molecular systematics of medicinal plants is an important part of the molecular systematics of plants. The difference between the two is that the objective of molecular systematics of medicinal plants is the abundant group of molecular systematics of plants, and the classification grade is mainly dominated by species and genera [4].

3.3.1 Development of Medicinal Plant Systematics

Studies on medicinal plant systematics go through three stages: the artificial system stage, the pre-evolutionary natural system stage and the phylogenetic system stage [5].

3.3.1.1 Artificial System Stage

The 'Book of Odes' in 700 B.C. was the first book to clearly record plant species. It included 160 classes of plants, of which 10 corresponded to genera and 112 corresponded to particular species. At the same time, the father of botany, Theophrastus, published 'Historian Plantarum', which included 480 plants. This book referred to 'arbor', 'shrub' and 'herb'; herbs were divided into annual plants, biennial plants and perennial plants.

'Shen Nong's Herbal', the first-documented monograph in existence, recorded 365 drugs, which were divided into three classes: the superior, the middle and the inferior. 'Compendium of Materia Medica', written by Li Shizhen, is the most important book in the history of medicinal plant study; it recorded 1892 plants, of which 1195 were plant medicines, divided into five groups: straw, corn, vegetable, fruit and wood.

The advent of microscopy brought plant systematics into a new stage of development; after this, anatomy, embryology, palynology, etc. formed one after another. Plant systematics became a special science, resulting in many artificial systems. The 'sexual system', proposed by Carl Linnaeus, also called the 'artificial system', is the most representative. In this system, plants were divided into 24 classes, mainly according to the presence or absence, amount and consolidation of their stamens. The opinion at that time was that when earth was first separated from heaven, the almighty God created a certain number of species, and that was how many species existed. Therefore, this system had no idea of species evolution.

3.3.1.2 Natural System Stage

After the second half of the nineteenth century, people gathered increasing amounts of knowledge about plants, and many scholars made efforts to search for the genetic relationships and development laws of plant in nature; thus, the natural system stage was established. These systems, such as the Bentham and Hooker system, Jussieu system and de Candolle system, take morphologic characteristics as a basis for categorisation and do not reflect the affinity and evolutionary relationships. Hence, this stage is called the natural system stage.

3.3.1.3 Phylogenetic System Stage

After the theory of biological evolution was proposed by Darwin in 'On the Origin of Species', taxonomists reappraised the established system and tried to establish phylogenetic classification systems.

The first system of this kind was published in 'Species and Medicine Botany Seminar' by W. Eichler, which constituted the basis for the Engler classification system. In his system, gymnosperm was placed in front of angiosperm, and the monocotyledon was arranged before the dicotyledon. The Engler classification system was first published in 'Die Naturichen Ptlanzen-familien' by A. Engler and Prantl in 1897, and it was the first relatively complete natural classification system in the history of plant classification. The Hutchinson classification system is an angiosperm classification system raised by J. Hutchinson in 'The Family of Flowering Plant' in 1926 and was further revised in 1959 and 1973, respectively. Soviet scholar A. Takhtajian published the Takhtajian system in 1954, which first broke through conventional concepts and established the 'superorder' classification system. American scholar A. Cronquist published the Cronquist system in 1958 and revised this system in 1981.

With the development of molecular biology, plant molecular systematics was born. The angiosperm phylogeny group (APG) I and II systems are the most influential molecular systems at the present time.

3.3.2 Research Methods in Medicinal Plant Systematics

The earliest research methods of plant systematics took the morphological characteristics of plants, especially those of flowers and fruits, as classification evidence, and on the basis of this, botanists established many categorisation systems. However, fossilised flowers are difficult to find, and so this system is difficult to establish. With the progression of science and inter-scientific penetration, plant systematics has rapidly developed during the past decades, resulting in many new research methods. The application of these methods in plant systematics has enabled plant classification systems to become more rational and closer to the objective reality.

3.3.2.1 Morphological Taxonomy

Morphological taxonomy uses the characteristics of herbarium specimens, such as the leaves, flowers, fruit, branches, roots and other external morphological characteristics, to classify and name plant species. In addition to museum specimen work, morphological taxonomy still needs to investigate, collect and record specimens and their actual degrees of variation in nature. This approach is characterised by simple manipulation equipment and practicability in application. However, it has the following shortcomings: (1) informative taxonomy with a long history requires taxonomists to memorise past articles, (2) it is easy to form understandings that break away from ecology and heredity and to propose new species that does not exist at all in nature and (3) there is too much emphasis on morphological characteristics.

In using morphology in taxonomy, it is very important to know the range of variability in a particular character under different conditions. Morphometric, allometric and morphological characters tend to vary under the influence of geographical and ecological conditions, which results in ecotypic and host-specific populations. Such populations are sometimes described as new species, although the differences between them are intra-specific variations.

3.3.2.2 Anatomy Taxonomy

Anatomy taxonomy is a method that uses optical microscopy to observe the construction of plants and provide the basis for plant classification. For example, some characteristics of Rutaceae, such as morphology, located position and density, are helpful for the analysis of the relationships between the subfamily and genus.

3.3.2.3 Embryo Taxonomy

Embryo taxonomy mainly focuses on the ovule shape and male and female gametophytes. The collection of experimental material is difficult, and the results have a high similarity in a low number of taxa. Therefore, this method is limited in the phylogenetic study of medicinal plants.

3.3.2.4 Cell Taxonomy

Chromosome data play a special role in taxonomy, and chromosomes often vary to some degree at the intra-genus, intra-species and intra-specific levels; this provides an important basis for investigating evolutionary relationships between genus, species and intra-specific variation patterns. The study of cell taxonomy comprises chromosome numbers, karyotype analysis, the analysis of reduction division, banding technique in chromosomes, in situ hybridisation of genome, etc., and these data can be applied to investigate the relationships between inter-genus, inter-species and intra-specific variation and to reveal the origin of species. The live experimental material of cell taxonomy is difficult to collect, so its application is limited.

3.3.2.5 Chemical Taxonomy

The morphological characteristics of plants are controlled by genes, which are the result of long-term natural selection. Thus, it has important reference value to use chemical methods to study the individual and systematic development of plants. Currently, chemical taxonomy mainly uses plant chemistry and biochemistry to study some compounds with finite distribution and their biosynthetic pathways and provide a more objective basis for measuring the similarities and relationships among plant taxa. Chemical taxonomy is divided into serum taxonomy and organic chemical taxonomy. Chemical taxonomy could solve classification problems from the subspecies to the present order. Species that have differences in their chemical constituents cannot be discovered or valued by classical taxonomy but are a concern of medicinal botanists. The differences between chemical constituents are not only material and the basis for drafting medicine production divisions and medicinal plant breeding programs but also one of the reasons for the formation of 'Dao-di herbs'.

3.3.2.6 Numerical Taxonomy

Numerical taxonomy is a classification system in biological systematics that deals with the grouping of taxonomic units by numerical methods based on their character states. It aims to create a taxonomic system using numeric algorithms, such as cluster analysis. The concept was first developed by Robert R. Sokal and Peter H.A. Sneath in 1963 and was later elaborated by the same authors. Phenetics is a closely related discipline and draws heavily from the methods of numerical taxonomy. Although intended as an objective classification method, in practice, the choice and weighing of morphological characteristics is often guided by available methods and research interests. Furthermore, the general consensus has become that taxonomic classification should reflect evolutionary (phylogenetic) processes. Some connections between phylogenetic trees and the spectral decomposition of the variance–covariance matrix of quantitative traits subject to Brownian motion over time have been established, providing a theoretical link between phylogenetic methods and numerical taxonomy. The specific phenetic algorithms proposed in numerical taxonomy, however, often fail to properly reconstruct the phylogenetic history of organisms. Numerical taxonomy remains useful in cases where biological species concepts cannot be applied, e.g. clonal evolution, as seen in apomictic micro-species like blackberry.

3.3.2.7 Branch Taxonomy

Branch taxonomy, also called cladistics, was created by the German entomologist W. Henning. Cladistics is a method of classifying species of organisms into groups called clades, which consist of all the descendants of an ancestral organism and the ancestor itself. Cladistics can be distinguished from other taxonomic systems, such as phenetics, by its focus on shared derived characters (synapomorphies). Previous systems usually employed overall morphological similarities to group species into genera, families and other higher level classifications; cladistic classifications (usually trees called cladograms) are intended to reflect the relative recency of common ancestry or the sharing of homologous features. Cladistics is also distinguished by its emphasis on parsimony and hypothesis testing (particularly falsificationism) rather than the subjective decisions that some other taxonomic systems rely upon.

3.3.2.8 Molecular Systematics

The aim of molecular systematics is to use the structure of molecules to gain information on an organism's evolutionary relationships. The result of a molecular phylogenetic analysis is expressed in a phylogenetic tree.

In short, various types of evidence can be used to establish plant classification systems; the research methods, of which the morphological method is the basis, are evolving. Many methods or techniques, such as plant anatomy, embryology, cytology, chemistry, numerical taxonomy, cladistics and molecular biology, have their particular uses. These methods have their own characteristics, applications and limitations. Regarding practical use, several methods are combined, and the most natural/effective relationship and evolution analysis of medicinal plants could be obtained by comparing the results of various methods.

3.4 Molecular Systematics of Medicinal Plants

Medicinal plant molecular systematics is an interdisciplinary science developed by the interaction between molecular biology and medicinal plant systematics. It uses various experimental methods of molecular biology to obtain a variety of molecular characters for discussing the classification of medicinal plants, phylogenetic relationships between groups, and evolutionary processes and mechanisms. Except for DNA sequences, molecular characters include structure features of the genome, characters of proteins, DNA fingerprint characteristics, etc. In practice, according to the different taxa, we can select different molecular characters to obtain the best and the largest amount of information on the phylogenetic relationships of specific groups. As the genetic material of plants, DNA is stable, reliable and free from outside influence, and thus it can be used in the study of the phylogenetic relationships of medicinal plants. Plant molecular systematics is actually cladistics using DNA sequence data.

Methods commonly used in plant molecular systematics include RFLP, SSR, SNP and DNA barcoding method. At present, the most commonly used method is sequence analysis.

3.4.1 *Research Method of Molecular Systematics of Medicinal Plants*

We know that life is a complex system going through a lengthy evolution and succession. In the process of biological evolution, the evolution of macromolecules is collectively called molecular evolution.

Nuclear DNA (nDNA), chloroplast DNA (cpDNA) and mitochondrial DNA (mtDNA) exist in plant cells. The evolutionary rates of plants are different in regard to their genome structure and function. In general, nDNA evolve the quickest, about twice as fast as cpDNA; mtDNA evolves the slowest, one third as fast as cpDNA. Generally, due to less restrictions in the sequence, non-coding regions show faster evolutionary rates than coding regions; the variations and rates are vastly different among the coding or non-coding sequences within the same genome. These differences in evolutionary rates among sequences provide alternative and diverse sources of characters for the phylogenetic study of different taxa. It is now clear that the selection of the appropriate molecular fragments relative to a particular problem in systematics is the most fundamental and crucial step in molecular systematic studies [6].

When selecting a sequence for phylogenetic analysis, we should consider the following points: (1) the sequence should be long enough to provide sufficient nucleotide sites with phylogenetic information, and the difference percentage of the selected sequence should be suitable to solve the system problem; (2) the sequence must be easy to sort, which is important for the correct assessment of character homology; and (3) the sequence must be orthologous.

3.4.1.1 nDNA

The current study of nDNA sequences focuses on the nrDNA encoding rRNA. Ribosomes are made from complexes of RNAs and proteins. Ribosomes are divided into two subunits, one larger than the other. The smaller subunit binds to the mRNA, while the larger subunit binds to the tRNA and the amino acids. Genes encoding rDNA in the plant nucleus are a multi-gene family composed of highly repetitive sequences. The 18S encodes small subunit rDNA in the ribosome, 5.8S and 26S comprise a transcription unit, and ITSs between 18S and 5.8S and 5.8S and 26S are called ITS1 and ITS2, respectively.

1. 18S rRNA: the 18S rRNA in most eukaryotes is in the small ribosomal subunit. The 18S rRNA sequence can provide more information for the phylogenetic study of higher level classification within the angiosperms. Its sequence variation is suitable for discussing the relationship between deep phylogenetic branches within angiosperm and seed plants. Due to the differences in the variation of sequences in different groups, 18S can also be used for the re-establishment of relationships between subfamily and species.

2. Internal transcribed spacer: the ITS between 18S and 26S nrDNA is divided into two parts (ITS1 and ITS2) by the 5.8S rRNA gene.

The variations in ITS regions in gymnosperms are complex, and only the length variation in ITS1 region may vary by several kilobases. Thus, ITS sequence analysis is one of the most widely used molecular sequences for the molecular systematic study of gymnosperms. In contrast, in angiosperms, the ITS region has both a high degree of nucleotide sequence variation and high conservation in length, which indicates that these sequences in the septal area are easy to sort in sibling groups. Rich variation in the lower taxa can solve the problem of plant phylogeny.

Each transcription unit contains an external transcribed spacer (ETS), 18S, 5.8S and 26S. Non-transcribed spacer (NTS) regions exist between genes.

3.4.1.2 cpDNA

cpDNA in plants is closed double-stranded DNA, accounting for about 10%–20% of the total DNA. The length of cpDNA is usually 120–220 kb, and its length variation is mainly caused by two inverted repeats (IR). The length of the two inverted repeats is 22–25 kb, and it divides the total cpDNA into a large single copy region and a small single copy region. cpDNA is exposed and does not form complexes with histones, similarly to *E. coli* DNA.

3.4.1.3 mtDNA

mtDNA is the DNA located in organelles called mitochondria, structures within eukaryotic cells that convert the energy from food into a form that cells can use. In humans (and probably metazoans in general), 100–10,000 separate copies of mtDNA are usually present per cell (egg and sperm cells are exceptions). In mammals, each double-stranded circular mtDNA molecule consists of 15,000–17,000 bp.

The low effective population size and rapid mutation rate (in animals) makes mtDNA useful for assessing the genetic relationships of individuals or groups within a species and also for identifying and quantifying the phylogeny (evolutionary relationships; see phylogenetics) among different species, provided they are not too distantly related. To do this, biologists determine and then compare the mtDNA sequences in different individuals or species. Data from the comparisons are used to construct a network of relationships among the sequences, which provides an estimate of the relationships among the individuals or species from which the mtDNA are taken. This approach has limits that are imposed by the rate of mtDNA sequence change.

3.4.1.4 Sequence Homology Alignment

Sequence alignment refers to the alignment of two or more nucleic acid or amino acid sequences using certain algorithms; column-by-column base similarities and differences; judging the degree of similarity and homology between them; and presuming their structure, function and evolution. Sequence alignment is not only a commonly used sequence analysis method for database searches, gene comparison and new gene discovery but also used as secondary biological data for the establishment of phylogenetic trees, protein structure and function predictions, and treatment of genetic diseases. Many biological studies, such as new drug designs, provide valuable information. Subsystems treat each base site as a trait, and the base composition at that position is regarded as a trait state. The similarity of each base site of these sequences is called the positional homology. Homologous sequence alignment is one of the most fundamental and important steps in phylogenetic analysis.

3.4.1.5 Molecular Evolution Modelling

The easiest way to compare the differences between two sequences is to compare the differences in their nucleotide composition. For example, if there are 10 loci differences on two nucleotide sequences of 100 bp in length, the genetic difference distance is $P = 10\% = 0.1$. Nucleotides are composed of only four bases, A, T, C and G, so there are only three possibilities when base substitution occurs at a certain site. When the rate of substitution of a sequence is small, the probability of one or more substitutions at any one site is negligible and the number of differences observed between the two sequences will be close to the actual number of substitutions. If a site replacement rate is high, there is a high probability of multiple substitutions or 'multiple hits', and these mutations will mask the previous mutation information. What we see is only the state of the last mutation. This means that if multiple mutations occur at the same site, we will lose the information on the previous mutations and the observed number of differences will be less than the actual number of substitutions. For example, a nucleotide at a locus first changes from A to C to T in a sequence and in another sequence changes from A to T. Although three substitutions have taken place, the two sequences are identical at that site, so it is difficult to estimate how many changes have actually occurred for a site replacement.

3.4.2 Application of Medicinal Plant Molecular Systematics

Under the influence of today's trend of 'back to nature', the development and application of medicinal plant resources have aroused widespread interest, so pharmaphylogeny has developed grounded in medicinal plant systematics. Pharmaphylogeny focuses on the phylogenetic correlation between the phylogeny, chemical constituents and pharmaceutical effects of medicinal plants. Like many other contemporary fields of science, pharmaphylogeny represents a highly interdisciplinary science that is one of the areas of pharmaceutical education. Its scope includes the studies of phylogeny, botanical taxonomy and the chemical, biochemical and biological properties of natural medicinal plants, as well as the search for new drugs from natural sources.

Studies on rhubarb are typical in pharmaphylogeny. Rhubarb is a commonly used Chinese herbal medicine, but its quality in some areas is unstable. A study on the pharmaphylogeny of seven groups with 27 species (44 samples) of domestic rhubarb showed that authentic rhubarb with obviously genuine cathartic effects had leaves that were divided into palmate sections. They contained sennoside and free rhein but no rhaponticin; as to morphology, the leaves have various degrees of palmate division. The results of numerical taxonomic and multivariate analysis indicated that the division of leaves was closely related to sennoside and cathartic contents, as well as purgative effect, which thus provided reliable scientific evidence for the quality control and production and utilisation of rhubarb resources. This proves the importance of medicinal plant systematics.

At present, as an important part of medicinal plant systematics, the study of medicinal plant molecular systematics has become one of the hotspots of pharmacognosy. Its application mainly focuses on the following two aspects.

3.4.2.1 Identification and Classification of Medicinal Plants

When studying each discipline of medicinal plants, one must ensure the correct identification and classification of experimental materials. Without correct identification, the experimental results would be difficult to verify. In practice, taxonomists often apply the concept of morphology based on visible characteristics, which have strong direct observable effects, and can meet the needs of multi-purpose classification. However, situations that can only be understood and not explained inevitably appear. Specific groups allow more in-depth research; comparative anatomy, palynology, cytology and other means can be used to define species more clearly; however, for experts in non-species groups, it is still difficult to identify plant species. The rise of molecular systematics makes it possible to change from making classifications according to morphological features to making them according to DNA sequences; DNA barcoding is a useful method for this [7, 8].

DNA barcoding is a taxonomic method that uses a short genetic marker in an organism's DNA to identify it as being a particular species. It differs from molecular phylogeny because the main goal of it is not to determine the classification but to identify an unknown sample in terms of a known classification. Although barcodes are sometimes used in an effort to identify unknown species or assess whether species should be combined or separated, such usage, if possible at all, pushes the limits of what barcodes are capable of. Applications include, for example, identifying plant leaves even when flowers or fruit are not available, identifying the diet of an animal based on stomach contents or faeces and identifying ingredients in commercial products (for example herbal supplements or wood).

3.4.2.2 Search for New Medicinal Plants

In the study of active constituents in medicinal plants, especially by family and genus, certain types of chemical components are found to be distributed in certain families and genus. According to this discipline, we can search for new medicinal plants. For example, most gentian plants contain bitter glycosides; *Ephedra* branch, Ranunculaceae, Papaveraceae, Menispermaceae, Rubiaceae, Loganiaceae, *Lilium* and others mainly contain alkaloids; Labiatae, Rutaceae, Umbelliferae, Lauraceae and others contain different types of volatile oil; and Polygonaceae mainly contain anthraquinone glycosides. Colchicine (used to treat tumours) was initially found in foreign species of *Colchicum* (Liliaceae), and in recent years it was also found in *Urginea indica* Kunth. in the same family in China. Plants in the same genus may contain the same types of chemical components; for example, *Ephedra* spp. contain ephedrine, and *Strychnos* spp. contain strychnine and brucine. These results indicate that the genetic relationships between medicinal plants provide clues when searching for new medicine sources. Molecular systematics can make these clues more accurate. According to the phylogenetic correlation between the phylogeny, chemical constituents and pharmaceutical effects of medicinal plants, searching for new medicinal resources by analysing the relationships between species and genera is the main application of studying medicinal plant molecular systematics [9, 10].

3.4.2.3 Protection of Medicinal Plants

The call for ‘return to nature’ around the world is growing, and the demand for botanicals is gradually increasing. Wild medicinal plant resources are rapidly declining due to environmental changes, habitat loss and uncontrolled excavation for the purpose of acquiring resources. Among the 398 endangered plants listed on the ‘Chinese Plant Red List’, there are as many as 168 medicinal plants, accounting for more than 42% of the total. The protection of medicinal plant biodiversity has become a hot topic in the international community. Biodiversity conservation is a complex project. Under limited financial resources, we need to know which populations should be protected first and the ways to protect them more effectively. This requires data on the status of the species population and evolutionary process. The differentiation and expansion of intra-species groups is essentially the same process as the phylogenetic evolution of species. If we can reveal the phylogenetic history between groups, we will be able to evaluate the evolutionary status of each group and thus identify the key units that need protection. In addition, by tracing the dynamic structure of the population and the process of genetic evolution, the genetic differentiation of wild populations with different geographical distributions can be effectively revealed; from this, the causes of endangerment can be elucidated so that we can more effectively protect, manage and predict the possible outcomes of conservation. Avise Slatkin and Moritz et al. explain the process of species differentiation and expansion from the perspective of population genetics and origin geography. Moritz emphasises the introduction of molecular systematics into the direction of biological protection. This would allow molecular systematics to identify protection units and better understand the historical process of the group, which would promote the development of conservation biology [11, 12].

References

1. Zhang N, Zeng L, Shan H, et al. Highly conserved low-copy nuclear genes as effective markers for phylogenetic markers in angiosperms. New Phytol. 2012;195(4):923–37.
2. Tonnabel J, Olivier I, Mignot A, et al. Developing nuclear DNA phylogenetic markers in the angiosperm genus Leucadendron (Proteaceae): A next-generation sequencing transcriptomic approach. Mol Phylogenet Evol. 2014;70:37–46.
3. Pridgeon AM, Cribb PJ, Chase MW, et al. Genera orchidacearum, Epidendroidea (Part 3), vol. 6. Oxford: Oxford University Press; 2014.
4. zong Y, Kang H t, Fang Q, et al. Phylogenetic relationship and genetic background of blueberry (Vaccinium spp.) based on retrotransposon-based SSAP molecular markers. Sci Hortic. 2019;247:116–22.
5. Schreiber M, Himmelbach A, Börner A, et al. Genetic diversity and relationship between domesticated rye and its wild relatives as revealed through genotyping-by-sequencing. Genomics Domest. 2019;12(1):66–77.
6. Fernanda A-S, Christian S, Castilho Carolina L, et al. Molecular identification of shark meat from local markets in Southern Brazil based on DNA barcoding: evidence for mislabeling and trade of endangered species. Front Genet. 2018;27(9):138.
7. Jiao L, Min Y, Wiedenhoeft AC, et al. DNA barcode authentication and library development for the wood of six commercial Pterocarpus species: the critical role of Xylarium specimens. Sci Rep. 2018;8:1945.
8. Banchi E, Ametrano CG, Stanković D, et al. DNA metabarcoding uncovers fungal diversity of mixed airborne samples in Italy. PLoS One. 2018;13(3):e0194489.
9. Raclariu AC, Heinrich M, Ichim MC, et al. Benefits and limitations of DNA barcoding and metabarcoding in herbal product authentication. Phytochem Anal. 2018;29:123–8.
10. De YW, Wang Q, Wang YL, et al. Evaluation of DNA barcodes in Codonopsis (Campanulaceae) and in some large angiosperm plant genera. PLoS One. 2017;12(2):e0170286.
11. Yuliya G, Saule A, Aibatsha Z, et al. Morphological description and DNA barcoding study of sand rice (Agriophyllum squarrosum, Chenopodiaceae) collected in Kazakhstan. BMC Plant Biol. 2017;17(S1):177.
12. Liu ZF, Ci XQ, Li L, et al. DNA barcoding evaluation and implications for phylogenetic relationships in Lauraceae from China. PLoS One. 2017;12(4):e0175788.

Chapter 4
Phylogeography of Medicinal Plant

Dan Jiang, Xiao-Lei Jin, Jia-Hui Sun, Qing-Jun Yuan, Yu-Chung Chiang, and Zhi-Yong Zhang

Abstract This chapter introduces the background, the basic theories, research methods in phylogeography and their application in biological evolution research, including the definition and formation of species, the evolutionary history of species, the identification of priority protected areas, and the origin of domestic animals and cultivated plants. In addition, the application of phylogeography in the evolution of medicinal plants is described, including research on the genetic basis of medicinal plants and the origin of cultivation. Case studies were also carried out.

4.1 Background of Phylogeography

Phylogeography was established in the early 1980s and emphasizes animal mtDNA [1, 2]. This is a field of study involving the principles and processes for the geographic distributions of genealogical lineages and those within and among closely related species [3, 4]. The analysis and interpretation of lineage distributions usually requires extensive input from molecular genetics, population genetics,

---

Dan Jiang, Xiao-Lei Jin and Jia-Hui Sun are equally contributed for this chapter.

D. Jiang
School of Chinese Materia Medica, Beijing University of Chinese Medicine, Beijing, China
e-mail: jiangdan1027@163.com

X.-L. Jin · Y.-C. Chiang
Department of Biological Sciences, National Sun Yat-sen University, Kaohsiung, Taiwan
e-mail: xiaolei@mail.nsysu.edu.tw; yuchung@mail.nsysu.edu.tw

J.-H. Sun · Q.-J. Yuan (✉)
National Resource Center for Chinese Materia Medica, China Academy of Chinese Medical Sciences, Beijing, China
e-mail: sunjh_2010@sina.com; yuanqingjun@icmm.ac.cn

Z.-Y. Zhang
School of Agricultural Sciences, Jiangxi Agricultural University, Nanchang, Jiangxi, China
e-mail: pinus-rubus@163.com

L.-q. Huang (ed.), *Molecular Pharmacognosy*,
https://doi.org/10.1007/978-981-32-9034-1_4

ethology, demography, phylogenetic biology, paleontology, geology, and historical geography. It pioneered the introduction of the macroevolution concept of phylogeny into the microevolution research, considering both the spatial distribution of the pedigree and the phylogenetic relationship (representing time) between the pedigrees, completely changing the classical population genetics practices of artificially defined populations as operational taxonomic units (OTUs), instead using naturally occurring individuals or alleles as operating units. Therefore, phylogeography is the combination of macroevolution and microevolution and has become one of the most important research fields in modern evolutionary biology [5].

Phylogeography is a branch of biogeography but differs from traditional biogeography in that the latter tends to focus only on the geographical distribution of higher taxa, such as marsupials or cycads. In contrast, phylogeography includes the study not only of high-order taxa but also of related species (emphasis on the history of their common origin) and intraspecies populations. Therefore, the study of phylogeography is a bridge between population genetics or microevolution and speciation or macroevolution.

Phylogeography has in recent years been gradually valued by the bioscience community, and it is expected to become one of the development trends of biogeography in the future. In 1998, Molecular Ecology, an internationally renowned publication, invited famous animal and plant evolution scholars such as Smith H. of the United Kingdom, Avise JC of the United States, Templeton AR, and Schaal BA to write essays on the principles and methods of phylogeography, and the papers, titled "Phylogeography" y different publications, have gradually become mainstream. In 1999, the 16th International Botanical Congress held in St. Louis, USA, inheriting the research and achievements of the phylogeny, added a group of phylogeography, and its group chairman, Michal Crunzan (Kentucky University), emphasized the importance and advancement of phylogeography research.

International research on phylogeography has become a hotspot and highlight of population genetics research. Chinese research has also begun to focus on evolution history, the protection of rare and endangered plants, and the origin of cultivated crops. For example, Ge et al. investigated the genetic structure and evolutionary history of the endangered plant *Dunnia sinensis* [6] and the wild banana (*Musa balbisiana*) [7], and proposed corresponding protective measures first applying the principles and methods of phylogeography. According to Wang et al., phylogeography analysis of the single relictual species *Cathaya argyrophylla* was carried out to elucidate its endangered mechanism [8]; Gao et al. studied the relationship between the phylogeography pattern of *Taxus wallichiana* and that of the flora of East Asia [9]. Yuan et al. revealed that the uplift of the Qinghai-Tibet Plateau leads to highly differentiated phylogeography pattern among populations of the vulnerable plant *Dipentodon sinicus*, was the result of the uplift of the Qinghai-Tibet Plateau [10]. Tian et al. analyzed the migration route and refuge of *Pinus kwangtungensis* in the glacial period [11]. Xu et al. constructed the phylogeography relationship in the middle of wild and cultivated *Zizania latifolia*, and explored the origin of cultivation [12]. However, the application of phylogeography in the evolution of medicinal plants has rarely been reported.

4.2 The Basic Theories and Methods of Phylogeography

4.2.1 The Basic Theories of Phylogeography

Traditional population genetics takes population as the basic unit of research. The variables analyzed are usually the gene frequency and genotype frequency of the population. The research perspective is forward-looking, that is, predicting the genetic and genotype frequencies of the population when they reach an equilibrium under the influence of a genetic drift. Phylogeography focuses on the genealogical relationship of genes, and the basic unit of study is the individual or allele (i.e., haplotype). The research perspective is retrospective, and the historical inheritance relationship between genes (haplotypes) is established. The historical inheritance of the gene lineage and the historical events that are pushed back to the common ancestor are the concerns of the coalescent theory. Therefore, the coalescent theory contributes to a significant theoretical frame for phylogeography.

The original idea of the coalescent theory was "common ancestor" and "identity by descent", which were first proposed by S. Wright and G. Molécot in the 1940s. In the following years, two theoretical population geneticists, F. Tajima and R. R. Husdon, made significant contributions. In the early 1980s, J. F. C. Kingman wrote two papers on the coalescent, stochastic processes and their application and the genealogy of large populations, which marked the birth of the theory of ancestor.

The basic concept of the coalescent theory is not complicated, which involves the following: a limited population, a genetic drift, every generation that had been replaced, a gene in the gene bank disappearing randomly, and the genes in the gene bank continuing to accumulate and produce new genes. As the number of generations increases, only one gene and its derived genes remain in the population. All genes in the current population have a most recent common ancestor (MRCA), if traced back from the offspring, which is a coalescent event [13] (Fig. 4.1). The rate of gene coalescent event depends on sampling strategy, population size, and gene mutation rate; it is affected by various evolutionary forces, such as genetic drift, natural selection, and gene flow.

4.2.2 The Study Methods of Phylogeography

4.2.2.1 Descriptive Phylogeography

The first study of molecular phylogeography is driven by PCR and first-generation DNA sequencing technology. In this wave, researchers typically sequence individual gene fragments (animals are usually mitochondrial DNA, plants are usually chloroplast DNA) from a population of different species and then qualitatively describe the obtained haplotype. The main description objects include the following:

(a) Phylogenetic relationships and phylogeography relationships of haplotypes: the former appears as a phylogenetic tree, while the latter appears as a network map.

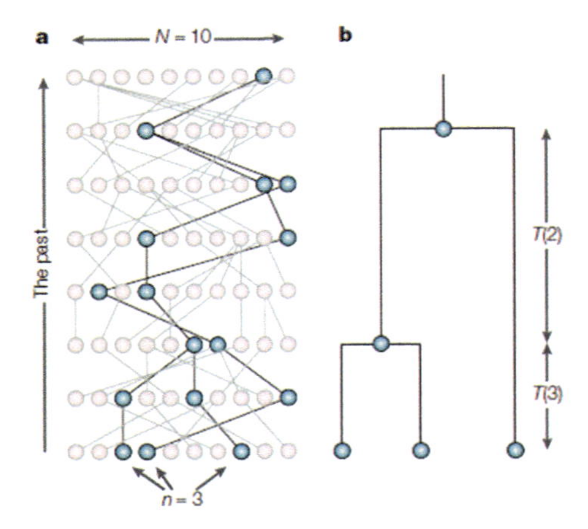

Fig. 4.1 The basic principle of the theory of ancestor. (**a**) The complete genealogy for a population of ten haploid individuals is shown (diploid populations of *N* individuals are typically studied using a haploid model with 2N individuals). The black lines trace the ancestries of three sampled lineages back to a single common ancestor. (**b**) The subgenealogy for the three sampled lineages. In the basic version of the coalescent, it is only necessary to keep track of the times between coalescence events (T(3) and T(2)) and the topology—that is, which lineages coalesce with which. *N* number of allelic copies in the population, *n* sample size. (Figure form Ref. [13])

(b) Description of the geographical distribution of haplotypes, typically using GIS software (such as ArcGIS) to map different haplotypes on the map (Fig. 4.2): the premise of this is that the sampled information must contain the latitude and longitude of the sample point.
(c) Estimation of the time of differentiation of the gene: the current favorite method is to use Bayesian to estimate the divergence time. The commonly used software is Beast. Before estimating the gene divergence time, it is necessary to confirm the evolution rate of the detected gene locus. In general, the estimation results of the same gene locus of the same population or related population reported by the predecessors are used. In many cases, predecessors did not report on the evolution rate of selected populations and genes, which must be time calibrated with fossil data. The fossil data used generally come from the outgroup of the research population.
(d) Using some common population genetics software, such as DnaSP, Arliquin, etc.: this is for calculating the population genetic parameters of the study population (such as genetic diversity, genetic differentiation).
(e) Using mismatch analysis, skyline analysis, and some parameters: this is for the detection of population size dynamics (demography).

Templeton et al. [15] initiated a nested clade phylogeographic analysis (NCPA) that integrates gene lineage and haplotype frequencies and geographical location information to change the drawbacks of judging the subjective nature of the phylogeography pattern and used permutation to statistically test phylogeography pattern, which avoids the subjectivity of the phylogeography pattern inference, although this method is still a descriptive phylogeography method.

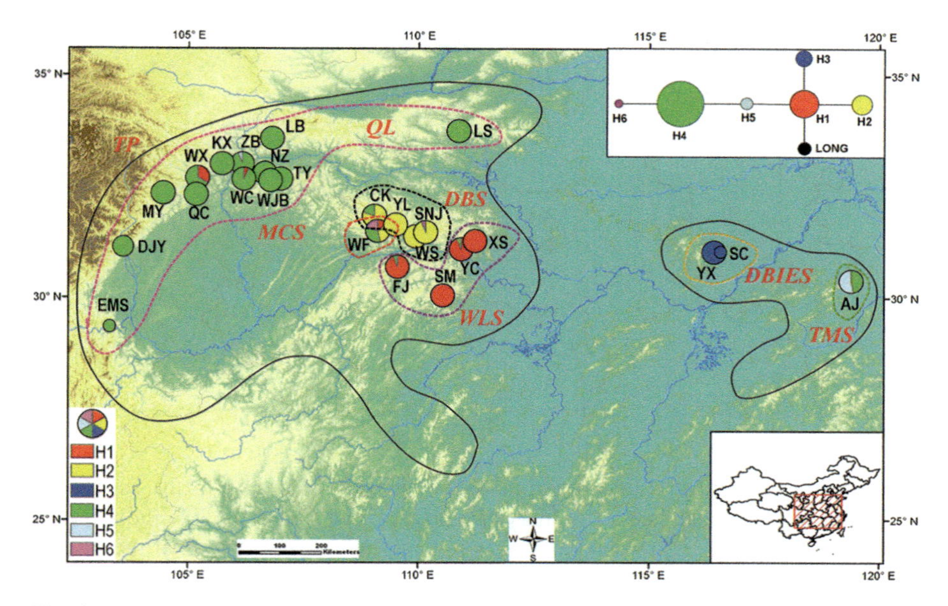

Fig. 4.2 Network map (top right) and geographic distribution of six haplotypes within and among populations of *Fagus engleriana*. Acronyms in red italics: *TP* Tibetan Plateau, *QL* Qinling Mountains, *MCS* Micangshan Mountains, *DBS* Dabashan Mountains, *WLS* Wulingshan Mountains, *DBIES* Dabieshan Mountains, *TMS* Tianmushan Mountains. (Figure form Ref. [14])

The NCPA method firstly connects the nearest haplotypes to form the lowest primary branch (I-level clade) [10]. When all the haplotypes are in the I-level branch, all the I-level subbranches are regarded as one unit. Then, according to the same method, the network is divided into a level-II branch and so on until all the haplotypes form a single branch (Fig. 4.3). Then, the NCPA uses a displacement test to test the null hypothesis of "a random distribution of haplotype". If the null hypothesis is rejected, it is identified specific reasons for phenotypes with significant geographic patterns, such as range expansion, isolation by distance, fragmentation according to a set of inference keys designed by Templeton et al. [15].

4.2.2.2 Statistical Phylogeography

shortcomings of overinterpreted data in the search tables designed by Templeton et al. [15], the method was gradually eliminated. Instead, people are increasingly inclined to use model-based statistical inference methods, the so-called statistical phylogeography. The most famous of these is the approximate Bayesian computation (ABC). Since historical events cannot be repeated, direct observation and repeated experiments are often incapable of evolutionary biology. Therefore, the strategy adopted by statistical phylogeography is to compare various possible hypotheses and then determine the most likely hypothesis based on available data.

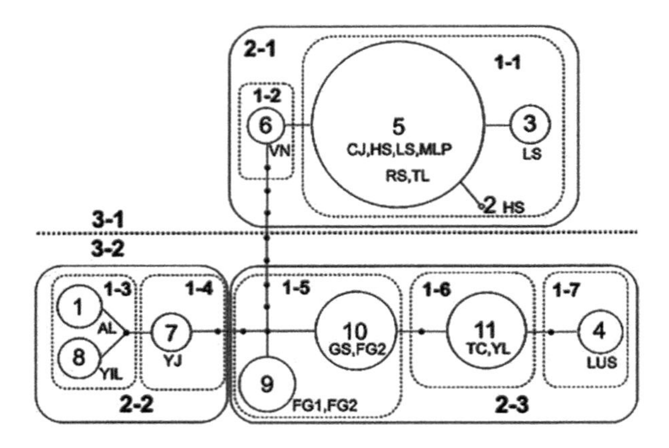

Fig. 4.3 Nested cladogram of 11 chloroplast haplotypes in *Dipentodon*. (Figure from Ref. [10])

The specific operation is first to simulate the data set according to the set model (hypothesis), then calculate a summary of statistics, and obtain the likelihood value of the statistics under each model (that is, get the inductive statistics of the probability observation); the larger is the likelihood value, the higher is the possibility of this hypothesis.

4.3 Application in Biological Evolution of Phylogeography

4.3.1 Species Definition and Species Formation

Species are the most basic unit in the biological world, and the concept of species is one of the most important concepts of biology. However, the definition of species is a problem in biology. First, scholars differ in the understanding of species. Mayden [16] has listed 24 different definitions of species, and there are naturally huge differences in every species defined under the different frameworks for species. Second, within the framework of the same species, the definition of species is difficult to unify due to the complexity of living things. For example, the most commonly used "biological species concept" emphasizes the importance of reproductive isolation; this concept is efficient in the animal kingdom, especially in higher animals, but in the plant world, this concept has met with significant challenges. For example, many higher plants (such as bamboo) have long-term vegetative reproduction, and they often have only one flower and fruit in their lifetime. For this group of organisms, it is hard to verify the existence of hybridization among different species, so the species definition of such organisms must rely on morphological differences. In addition, even in groups with sexual reproduction (such as Fagaceae), it is difficult to define species using the criteria of reproductive isolation due to extremely many crosses.

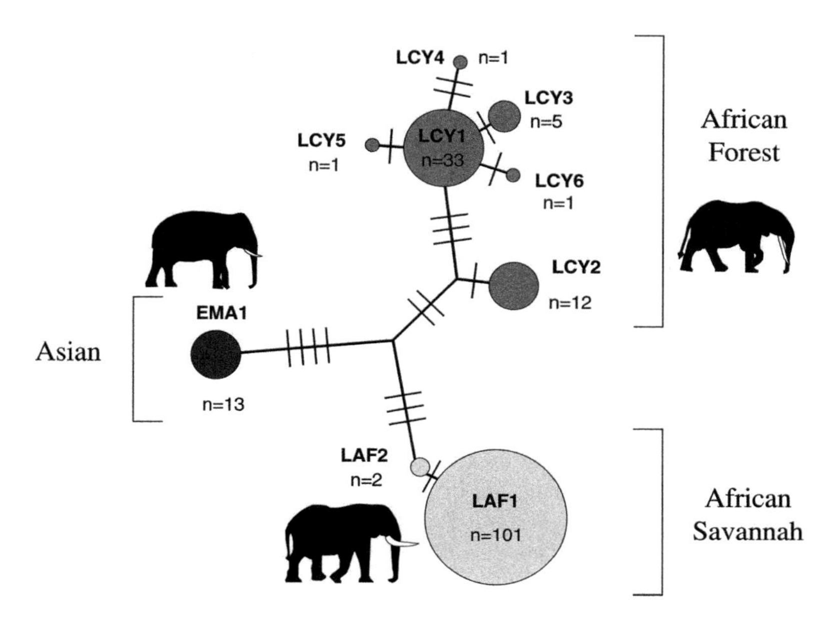

Fig. 4.4 Minimum spanning network depicting relationships among nine haplotypes observed for the X-linked BGN gene for Asian, African forest, and African savannah elephants. Each circle represents a haplotype. (Figure form Ref. [18])

Since phylogeography is a bridge connecting intraspecific and interspecies evolution, it is very suitable for exploring the process and dynamics of species formation. Therefore, from the day of its birth, phylogeography has become a powerful tool for species definition. Early animal phylogeography used mitochondrial DNA to investigate on issues involving species. Slade and Moritz [17] studied *Bufo marinus* and found that the populations on the east and west sides of the Eastern Andean Cordillera showed huge genetic differences by using mitochondrial DNA. On the contrary, the related species *Bufo paracnemis* is in the eastern branch, so the authors questioned the definition of the species of *Bufo marinus* and its related groups [17]. In recognition of the fact that different genes have coalescent stochasticity, which leads to gene trees being different from species trees, in recent years, multiple gene loci have been used in phylogeography to examine the relationship between species and their formation. Roca et al. [18] took four nuclear gene fragments to conduct a phylogeography study of African elephants and found that African elephants can be divided into African forest elephants and African savannah elephants. The genetic difference between the two branches is equivalent to 58% of the difference between *Elephas* and *Loxodonta*; considering the larger morphology and habitat differences between the two African elephants, the authors suggest that the two branches be raised to species-level units [18] (Fig. 4.4).

4.3.2 Evolutionary History of Species

The most significant difference between phylogeography and traditional population genetics is the pedigree relationship introduced in the species, and the gene pedigree accumulates mutations over time. These genetic variations usually appear at a rather consistent rate (molecular clock). Thus, the specific amount of variation can be converted to time, providing a time scale for the evolution of species. In addition, with the continuous development of the coalescent theory, phylogeographers can mathematically simulate the historical events involving the population, such as population expansion and contraction, population differentiation, gene flow, and natural selection, and then more clearly trace the evolutionary history of the species.

Pinus densata is a typical homoploid hybrid and widely circulated in the south and east, which is very suitable for the alpine climate of the Qinghai-Tibet Plateau. Gao et al. [19] used eight nuclear loci to study the population differentiation history of 19 populations to understand the history of isolation and the differentiation of various populations of *Pinus densata*. The study found that *Pinus densata* originated in the Miocene, and all populations could be clustered into three groups [19]. The differentiation of these three groups is associated with the rapid rise of the Qinghai-Tibet Plateau in the evening and the Quaternary glacial period (Fig. 4.5). Thus, the evolutionary history of *Pinus densata* is very clearly stated [19].

4.3.3 Determining Priority Protection Areas

Genetic diversity has a significant role in the survival and evolution of species and the preservation of ecosystem functions. Therefore, genetic diversity is one of the most significant issues in conservation biology. Due to geological history, some regions tend to have genetic diversity or genetic differentiation far higher than other regions, such as southern Europe (Iberia), Italy and the Balkans, which are the Quaternary ice age refuge for many thermophilic plants and animals. The are rich in genetic variation and have large genetic differentiation among regions. There are also some areas, such as the Hengduan Mountains in southwestern China, that, due to their complex terrain and variable climate (large ecological gradient), provide a variety of habitats for biological evolution. The genetic diversity and genetic differences of species within the region are much higher than in other regions. These areas with high genetic diversity or genetic differentiation are known as evolutionary hotspots [20] or phylogeographical hotspots [21]. Since different species in the same region have a similar evolutionary history or ecological processes, comparative phylogeography studies of different species may make it possible to identify these evolutionary hotspots and incorporate them into nature conservation plans [22–24]. Therefore, identifying evolutionary hotspots or phylogeographical hotspots is one of the most important scientific issues in phylogeography.

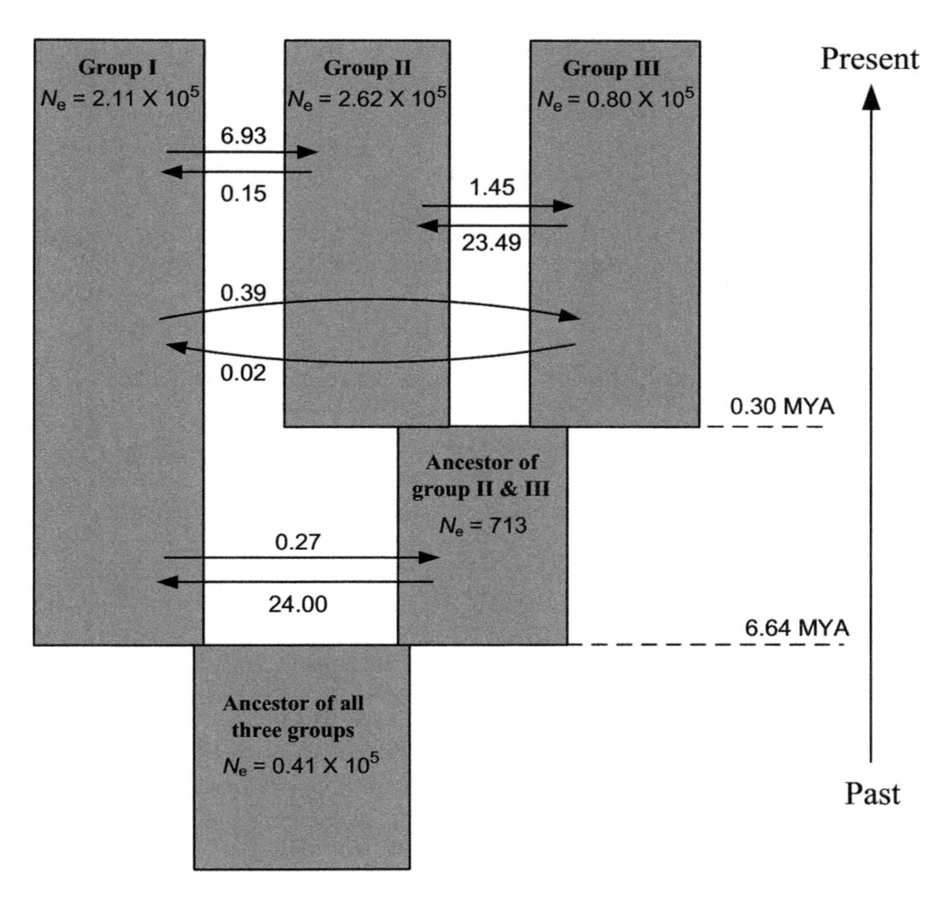

Fig. 4.5 IMa (Isolation with Migration) method for calculating the differentiation time of three *Pinus densata* groups. (Figure from Ref. [19])

Since the end of the Third Age, especially in the Quaternary, the Earth's climate has experienced large-scale glacial-interglacial climate fluctuations. During the cold Quaternary glacial period, different plants and animals tend to retreat to areas with suitable climates, forming a so-called "refuge." These refuges often retain many genetic variations. In the same period, due to the isolation between glacial refuges, these areas also have strong genetic differentiation. Therefore, glacial refuges are often phylogeographical hotspots, and the identification of glacial refuges becomes one of the most important goals in learning phylogeography. As mentioned earlier, most of the flora and fauna retreat to the three southern peninsulas of Europe, namely, Iberia, Italy, and the Balkans, which are becoming the famous "Southern Refuge" due to the large-scale continental glaciers that developed in these regions during the Quaternary ice age. This pattern is also more apparent in North America. China is in the eastern part of Eurasia, and it has complex terrains and diverse climate types. Due to the impact of the Asian monsoon climate, most of the Chinese regions are not covered by continental glaciers. Therefore, the pattern of Chinese biological ice age refuges is quite distinctive from Europe and North America. Recently, findings related to plant phylogeography showed that most of the plants in eastern China have multiple refuge patterns, and the refuges of different plants have individual characteristics. For example, Zhang et al. [25] conducted a comparative phylogeography study on two subtropical *Fagus* L. plants in China, showing that *F. longipetiolat*a and *F. Lucida* have multiple glacial shelters. However, although the two are in the same domain, the refuge pattern is entirely different (Fig. 4.6).

4.3.4 The Origin of Domestic Animals and Cultivated Plants

Since Darwin's theory of evolution, evolutionary biologists have attached great importance to the domestication of animals and plants. Darwin has personally participated in the breeding of pigeons. Domesticated animals and cultivated plants have undergone tremendous changes in morphology, physiology, and behavior, and domestication has become one of the focuses of evolutionary biology research. From the perspective of evolutionary biology, the domestication of animals and the cultivation of plants are equivalent to a process of population differentiation and even species formation under artificial selection pressure, and describing this process is precisely the strength of phylogeography. Thus, it has performed a crucial part in the study of the origin of home animals and cultivated plants.

The dog is a friend of human beings and the first animal to be domesticated by humans. Due to long-term artificial breeding and selection, the breeds of dogs are numerous and varied. FCI (Fédération Cynologique Internationale, World Dog Organization) officially recognized 332 conventional varieties and 11 tentative varieties. It is natural to ask whether these varieties originate from the same species. To solve this problem, Vilà et al. [26] collected samples of 162 wolves, 67 breeds of 140 dogs, and other wolves from around the world. Using mitochondrial DNA for phylogeography, they found that the gray wolf is the sole ancestor of modern dogs [26]. Today, the domestication of dogs is very clear [27] (Fig. 4.7).

Cassava (*Manihot esculenta*) is an important potato crop in the global tropics. Because of its high root starch content, it can be used as a ration and as an energy plant. Olsen and Schaal [28] used DNA glyceraldehyde 3-phosphate dehydrogenase (G3pdh) to sequence DNA from 212 cultivated and wild individuals, and 28 haplotypes were detected. It was found that cultivated cassava originated from the southern margin of the Amazon basin; the original type was *M. esculenta* subsp. *flabellifolia*.

Rice is one of the essential rations of human beings. It is generally understood to originate from the east of Asia and southeast of Asia, but it has always been a mystery where it originated. Some scholars believe that cultivated rice has many origins, and the evidence includes the considerable genetic distance between *Oryza sativa* L. subsp. indica Kato and subsp. japonica Kato in Journ. Molina et al. [29] examined the genetic variation of 630 nuclear genes in cultivated rice and wild rice

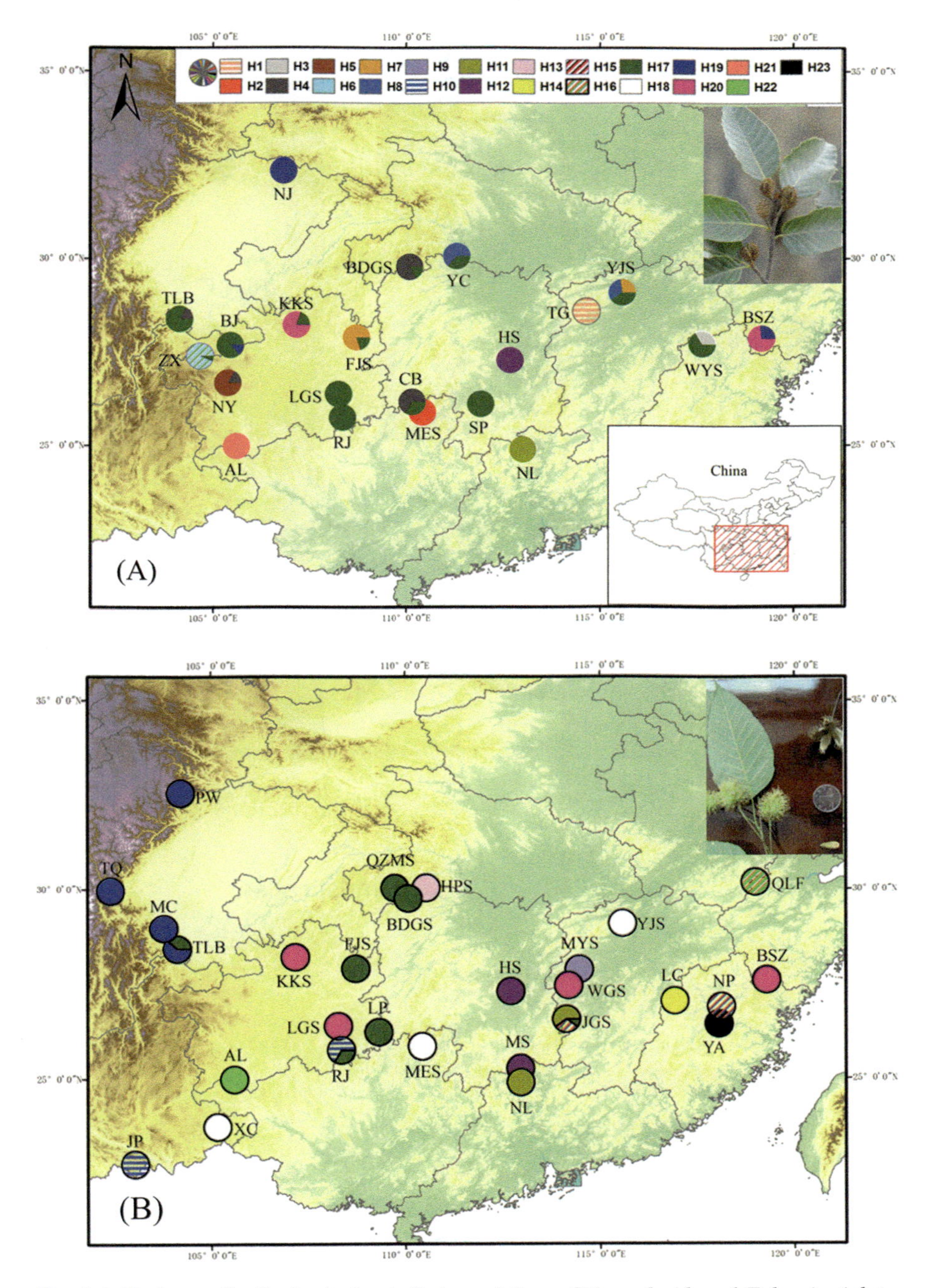

Fig. 4.6 Haplotype distribution in the studied populations of *Fagus lucida* and *F. longipetiolata*, respectively. The insets in (**a**) and (**b**) show the branches with capsules of each species. (Adapted from Ref. [25])

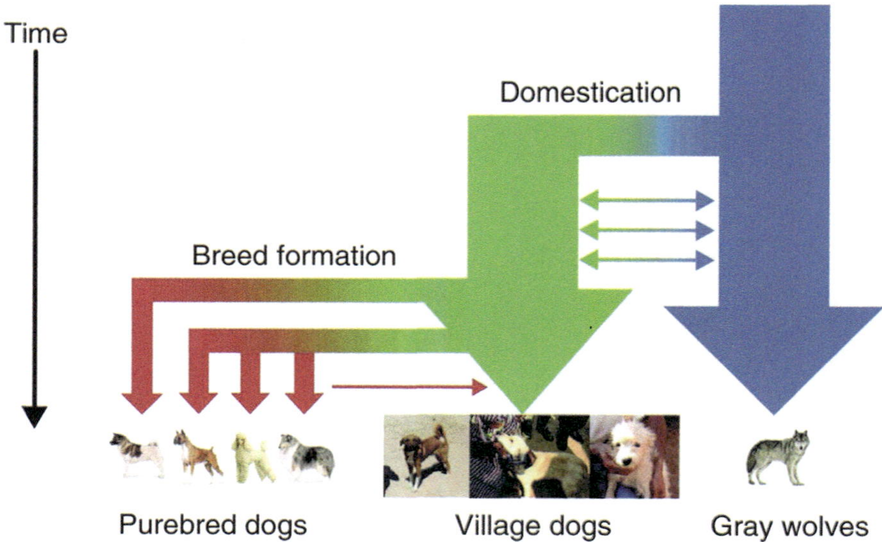

Fig. 4.7 Phenotypic diversity and evolutionary history of dog. (Adapted from Ref. [27])

(*Oryza rufipogon*) and found that cultivated rice originated in the Yangtze River basin around 8200–13,500 years ago. However, recent studies have suggested that cultivated rice originated in the Pearl River basin [30].

4.4 Application in Medicinal Plants of Phylogeography

4.4.1 Basic Research on the Genetics of Medicinal Plants

The genetic basis for identifying the geoherbalism of medicinal plants must be clarify their phylogeographic pattern of genetic differences. Genetic differentiation among populations is the result of gene flow blocked between populations that result in genetic differences. The degree of gene flow blocked among populations can be altered by allele frequencies or haplotype frequencies in different populations. Templeton et al. [15] proposed that three main elements can cause changes in the distribution of allele frequencies or haplotype frequencies, resulting in three major population genetic differentiation models, and phylogeographic analysis can provide a quantitative inference of the genetic differentiation patterns of these three populations. According to the phylogeographic pattern of different populations of different medicinal plants, the degree of genetic differentiation of a plant population can be obtained, and further analysis of geoherbalism and correlation can be done.

Mode 1: Allopatric fragmentation

In this model, gene flow between populations is almost blocked or completely blocked, and allele frequencies or haplotype frequencies between populations vary widely or are entirely different. In this model, there should be a sizeable genetic differentiation between the Daodi population and the non-Daodi population. The Daodi and non-Daodi populations are genetically distinct. The formation of the geoherbalism may be directly related to the genetic differentiation between the populations.

Mode 2: Restricted gene flow with isolation by distance

In this model, the gene flow between populations decreases with the increase of geographical distance, and the allele frequency between populations or the difference in haplotype frequency gets bigger with the increase of geographical distance. The genetic differentiation between the Daodi and non-Daodi populations is inevitably a continuous variation process in this model. The formation of the geoherbalism is partially related to the genetic differentiation between populations.

Mode 3: Range expansion

This model is divided into two types

Long-distance transmission: this model refers to the fact that a population of seeds spreads through a particular medium (such as birds) to a place far from the distribution area to form a new population. In this case, two populations with long distance may have very similar allele frequencies or haplotype frequencies, while adjacent populations have relatively large genetic differentiation. If this happens in Daodi population, and the geographical location of the allele frequency or the haplotype frequency is similar to the distribution of a drug multi-Daodi. It is evident that geoherbalism formation is directly related to genetic differentiation. If there is no or only one Daodi area, the formation of geoherbalism is not related to genetic differentiation.

Rapid expansion of adjacent areas: this model refers to the rapid expansion of the distribution of species (such as invasive plant water hyacinth, *Eupatorium adenophorum*, etc.) in a short period of time, in which case the gene flow between populations is almost unimpeded and allele frequencies or haplotype frequencies between populations are almost identical, without genetic differentiation. There are very few medicinal plants in this mode, and it is difficult to form Daodi herbs even if they are present.

Phylogeographic analysis quantitatively reveals the phylogeographic pattern of population genetic differentiation. By reanalyzing its correlation with the formation of geoherbalism, it is possible to accurately classify the correlation between them into different levels. This is not solved by the previous methods of genetic diversity analysis. Phylogeographic analysis is a breakthrough in theoretical basis and research methods and opens new avenues for research on the formation of Daodi herbs.

4.4.2 Study on the Origin of Cultivated Medicinal Plant

Wild Chinese traditional medicines are being overutilized as several governments and agencies recommend that wild medicinal species should be brought into cultivation systems. The cultivation of medicinal plants is about 2600 years old in China, and most cultivation methods have risen rapidly in recent decades. From the perspective of evolutionary biology, the cultivation history of medicinal plants has the following characteristics:

(a) Compared with the cultivation history of crops, the cultivation history of medicinal plants is short, and it involves the early stages of cultivation and domestication—that is, domestication and selection breeding.
(b) Due to the short cultivation history of medicinal plants, selected traits of medicinal plants (such as medicinal ingredients, drug effects) are not easy to observe. The direction of artificial selection is not apparent. Therefore, the artificial selection pressure of medicinal plants is smaller, the cultivation traits are not typical, the wild type and cultivation type are difficult to distinguish, and the cultivation type can often survive under natural conditions.
(c) Medicinal plants are cultivated through a variety of ways; every cultivation area is small, but the biological characteristics of each area are different. Medicinal

plants are often cultivated in a semi-wild manner, and there is substantial gene flow between the wild and cultivated types. These characteristics signify that the origin of cultivated medicinal plants is very different from that of crops.

However, the study of the origin of medicinal plant cultivation not only has important theoretical significance but also has practical significance. The cultivation origin of crops is one of the hotspots of plant evolution research in the world today. The cultivation and domestication of most crops occurred in the ancient Neolithic Age 10,000 years ago. It has entered the late stage of cultivation and domestication, and can only speculate on the events that cause genetic variation in the early stage of crop cultivation and domestication. However, the cultivation and domestication history of medicinal plants is no more than 2600 years, and most of it occur in recent decades, at the beginning of cultivation and domestication. The process of changing the evolutionary process of plants due to human intervention is underway. Therefore, medicinal plants are the corking materials for artificially inducing genetic diversity and genetic structure changes in the early stages of cultivation and domestication, but this period of crop cultivation and domestication has ceased to exist. At this point, the origin cultivated research of medicinal plants have special value. In practice, the original species, origin and origin time of the cultivated medicinal plants have a strong guiding role for the protection of the source, the safe drug, and the medicinal plant resources. In addition, the origin of medicinal cultivated plants provides first-hand evidence for the history of Chinese medicine development.

4.5 The Case Study of Chinese Medicinal Plant [31]

4.5.1 Background of Scutellaria Baicalensis Georgi

Scutellaria baicalensis Georgi (Huangqin or Chinese skullcap) is a plant commonly used in traditional Chinese medicine; it was first discovered in *Shen Nong Ben Cao Jing* in *ca.* 100 BC [32]. The roots of this herb have been widely used in the treatment of hepatitis, jaundice, tumor, leukemia, hyperlipaemia, arteriosclerosis, diarrhea, and inflammatory diseases; it is one of the most important heat-clearing and dampness-drying drugs. It is an important ingredient in many traditional Chinese medicine formulas, such as Shuanghuanglian oral liquid, Compound Huangqin granule, and Yinhuang tablet, which are used for detoxication and for the relief of fever. Previous phytochemical research discovered that the main bioactivities of this herb are attributed to baicalin and other flavone glucuronide analogues. In recent years, *Scutellaria* flavonoids have received worldwide attention due to their several interesting pharmacological effects, which include hepatoprotection, neuroprotection, anti-inflammation, anti-cancer, anti-HIV, and anti-hepatitis B virus activities [33–35].

S. baicalensis is a perennial herb of the family Lamiaceae, which is widely found in North China, Russia, Korea, and Japan. The wild herb can be found in Inner Mongolia, Shanxi, Shaanxi, Hebei, and Shandong provinces of China, where it mainly originated. The commercial product, on the other hand, is famous for its high quality and therapeutic potency, and it can be found in an area around Cheng De of Hebei province, which is well known as "Rehe Huangqin". Quality control is becoming one of the critical problems in the modernization of Chinese medicine and the phytopharmaceutical industry. Nevertheless, the current traditional methods of authentication are ineffective for distinguishing materials of *S. baicalensis by morphological and histological methods*. Chemical characterization is also not always reliable because of the difference in chemical profiles due to a lot of nongenetic factors such as age and post-harvest conditions. Therefore, it is crucial to develop a reliable genetic marker system that could be used to affordably and accurately locality identify of *S. baicalensis*.

4.5.2 Sampling and Methods

4.5.2.1 Sampling

An overall of 602 and 451 individuals of *S. baicalensis* representing 28 wild and 22 cultivated populations were found in Northeast to Northwest China, respectively (Table 4.1). Each population had samples of 17–24 individuals. Twenty-Two *Scutellaria rehderiana* Diels individuals as an outgroup were sampled from a wild population of Weiyuan County in Gansu province (WYW). The leaves of *S. baicalensis* for DNA extraction were dried with silica gel. Voucher specimens were kept in the herbaria of the Institute of Chinese Materia Medica (CMMI) and China Academy of Chinese Medical Sciences. The geographic location of each sampled population was recorded by using a Garmin GPS unit (Table 4.1).

4.5.2.2 DNA Extraction

The dried leaves were milled by using a RETSCH mixer mill (MM301). A modified cetyltrimethylammonium bromide (CTAB) protocol [36] was conducted for Genomic DNA extraction.

The leaf tissue was homogenized in 900 μl of 2 × CTAB by using a hand-operated homogenizer with a plastic pestle for 15~20 s. Later, 6 μl of Mercaptoethanol was added and vortexed for 30 s. The specimens were incubated at 65 °C for 1.5–2 h for cell lysis. Then the incubated material was removed and placed at room temperature. A same volume of chloroform/isoamyl alcohol (24:1)

Table 4.1 Details of sample locations, sample sizes in 28 wild and 22 cultivated populations of *S. baicalensis* and one wild population of *Scutellaria rehderiana*

Province	County	P	Lat.(N)	Long(E)	Alt.(m)	N
Neimenggu	Eerguna	EGW	50.42°	119.51°	564.7	20
Neimenggu	keshiketeng	KKW	43.28°	117.23°	1399	24
Neimenggu	linxi	LXW	44.05°	117.76°	1212	22
Neimenggu	kalaqin	KQIC	42.90°	118.76°	691.2	24
Neimenggu	kalaqin	KQLC	42.90°	118.50°	691.2	20
Neimenggu	Guyang	GYW	41.20°	110.60°	1854	24
Heilongjiang	huma	HMW	51.93°	126.43°	288.3	24
Heilongjiang	Duerbote	DMW	46.51°	124.59°	146.2	24
Heilongjiang	luobei	LBC	47.91°	130.72°	184.2	17
Jilin	Baicheng	BCW	45.90°	122.42°	240.8	19
Jilin	Changchun	CCC	43.87°	125.27°	241	21
Jilin	Yanji	YJW	42.92°	129.60°	302.1	19
Liaoning	Yixian	YXC	41.53°	121.23°	61	21
Liaoning	Jianchang	JCW	40.82°	119.78°	362	19
Liaoning	Jinzhou	JZW	39.13°	121.70°	12	22
Beijing	Yanqing	YQ1W	40.47°	115.97°	900	20
Beijing	Yanqing	YQ2W	40.52°	115.78°	1300.6	22
Beijing	Yanqing	YQC	40.52°	115.78°	600	18
Hebei	Chicheng	CCW	40.92°	115.82°	1100	17
Hebei	Kuancheng	KCW	40.62°	118.47°	304	23
Hebei	Kuancheng	KCC	40.61°	118.49°	304	21
Hebei	Luanping	LPW	40.95°	117.53°	523	17
Hebei	Luanping	LPC	40.93°	117.33°	523	19
Hebei	Chengde	CD1W	41.20°	117.94°	58	18
Hebei	Chengde	CD1C	41.20°	117.94°	58	20
Hebei	Chengde	CD2W	40.20°	117.94°	60	22
Hebei	Chengde	CD2C	40.20°	117.73°	60	19
Shandong	Yantai	YTW	37.55°	121.52°	25	22
Shandong	Jinan	JNC	36.64°	117.36°	200	23
Shandong	Juxian	JUXC	35.90°	118.97°	179	24
Jiangsu	Jurong	JRC	31.87°	119.22°	22	20
Henan	Huixian	HXW	35.46°	113.77°	863	19
Henan	Songxian	SXW	34.22°	111.91°	864.5	24
Henan	Songxian	SXC	34.18°	111.97°	539.3	22
Hubei	Fangxian	FXC	33.06°	110.07°	918	19
Shanxi	Wutai	WTW	38.83°	113.36°	1148	21
Shanxi	Fenyang	FYW	37.42°	111.65°	1588	21
Shanxi	Fenyang	FYC	37.35°	111.77°	947	24
Shanxi	Lingchuan	LCW	35.98°	113.49°	1406	23
Shanxi	Lingchuan	LCC	35.95°	111.72°	1363	21
Shanxi	Jiangxian	JXW	35.39°	111.61°	854.5	21
Shanxi	Jiangxian	JXC	35.47°	111.46°	545	22

(continued)

Table 4.1 (continued)

Province	County	P	Lat.(N)	Long(E)	Alt.(m)	N
Shaanxi	Huanglong	HLW	35.59°	109.88°	1218	23
Shaanxi	Huanglong	HLC	35.59°	109.88°	1160	20
Shaanxi	Shanyang	SYW	33.56°	109.89°	949.0	24
Shaanxi	Taibai	TBW	34.06°	107.30°	1682	18
Shaanxi	Taibai	TBC	34.04°	107.30°	1565	19
Gansu	Heshui	HSW	36.12°	108.67°	1121	24
Gansu	Weiyuan	WYW	35.15°	104.21°	2115	22
Gansu	Weiyuan	WYC	35.15°	104.21°	2090	21
Gansu	Zhangxian	ZXC	34.60°	104.60°	2065	20

Adapted from Ref. [31]

was added to the samples and vortexed for 5–10 min, then centrifuged at 12,000 g for 5–10 min at 4 °C. The supernatant was transferred to a fresh tube, then a double volume of isopropanol was added to each sample and incubated at −20 °C for 30 min. The incubated samples were then centrifuged at 12,000 g for 5–10 min at 4 °C. The pellet was washed with 70% ethanol, dried, and resuspended in sterile dH_2O containing 0.1 × TE100μL and 20 μg/ml DNase-free RNase A. Concentration and purity were recorded from the A260/A280 ratio by using a spectrophotometer then analyzed on 1.0% agarose gels using 15 μl aliquots of the reaction mixture.

4.5.2.3 Amplification and Sequencing

Amplification reactions were carried out in a volume of 20 μL containing 2.0 mm/L MgC12, 0.5 μm/L dNTP, 10 × buffer, 2.5 μm/L primer, 1 U Taq DNA and 20 ng DNA template.

PCR amplification was carried out in a TC-512 thermocycler (Techne, England) programmed for an initial 240 s at 94 °C, followed by 30 cycles of 45 s at 94 °C, 30 s at 55 °C (*atpB-rbcL*), 54 °C (*trnL*-trnF) or 58 °C (*psbA-trnH*), 90 s at 72 °C, and a final 240 s at 72 °C.

Sequencing reactions were performed with the forward or reverse primers of the PCR using the DYEnamic ET Terminator Kit (Amersham Pharmacia Biotech) according to the manufacturer's protocol. Sequencing was conducted on a HITACHI 3130 Genetic Analyzer (Hitachi High-Technologies Corporation, Tokyo Japan) after the purification of the reaction product through precipitation with 95% ethanol and 3 M sodium acetate (pH 5.2).

Table 4.2 Screened cpDNA fragments and primers

Locus	Length (bp)	Number of variable site	Primer	5'–sequence–3'
atpB-rbcL	765–767	13	*atpB*	ACATCKARTACKGGACCAATAA
			rbcL	AACACCAGCTTTRAATCCAA
			atpB[a]	ATCCCTCCCTACAACTCATG
			rbcL[a]	TTTTTTCAAGCGTGGAAGCC
*trnL-trn*F	781–795	7	*trnL*	CGAAATCGGTAGACGCTACG
			trnF	ATTTGAACTGGTGACACGAG
psbA–trnH	361–458	14	*psbA*F	GTTATGCATGAACGTAATGCTC
			*trnH*R	CGCGCATGGTGGATTCACAATCC

[a]Primers designed for this study. Adapted from Ref. [31]

4.5.2.4 Primers Screening and Design

The universal primers reported in Hamilton [37] and Sang et al. [38] were used for screening polymorphic cpDNA fragments. After a preliminary screening of eight chloroplast fragments, *atpB-rbcL*, *trnL-trn*F, and *psbA-trnH* intergenic spacers were chosen for the full survey as they contained the most polymorphic sites.

The necessary information and amplification primers of these fragments are shown in Table 4.2. *PsbA-trnH* has the most substantial variation, with 14 polymorphic loci; followed by *atpB-rbcL* with 13 polymorphic loci, *trnL*-trnF has the smallest variation and only seven polymorphic loci. The three segments contained a total amount of information sufficient for population genetic analysis, so these three segments were finally selected for the geographic pedigree analysis of *S. baicalensis*. *PsbA–trnH* (annealing temperature 58 °C) and *trnL*-trnF (annealing temperature 54 °C) were amplified using the reported primers, which have proper amplification. *AtpB-rbcL* was amplified using the reported primers (annealing temperature 48 °C), and only a few individuals could be amplified. So specific amplification primers (annealing temperature 55 °C) were redesigned (Table 4.2), and a good amplification effect was obtained.

4.5.2.5 Data Analysis

4.5.2.5.1 Sequence Analysis

Sequences were arranged using Clustal_X version 1.81 [39], and all indels and gaps were coded as substitutions by Caicedo and Schaal [40]. The entire sequences were saved in a PHYLIP or a NEXUS file format for later analysis.

4.5.2.5.2 Population Genetic Diversity and Genetic Structure Analysis

The genetic diversity and genetic structure of the population are obtained by calculating and comparing the values of G_{ST} and N_{ST}. Nei [41] refers to G_{ST} as "coefficient of gene differentiation," according to the concept of Nei [41–43]; for any polymorphic locus, the total genetic diversity (h_T) of all populations of a "species" includes genetic diversity within each population (h_S) and genetic diversity among populations (d_{ST}); the relationship among them can be expressed as:

$$h_T = h_S + d_{ST}$$

The ratio of genetic diversity that exists between populations can be calculated using the following formula:

$$G_{ST} = d_{ST}/h_T = (h_T - h_S)/h_T$$

The value of G_{ST} can range from 0 to 1. When the value of G_{ST} is close to 0, $h_T \approx h_S$, there is almost no differentiation between populations and the total genetic diversity mainly exists in the population; when the value of G_{ST} is close to 1, $h_T - h_S \approx h_T$, the genetic differentiation within the population is almost zero and the total genetic diversity mainly exists among the populations. The larger is the value of G_{ST} from 0 to 1, the greater is the relative amount of genetic differentiation among populations. The following is the calculation process of G_{ST}:

- Genetic diversity within each population (h_S):

  First calculate the genetic diversity of a locus in each population(h_{Si}):

$$h_{Si} = 1 - \sum_{j=1}^{m} q_j^2$$

m: the number of alleles at this locus, q_j: the frequency of the *j*-th allele in this population at this locus.

$$h_S = \sum_{i=1}^{n} h_{Si}/n = \sum_{i=1}^{n}\left(1 - \sum_{j=1}^{mi} q_{ij}^2\right)/n$$

n: the total number of populations determined, q_{ij}: the frequency of the *j*-th allele at the *i*-th population at this locus, *mi*: the number of alleles at the *i*-th population at this locus.

- Total population genetic diversity(h_T):

$$h_T = 1 - \sum_{j=1}^{m} r_j^2$$

m: the number of alleles at this locus, r_j: the average frequency of the j-th allele at this locus in all populations.

- Genetic diversity among populations (d_{ST}):

$$d_{ST} = h_T - h_S$$

- Gene differentiation coefficient(G_{ST}):

$$G_{ST} = d_{ST}/h_T = (h_T - h_S)/h_T$$

The above is the calculation of one locus. For all loci, the h_S, h_T, and G_{ST} of the average genetic diversity can be obtained by calculating the arithmetic mean of the $h_{S\ I}$, $h_{T\ i}$, and $G_{ST\ i}$ of all loci.

From the above calculations, the G_{ST} of Nei's only considers one factor of the allele frequency when detecting the genetic differentiation of the population. Pons and Petit [44, 45] proposed the concept of N_{ST}, which further considers the genetic distance between haplotypes (alleles) based on the allele frequency. The following briefly introduces the calculation method of N_{ST}:

Let p_i and p_j be the frequencies of the i-th and j-th haplotypes in the total population, p_{ki} and p_{kj} be the frequencies of the i-th and j-th haplotypes in the k-th population, and $c_{i\,j}$ be p_{ki} and p_{kj} sum of the differences in all populations, π_{ij} be the genetic distance between the i-th haplotype and the j-th haplotype, $\pi_{ii} = 0$, and $\pi_{i\,j} = \pi_{j\,i}$. Then the genetic diversity of the k-th population is v_k:

$$v_k = \sum_{ij} \pi_{ij} p_{ki} p_{kj}$$

The genetic diversity within the average population (v_S) is the expected value of the genetic diversity of all populations:

$$v_S = \sum_{ij} \pi_{ij} \left(p_i p_j + c_{ij} \right)$$

Total population genetic diversity (v_T):

$$v_T = \sum_{ij} \pi_{ij} p_i p_j$$

Replacing the overall genetic diversity h_T in the gene differentiation coefficient G_{ST} of Nei's and the genetic diversity h_S in each population by v_T and v_S:

$$N_{ST} = (v_T - v_S)/v_T = \frac{\sum_{ij}\pi_{ij}c_{ij}}{v_T}$$

Since the value of G_{ST} is only determined by the frequency of the haplotype and the value of N_{ST} determines the frequency and genetic distance of the haplotype, in many cases they are not equal, and the geographical structure of the population is causing the difference between these two values. Therefore, according to the differentiation between G_{ST} and N_{ST}, the geographical relationship among the populations can be judged.

When $N_{ST} >> G_{ST}$, it is indicated that the haplotypes with similar genetic distances are distributed in the same or geographically similar populations, and the populations are increasingly genetically differentiated with the increase of geographical distance. There is an apparent geographic structure among the populations.

When $N_{ST} = G_{ST}$, it is shown that the genetic distances among the haplotypes are similar. They reflect similar among group relationships and the same geographical structure, no matter how they are distributed in the population.

When $N_{ST} << G_{ST}$, it is suggested that haplotypes with similar genetic distances are distributed in distant, geographically completely separate populations, or haplotypes with vast genetic distances are distributed in the same or similar populations, which may be some random factors. The result of the redistribution of haplotypes among populations does not conform to the law of geographical evolution of relatives.

4.5.2.5.3 Haplotype Network Depiction

A haplotype network showing the mutational relationships among distinct haplotypes was drawn according to the principle of parsimony by TCS version 1.13 [46, 47], which positing *S. rehderiana* as outgroup. This method uses the coalescent theory to estimate the kinship between haplotypes with the credibility of over 95% [45].

4.5.2.5.4 Haplotype Geographical Distribution Depiction

The geographical distribution of haplotypes was mapped on a map of China by using ArcMap 8.3 (ESRI, Inc.). The significant difference of haplotype frequencies between wild and cultivated populations was calculated using the x^2 test.

4.5.2.5.5 Mantel Test

The effect of geographical distance on genetic structure, correlations between pairwise genetic distances (Kimura 2-parameter distance generated with Kimura's 2-parameter model in MEGA 3), and pairwise geographic distances were examined using a Mantel test implemented by Isolation by Distance Web Service [48].

4.5.3 Analysis of Daodi Genetic Basis

4.5.3.1 Results

4.5.3.1.1 Sequence Characteristics of Three Chloroplast Intergenic Spacers

The 602 individuals from 28 wild populations of *S. baicalensis* and 22 individuals from a wild population of *S. rehderiana* were successfully amplified and sequenced, and three chloroplast DNA fragments (*atp*B-*rbc*L, *trn*L-*trn*F, and *psb*A-*trn*H) were aligned from these individuals.

The aligned sequences of *atpB-rbcL* spacers in *S. baicalensis* and *S. rehderiana* were 767 base pairs in length, and indel occurred at positions 202 and 213 to mutate the length of the sequence (Table 4.3). In terms of nucleotide composition, A/T is abundant, ranging from 69.32 to 69.75%, which is consistent with the nucleotide structure of most chloroplast DNA noncoding regions [49]. Nine substitutions and two indels result in a total of 7 haplotypes (Table 4.3).

The aligned sequences of *trnL*-trnF spacers in *S. baicalensis* and *S. rehderiana* were 790 base pairs in length, and the sequence length variation ranges from 781 to 788 bp; indel occurred at positions 169, 256, 454, and 493–499 to mutate the length of the sequence. The short sequence repeats (poly-C) between 493 and 499 bp of *trnL*-trnF were not treated as polymorphic sites and were removed from subsequent analyses because the sequencing poly-N regions could easily lead to homoplasies due to polymerase error. In the nucleotide composition, the A / T content is lower than that of *atp*B-*rbc*L, and the range is 62.31–62.87%. Two substitutions and three indels result in a total of seven haplotypes (Table 4.3).

The aligned sequences of *psbA-trnH* spacers in *S. baicalensis* and *S. rehderiana* were 481 base pairs in length, and the sequence length variation ranges from 361 to 458 bp; indel occurred at positions 151–163, 188–211, 212–235, 236–259, 260–283, 305, 358–363, and 414–423 to mutate the length of the sequence. In the nucleotide composition, the A / T content is the most abundant of the three fragments, ranging from 70.99 to 71.80% (Table 4.3).

The sequences of nine atpB-rbcL, eight trnL-trnF, and 13 psbA-trnH haplotypes have been placed in the GenBank databases [GenBank: GQ374124-GQ374155]. Thirty-five haplotypes of *S. baicalensis* (HapA-Y) and two haplotypes of *S. rehderiana* (HapZ1-Z2) were recognized when atpB-rbcL, trnL-trnF, and psbA-trnH sequences were merged (Table 4.3).

Table 4.3 Variable sites of the aligned sequences of three chloroplast DNA fragments in 32 haplotypes of wild *S. baicalensis* (HapA-Y) and 2 haplotypes of *S. rehderiana* (HapZ1-Z2)

	Nucleotide position																											
	*atp*B-*rbc*L								*trn*L-*trn*F								*psb*A-*trn*H											
Haplotype	60	195	202	213	309	380	502	506	602	655	722	169	256	454	475	717	73	151	188	212	236	260	305	345	358	414	426	430
HapA	C	T	T	–	C	G	A	C	T	G	T	–	A	–	C	C	G	–	–	–	–	–	–	G	∮	–	T	A
HapB	C	T	T	–	G	G	A	C	T	G	T	–	A	–	C	A	G	–	–	–	–	–	A	G	∮	–	T	T
HapC	C	T	T	–	C	G	A	C	T	G	T	–	A	–	C	A	G	–	–	–	–	–	A	G	∮	–	T	A
HapD	C	T	T	–	C	G	A	C	T	G	G	–	A	–	C	A	G	–	–	–	–	–	A	G	∮	–	T	T
HapE	C	T	T	–	C	G	A	C	T	G	T	–	–	–	C	A	G	–	–	–	–	–	A	G	∮	–	T	A
HapF	C	T	T	–	C	G	A	C	T	G	T	–	A	–	C	A	G	–	¶	¶	–	–	A	G	∮	–	T	T
HapG	C	T	T	–	C	G	A	C	T	G	T	–	A	–	C	A	G	–	–	–	–	–	A	G	∮	–	T	T
HapH	C	T	T	–	C	G	A	C	T	G	T	–	A	–	C	A	G	–	¶	¶	¶	¶	A	G	∮	–	T	T
HapI	C	T	T	–	C	G	A	C	T	G	T	–	A	–	C	A	G	–	–	–	–	–	A	A	∮	–	T	T
hapJ	C	T	T	–	C	G	A	C	T	G	T	–	A	–	C	A	G	–	¶	–	–	–	A	G	∮	–	T	T
HapK	C	T	T	–	C	G	C	C	T	G	T	–	A	–	C	A	G	–	–	–	–	–	A	G	∮	–	T	A
HapL	C	T	T	–	G	G	A	C	T	G	T	–	A	–	C	A	G	–	–	–	–	–	A	A	∮	–	T	T
HapM	C	T	T	–	G	G	A	C	T	G	T	–	–	–	A	A	G	–	–	–	–	–	A	G	∮	–	G	T
HapN	C	T	T	–	C	G	A	C	T	G	T	–	A	A	C	A	G	–	–	–	–	–	A	G	∮	–	T	T
HapO	C	T	T	–	G	G	A	C	T	G	T	–	A	–	C	A	G	–	–	–	–	–	A	G	∮	–	T	T
HapP	C	T	T	–	C	G	A	C	T	G	T	T	A	–	C	A	G	–	–	–	–	–	A	G	∮	–	T	T
HapQ	C	T	T	–	C	G	A	C	T	G	T	–	A	–	C	A	G	–	–	–	–	–	A	G	∮	–	G	T
HapR	C	T	T	T	C	G	A	C	T	G	T	–	A	–	C	A	G	–	–	–	–	–	A	G	∮	–	T	A
HapS	C	T	T	–	G	G	A	C	T	G	T	–	A	–	C	A	G	–	–	–	–	–	A	G	∮	–	T	A
HapT	C	T	T	–	C	G	A	C	T	G	T	–	–	–	A	A	G	–	–	–	–	–	A	G	∮	–	T	T
HapU	C	T	T	–	C	G	A	C	T	G	T	–	A	–	C	A	G	#	–	–	–	–	A	G	∮	–	T	T
HapV	C	T	–	–	C	G	A	C	T	G	T	–	A	–	C	A	G	–	–	–	–	–	A	G	∮	–	T	T
HapW	C	T	T	–	C	G	A	C	T	G	T	–	–	–	A	A	G	–	–	–	–	–	A	G	∮	–	G	T
HapX	C	T	T	–	G	G	A	C	T	G	T	–	–	–	A	A	G	–	–	–	–	–	A	G	∮	–	T	T
HapY	C	T	T	–	C	G	A	C	T	G	T	–	–	–	A	A	G	–	–	–	–	–	A	A	∮	–	T	T
HapZ1	T	C	T	T	C	A	A	T	C	A	T	–	–	–	A	A	T	–	–	–	–	–	A	G	–	£	G	T
HapZ2	T	C	T	T	C	A	A	T	C	A	T	–	A	–	C	A	T	–	–	–	–	–	A	G	–	£	G	T

Adapted from Ref. [31]

#: TTAGTAGTCTTTC; ¶: CTAGACTTATTTCTTTCCATTAAG; ∮: AAAATT; £: GAAAAAAATA

4.5.3.1.2 Genetic Diversity and Genetic Structure

The cpDNA diversity of 28 wild populations of *S. baicalensis* was moderate at the level of population ($G_{ST} = 0.701$, $h_S = 0.265$, $h_T = 0.888$). When considering the genetic distance among haplotypes, the differentiation of genetic structure was much higher ($N_{ST} = 0.742$, $v_S = 0.229$, $v_T = 0.889$) and the result of U-test is significantly larger than G_{ST} ($U = 0.50$, $P < 0.01$). This result indicates that haplotypes with similar genetic distances are distributed in the same or geographically similar populations, while haplotypes with larger genetic distances are distributed in populations with geographical distances, that is, between populations that have significant phylogeographic patterns.

4.5.3.1.3 Haplotype Network Phylogenetic Tree

The distribution frequency of 27 haplotypes in 28 wild Huangqin populations and one wild population of *S. rehderiana* is shown in Table 4.4, which was drawn according to the principle of parsimony by TCS version 1.13. In the haplotype network (Fig. 4.8), the 25 haplotypes of *S. baicalensis* and two haplotypes of *S. rehderiana* were varied by at least ten mutations. There is no shared haplotype, which indicates significant genetic differentiation between the two species, not genetic infiltration or hybridization between species.

The genealogical structure of 25 haplotypes of *S. baicalensis* gave a shallow gene tree with four apparent centers: HapG (owned by 78 individuals) is positioned at the center of the phylogenetic tree and is the most primitive haplotype, which differentiates 10 haplotypes (HapB, C, U, P, D, J, N, V, Q, I) by a single mutation, which together form the most central branch I; HapB (owned by 146 individuals) further differentiates from HapI (owned by 45 individuals) to form a subnetwork, subbranch II; HapG differentiates from another haplotype HapC (owned by 127 individuals); and HapJ (owned by 37 individuals) form subbranches III and subbranches IV, respectively. The central haplotypes of the centers or subbranches of these four phylogenetic trees are the most common haplotypes, with the number of individuals with them in the top four.

4.5.3.1.4 Geographical Distribution of Haplotypes

The geographical distribution of 27 chloroplast haplotypes was edited using ArcMap 8.3 (ESRI, Inc.) (Fig. 4.9). The results showed a noticeable geographical pattern of the geographical distribution of chloroplast haplotypes in the wild populations. The most primitive HapG and the most central branch I are mainly distributed in Chengde County in Hebei province and surrounding areas, the subbranches II evolved by HapB and HapI are mainly distributed in the northwest, and the

Table 4.4 The frequency of 27 chloroplast haplotypes and their distribution in 28 wild populations of *S. baicalensis* and one wild population of *S. rehderiana* (P: Population N: Sample)

P	N	cpDNA haplotype																										
		A	B	C	D	E	F	G	H	I	J	K	L	M	N	O	P	Q	R	S	T	U	V	W	X	Y	Z1	Z2
Scutellaria baicalensis																												
HMW	24	24																										
DMW	24		7	16	1																							
BCW	19		1	11		1	3	1	2																			
YJW	21			21																								
JZW	20										8	12																
JCW	22		3	6	1			3			1					2	4	1					1					
EGW	20	4		16																								
GYW	24										2								22									
KKW	22		18																	4								
LXW	24		5																			19						
YQ1W	19														19													
CCW	22			21				1																				
KCW	17					16														1								
CD1W	18			13				5																				
CD2W	23		23																									
LPW	18			5				12															1					
HXW	22		3		4			5		6													4					
SXW	19							6		12			1															
WTW	24							24																				
FYW	23		22							1																		
LCW	21							21																				
JXW	23		23																									
HLW	24		21							1															2			
SYW	19									1			1	1				1			1			11	1	2		
TBW	24									24																		
HSW	20		20																									
YQ2W	22			18							2									2								
YTW	24										24																	
Total	602	28	146	127	6	17	3	78	2	45	37	12	2	1	19	2	4	2	22	7	1	19	6	11	3	2		
Scutellaria rehderiana																												
WYW	22																										11	11

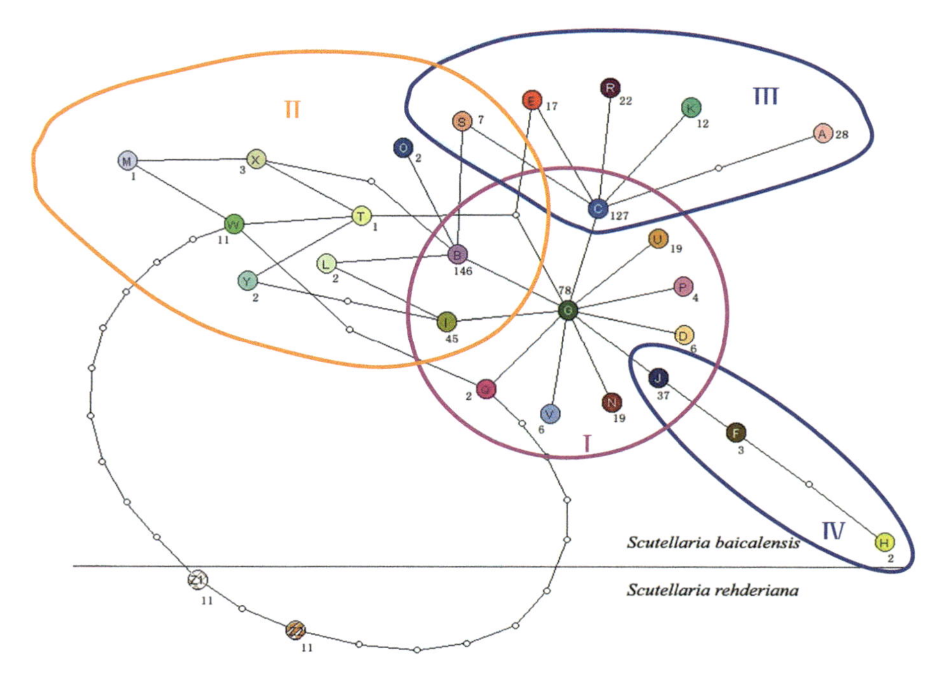

Fig. 4.8 Nested cladogram of 25 chloroplast haplotypes in *S. baicalensis* and two haplotypes in *S. rehderiana*. The bold and italic numbers besides haplotypes represent the number of wild and cultivated individuals with certain haplotype, respectively. Open small circles represent inferred interior nodes that were absent in the samples. Each branch indicates one mutation

subbranches III and subbranches IV evolved by HapC and HapJ mainly distributed in the northeast and the Shandong Peninsula, only HapR and HapJ are exceptionally distributed to the Guyang population in the west of Chengde. This might be the result of some human factors such as commercial exchanges that increase the exchange of Huangqin seeds. In addition, each region has its unique haplotypes, such as HapE, HapN, HapU, and HapS in Chengde County in Hebei province and their surrounding areas; HapI and HapW in the northwest; and HapA in the northeast.

4.5.3.1.5 Mantel Test

Mantel test analyses demonstrated that genetic distance was significantly correlated with geographical distance ($r = 0.4346$, $P < 0.0010$; Fig. 4.10) in the wild populations. The result is consistent with the results of the genetic structure showing a significant geographic structure and the clear structuring of the geographic haplotype distribution.

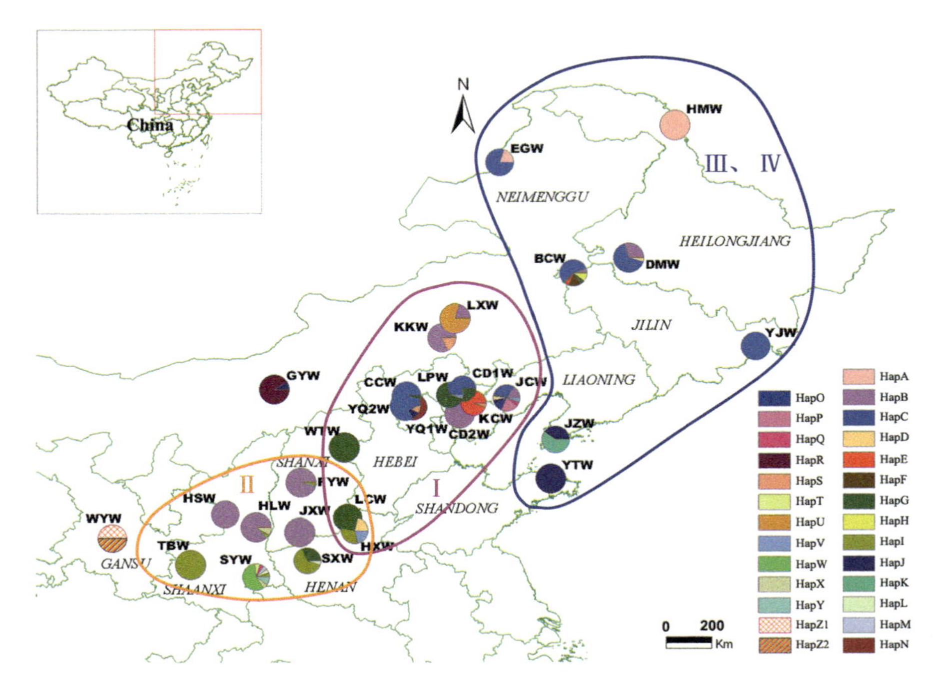

Fig. 4.9 Geographic distribution and frequencies of chloroplast haplotypes in wild *S. baicalensis* and *S. rehderiana* (WYW). Population abbreviations are the same as Table 4.1. The pie sizes of populations are proportional to their sample sizes. (Figure reproduced from Ref. [31])

4.5.3.2 Discussion

4.5.3.2.1 Genetic Diversity and Genetic Structure

Compared to most plants, wild *S. baicalensis* has a higher level of cpDNA diversity ($h_T = 0.888$), such as *Cedrela odorata* ($h_T = 0.700$) [50], *Juniperus przewalskii* ($h_T = 0.700$) [51], *Alnus glutinosa* ($h_T = 0.773$) [52], *Quercus sp.* ($h_T = 0.874$) [53], *Tilia cordata* ($h_T = 0.881$) [54], *Cyclobalanopsis glauca* ($h_T = 0.681$) [55], *Vouacapoua Americana* ($h_T = 0.87$) [56]. Only *Cycas taitungensis* ($h_T = 0.998$) [57] and *Cunninghamia lanceolate* ($h_T = 0.952$) [58] are higher than wild *S. baicalensis*. The average cpDNA diversity ($h_T = 0.67$) of 170 plants with different molecular markers was collected [59], and the cpDNA diversity of wild *S. baicalensis* was significantly higher than the average level of these plants. The genetic differentiation of *S. baicalensis* is because the distribution of *S. baicalensis* spans a wide geographical range. The diverse geographical and climatic conditions in the distribution area form a variety of vegetation types. The vegetation types from northeast to northwest include boreal deciduous forest, cold mixed forest, steppe and temperate deciduous forest. The different vegetation types undergo different genetic

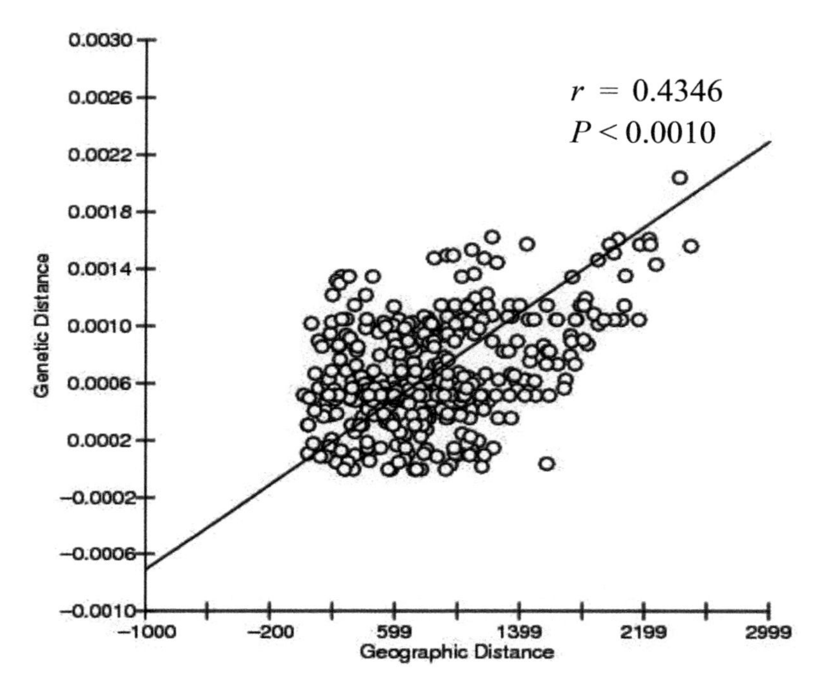

Fig. 4.10 Scatterplots of genetic distances (Kimura 2-parameter distance) against geographical distances (kilometers) separating each pairwise combination of populations within wild *S. baicalensis*

differentiation due to different geographical, climatic conditions and habitats, thus forming a higher genetic diversity of *S. baicalensis*.

Compared with the cpDNA diversity with higher species levels, there is also a certain degree of genetic diversity ($h_S = 0.265$) within the population of *S. baicalensis*, which makes the degree of genetic difference among populations relatively lower ($N_{ST} = 0.742$ and $G_{ST} = 0.701$), but the relative genetic differentiation between the populations of *S. baicalensis* is still at a medium level between other plant populations, such as *Cedrela odorata* ($N_{ST} = 0.980$, $G_{ST} = 0.960$) [50], *Juniperus przewalskii* ($N_{ST} = 0.834$, $G_{ST} = 0.772$) [51], *Alnus glutinosa* ($N_{ST} = 0.905$, $G_{ST} = 0.886$) [52], *Quercus sp.* ($N_{ST} = 0.850$, $G_{ST} = 0.830$) [53], *Tilia cordata* ($N_{ST} = 0.662$, $G_{ST} = 0.552$) [54], *Cyclobalanopsis glauca* ($N_{ST} = 0.612$, $G_{ST} = 0.702$) [55], *Vouacapoua Americana* ($G_{ST} = 0.890$) [56], *Cycas taitungensis* ($F_{ST} = 0.0056$) [57], *Cunninghamia lanceolate* ($\Phi_{ST} = 0.130$) [58]. The correlation between genetic distance and geographic distance also indicated that the variation of cpDNA was significantly structured in the geographical distribution of haplotypes. The four branches in the phylogenetic tree are distributed in three regions with distinct geographical climate and vegetation: branches III and IV are distributed in the northeastern region and the Shandong peninsula. This region is characterized by high latitude and low altitude, and the sampling population is at an elevation (below 600 m), close to the ocean; the climate is cold and humid, and the vegetation types are cold-leaved deciduous forest, cold-coniferous mixed forest, and meadow grassland. Under these conditions, the cold-wet haplotypes of branches III and IV are differentiated. Branch II is distributed in the northwestern region, characterized by low latitude and high altitude. The sampling population is between 800 and 1600m above sea level. It is far away from the ocean, the climate is warm and dry, and the vegetation types are temperate deciduous forests and mountain grasslands. Under these conditions, the warm-arid haplotypes of branch II are differentiated. Branch I is located between these two regions. The latitude, altitude, and distance from the ocean are between these two regions. The climate is warm and humid. The vegetation types are meadow grassland and temperate deciduous forest. The superior geographical and climatic conditions differentiate the warm-moist haplotype of branch I. The geographical climate and ecological environment in the distribution area of *S. baicalensis* have regular changes, which promotes the significant structural differentiation of *S. baicalensis* and finally forms the phylogeography structure of *S. baicalensis*.

4.5.3.2.2 The Daodi District Is the Center of Origin and Diversity of *S. baicalensis*

The doctors of the past generations had different understandings of the meaning of "Daodi." Huang et al. summarized the three explanations of "Daodi" in the past: ① "Daodi" refers to local specialties, which later evolved into good, high-quality and reliable pronouns. ② "Daodi" also refers to "Didao"; "Di" refers to geography, zone, topography, and landform; and "Dao" means a noun by geographical area. ③ "Dao" is a biological "population," an individual group of the same species linked by mating and kinship with a common gene bank, which is the result of a combination of genotype and environmental change. Huang Luqi et al. proposed a model hypothesis that "the more obvious the geoherbalism is, the more obvious the specialization of its genes" further expounds the relationship between geoherbalism and genetic differentiation. It is generally believed that *S. baicalensis* is a medicinal material of the wild in Chengde and its surrounding areas. Its roots are solid, with less hollow, yellow color and good quality, which is known as "Rehe Huangqin." The history of Rehe includes the current Chengde City, Kuancheng County, Luanping County, Jianchang County, Keshiketeng Banner, and Linxi County (http://www.xzqh.org/yange/chexiao/Rehe.htm); they should all be the Daodi Huangqin. In the Daodi area, the haplotype of the most central branch I is concentrated in the phylogenetic tree (Figs. 4.8 and 4.9). According to the coalescent theory, the haplotypes and branches at the center of the phylogenetic tree are the most primitive and are the ancestors of other haplotypes and subbranches [60]. The ancestor haplotypes concentrated distribution of the region is the origin of the species. From this, it can be inferred that the most primitive haplotypes are concentrated in the Daodi area of Huangqin, which is the origin of Huangqin. At the same time, the area of the Daodi area is small in the whole distribution area, but there are 12 kinds of haplotypes (HapG, B, C, D, E, J, O, P, Q, S, U, V), accounting for 48% of all haplotypes. Thus, the Daodi area is also the diversified center of Huangqin.

4.5.3.2.3 The Superior Environmental Conditions of the Glacial Shelter Have Formed the Geoherbalism of *S. baicalensis*

The glacial period (especially the Pleistocene glacial period) has had a profound influence on the spatial distribution patterns and the genetic structures of current animals and plants [61–64]. The Ice Age Refuge was in the four glacial periods of the Pleistocene, especially in the recent glacial period, which was not enclosed by the ice sheet and became a place for animals and plants to escape from the glacial period. They are the starting point and source pool of species redistribution after the glacial period [65]. Thus, the areas where the original haplotypes are concentrated and genetically diverse are often expressed as the center of origin and the center of diversity of the species. Mengel [66] believes that the repeated formation and ablation of Pleistocene glaciers separates the different populations of the species, that is, the formation of exotic populations, which eventually leads to the emergence of new species. Recently, Klicka and Zink [67] argued that the differentiation between the populations of the same species distributed in different shelters during the glacial period was exaggerated. The glacial shelters can only lead to intraspecific subspecies differentiation, but not enough to form new species. In any case, the glacial shelter has accelerated the differentiation between different populations of the species, providing conditions for the development of new species or new subspecies, which lays the foundation for the differentiation of medicinal plant quality and the formation of geoherbalism.

There are many mountain glaciers in the western mountains and the Qinghai-Tibet Plateau in China. In China, due to complex terrain conditions, and many mountains are moving east and west, the impact of the glacial climate on the organisms has been alleviated, and there are many glacial shelters. Yanshan Mountain is located in the north of Hebei Plain. The mountain is eastwest, between 39°40′–42°10′ north latitude and 115°45′–119°50′ east longitude. It is 500–1500 m over the sea level and is high in the north and low in the south. The south fell below 500 m and became a low hill. There are many basins and valleys in the mountains, such as Chengde, Kuancheng, and Yiping Valley. The east-west terrain of the Yanshan Mountains can effectively block the glaciers moving from north to south during the glacial phase, so Chengde and its surrounding areas have become glacial shelters for northern plants. When ice age came, the Huangqin, which can be found in the northeastern region, was affected by the Siberian continental ice sheet and began to grow southward. When it reached the refuge in the Chengde area, it survived because of the protection of the Yanshan Mountains.

Huangqin, which is distributed in the northwestern region, was attacked by the western alpine mountain glaciers and began to migrate to the low-altitude areas in the east. After moving to Chengde and its surrounding areas, the Taihang Mountains in the west were north-south, resisting the western mountain glaciers. So Huangqin in Chengde and surrounding areas further protected and survives well. When the temperature during the interglacial period rose, the Huangqin in the Chengde refuge began to gradually migrate to the northeast and northwest, forming different evolutionary branches that adapt to the type of climate in the northeast and northwest. It can be seen that Chengde and its surrounding areas in the Daodi District are the only glacial refuges of Huangqin. The excellent environmental conditions of the refuges have caused significant genetic differentiation in the Huangqin in these and other areas. Genetic differentiation further causes phenotypic changes, and phenotypic differentiation leads to the differentiation of medicinal quality. Eventually, the jaundice in this area becomes a local medicinal material, forming the geoherbalism of huangqin.

4.5.4 Analysis of Cultivation Origin

4.5.4.1 Results

4.5.4.1.1 Sequence Characteristics of Three Chloroplast Intergenic Spacers

The aligned sequences of *atp*B-*rbc*L, *trn*L-*trn*F, and *psb*A-*trn*H spacers in *S. baicalensis* and *S. rehderiana* had 768, 799, and 502 base pairs in length. The three spacers had 13, 7, and 14 polymorphic sites (including substitutions and indels) (Table 4.5) because the sequencing poly-N regions could easily lead to homoplasies due to polymerase error [28], and the short sequence repeats were removed from subsequent analyses. Sequence divergence was measured with Kimura 2-parameter algorithm, which varied from 0.000% to 0.262%, 0.000% to 0.257%, and 0.000% to 0.559% for *atp*B-*rbc*L, *trn*L-*trn*F, and *psb*A-*trn*H, respectively. Tajima's dcriterion (*atp*B-*rbc*L: $D = -1.53470$, $0.10 > P > 0.05$; *trn*L-*trn*F: $D = -1.31009$, $P > 0.10$; and *psb*A-*trn*H: $D = -1.38479$, $P > 0.10$) and Fu and Li's criterion (*atp*B-*rbc*L: $D* = -1.66525$, $P > 0.10$; $F* = -1.79736$, $P > 0.10$; *trn*L-*trn*F: $D* = -1.40980$, $P > 0.10$; $F* = -1.51361$, $P > 0.10$, and *psb*A-*trn*H: $D* = -1.28584$, $P > 0.10$; $F* = -1.48316$, $P > 0.10$). These results indicate that all three fragments are neutral. The sequences of nine *atp*B-*rbc*L, eight *trn*L-*trn*F, and 13 *psb*A-*trn*H haplotypes have been placed in the GenBank databases [GenBank: GQ374124-GQ374155]. Thirty-two haplotypes of *S. baicalensis* (HapA-Y and Hap1–7) and two haplotypes of *S. rehderiana* (HapZ1-Z2) were recognized when *atp*B-*rbc*L, *trn*L-*trn*F, and *psb*A-*trn*H sequences were merged (Tables 4.5).

Table 4.5 Variable sites of the aligned sequences of three chloroplast DNA fragments in 32 haplotypes of wild and cultivated *S. baicalensis* and two haplotypes of *S. rehderiana* (∗, #, ¶, ‡, §, and £ denote six indels)

	Nucleotide position																																	
	atpB-rbcL													*trnL-trnF*							*psbA-trnH*													
	6	1	1	2	2	2	3	3	5	5	6	6	7	2	1	2	2	4	4	7	4	7	1	1	2	2	2	2	3	3	3	4	4	4
	0	9	9	0	0	1	1	8	0	0	0	5	2	2	7	6	6	6	8	2	8	3	5	8	1	3	6	9	2	6	7	3	4	5
Haplotype		3	5	2	3	4	0	1	3	7	3	6	3		7	4	5	3	4	6			1	8	2	6	0	5	6	6	9	5	7	1
HapA	*C*	*C*	*T*	*T*	–	–	*C*	*G*	*A*	*C*	*T*	*G*	*T*	–	–	*A*	–	–	*C*	*C*	*C*	*G*	–	–	–	–	–	–		*G*	§	–	*T*	*A*
HapB	*C*	*C*	*T*	*T*	–	–	*G*	*G*	*A*	*C*	*T*	*G*	*T*	–	–	*A*	–	–	*C*	*A*	*C*	*G*	–	–	–	–	–	–	*A*	*G*	§	–	*T*	*T*
HapC	*C*	*C*	*T*	*T*	–	–	*C*	*G*	*A*	*C*	*T*	*G*	*T*	–	–	*A*	–	–	*C*	*A*	*C*	*G*	–	–	–	–	–	–	*A*	*G*	§	–	*T*	*A*
HapD	*C*	*C*	*T*	*T*	–	–	*C*	*G*	*A*	*C*	*T*	*G*	*G*	–	–	*A*	–	–	*C*	*A*	*C*	*G*	–	–	–	–	–	–	*A*	*G*	§	–	*T*	*T*
HapE	*C*	*C*	*T*	*T*	–	–	*C*	*G*	*A*	*C*	*T*	*G*	*T*	–	–	-	–	–	*C*	*A*	*C*	*G*	–	–	–	–	–	–	*A*	*G*	§	–	*T*	*A*
HapF	*C*	*C*	*T*	*T*	–	–	*C*	*G*	*A*	*C*	*T*	*G*	*T*	–	–	*A*	–	–	*C*	*A*	*C*	*G*	–	¶	¶	–	–	–	*A*	*G*	§	–	*T*	*T*
HapG	*C*	*C*	*T*	*T*	–	–	*C*	*G*	*A*	*C*	*T*	*G*	*T*	–	–	*A*	–	–	*C*	*A*	*C*	*G*	–	–	–	–	–	–	*A*	*G*	§	–	*T*	*T*
HapH	*C*	*C*	*T*	*T*	–	–	*C*	*G*	*A*	*C*	*T*	*G*	*T*	–	–	*A*	–	–	*C*	*A*	*C*	*G*	–	¶	¶	¶	¶	–	*A*	*G*	§	–	*T*	*T*
HapI	*C*	*C*	*T*	*T*	–	–	*C*	*G*	*A*	*C*	*T*	*G*	*T*	–	–	*A*	–	–	*C*	*A*	*C*	*G*	–	–	–	–	–	–	*A*	*A*	§	–	*T*	*T*
hapJ	*C*	*C*	*T*	*T*	–	–	*C*	*G*	*A*	*C*	*T*	*G*	*T*	–	–	*A*	–	–	*C*	*A*	*C*	*G*	–	¶	–	–	–	–	*A*	*G*	§	–	*T*	*T*
HapK	*C*	*C*	*T*	*T*	–	–	*C*	*G*	*C*	*C*	*T*	*G*	*T*	–	–	*A*	–	–	*C*	*A*	*C*	*G*	–	–	–	–	–	–	*A*	*G*	§	–	*T*	*A*
HapL	*C*	*C*	*T*	*T*	–	–	*G*	*G*	*A*	*C*	*T*	*G*	*T*	–	–	*A*	–	–	*C*	*A*	*C*	*G*	–	–	–	–	–	–	*A*	*A*	§	–	*T*	*T*
HapM	*C*	*C*	*T*	*T*	–	–	*G*	*G*	*A*	*C*	*T*	*G*	*T*	–	–	–	–	–	*A*	*A*	*C*	*G*	–	–	–	–	–	–	*A*	*G*	§	–	*G*	*T*
HapN	*C*	*C*	*T*	*T*	–	–	*C*	*G*	*A*	*C*	*T*	*G*	*T*	–	–	*A*	–	*A*	*C*	*A*	*C*	*G*	–	–	–	–	–	–	*A*	*G*	§	–	*T*	*T*
HapO	*C*	*C*	*T*	*T*	–	–	*G*	*G*	*A*	*C*	*T*	*G*	*T*	–	–	*A*	–	–	*C*	*A*	*C*	*G*	–	–	–	–	–	–	*A*	*G*	§	–	*T*	*T*
HapP	*C*	*C*	*T*	*T*	–	–	*C*	*G*	*A*	*C*	*T*	*G*	*T*	–	*T*	*A*	–	–	*C*	*A*	*C*	*G*	–	–	–	–	–	–	*A*	*G*	§	–	*T*	*T*
HapQ	*C*	*C*	*T*	*T*	–	–	*C*	*G*	*A*	*C*	*T*	*G*	*T*	–	–	*A*	–	–	*C*	*A*	*C*	*G*	–	–	–	–	–	–	*A*	*G*	§	–	*G*	*T*
HapR	*C*	*C*	*T*	*T*	–	*T*	*C*	*G*	*A*	*C*	*T*	*G*	*T*	–	–	*A*	–	–	*C*	*A*	*C*	*G*	–	–	–	–	–	–	*A*	*G*	§	–	*T*	*A*
HapS	*C*	*C*	*T*	*T*	–	–	*G*	*G*	*A*	*C*	*T*	*G*	*T*	–	–	*A*	–	–	*C*	*A*	*C*	*G*	–	–	–	–	–	–	*A*	*G*	§	–	*T*	*A*
HapT	*C*	*C*	*T*	*T*	–	–	*C*	*G*	*A*	*C*	*T*	*G*	*T*	–	–	-	–	–	*A*	*A*	*C*	*G*	–	–	–	–	–	–	*A*	*G*	§	–	*T*	*T*

(continued)

Table 4.5 (continued)

Haplotype	*Nucleotide position*																																
	atpB-rbcL													*trnL-trnF*						*psbA-trnH*													
	6	1	1	2	2	2	3	3	5	5	6	6	7	2	1	2	2	4	4	7	4	7	1	1	2	2	2	2	3	3	3	4	4	4
	0	9	9	0	0	1	1	8	0	0	0	5	2	2	7	6	6	6	8	2	8	3	5	8	1	3	6	9	2	6	7	3	4	5
Haplotype		3	5	2	3	4	0	1	3	7	3	6	3		7	4	5	3	4	6			1	8	2	6	0	5	6	6	9	5	7	1
HapU	*C*	*C*	*T*	*T*	–	–	*C*	*G*	*A*	*C*	*T*	*G*	*T*	–	–	*A*	–	–	*C*	*A*	*C*	*G*	#	–	–	–	–	–	*A*	*G*	§	–	*T*	*T*
HapV	*C*	*C*	*T*	-	–	–	*C*	*G*	*A*	*C*	*T*	*G*	*T*	–	–	*A*	–	–	*C*	*A*	*C*	*G*	–	–	–	–	–	–	*A*	*G*	§	–	*T*	*T*
HapW	*C*	*C*	*T*	*T*	–	–	*C*	*G*	*A*	*C*	*T*	*G*	*T*	–	–	–	–	–	*A*	*A*	*C*	*G*	–	–	–	–	–	–	*A*	*G*	§	–	*G*	*T*
HapX	*C*	*C*	*T*	*T*	–	–	*G*	*G*	*A*	*C*	*T*	*G*	*T*	–	–	–	–	–	*A*	*A*	*C*	*G*	–	–	–	–	–	–	*A*	*G*	§	–	*T*	*T*
HapY	*C*	*C*	*T*	*T*	–	–	*C*	*G*	*A*	*C*	*T*	*G*	*T*	–	–	–	–	–	*A*	*A*	*C*	*G*	–	–	–	–	–	–	*A*	*A*	§	–	*T*	*T*
HapZ1	*T*	*C*	*C*	*T*	–	*T*	*C*	*A*	*A*	*T*	*C*	*A*	*T*	–	–	–	–	–	*A*	*A*	*C*	*T*	–	–	–	–	–	–	*A*	*G*	–	£	*G*	*T*
HapZ2	*T*	*C*	*C*	*T*	–	*T*	*C*	*A*	*A*	*T*	*C*	*A*	*T*	–	–	*A*	–	–	*C*	*A*	*C*	*T*	–	–	–	–	–	–	*A*	*G*	–	£	*G*	*T*
Hap1	*C*	*C*	*T*	*T*	–	–	*C*	*G*	*A*	*C*	*T*	*G*	*T*	–	–	*A*	*A*	–	*C*	*A*	*C*	*G*	–	–	–	–	–	–	*A*	*G*	§	–	*T*	*T*
Hap2	*C*	*T*	*T*	*T*	–	–	*C*	*G*	*A*	*C*	*T*	*G*	*T*	–	–	*A*	–	–	*C*	*A*	*C*	*G*	–	–	–	–	–	–	*A*	*G*	§	–	*T*	*T*
Hap3	*C*	*C*	*T*	*T*	–	–	*G*	*G*	*A*	*C*	*T*	*G*	*T*	–	–	*A*	–	–	*C*	*A*	*T*	*G*	–	–	–	–	–	–	*A*	*G*	§	–	*T*	*T*
Hap4	*C*	*C*	*T*	*T*	–	–	*G*	*G*	*A*	*C*	*T*	*G*	*T*	*	–	*A*	–	–	*C*	*A*	*C*	*G*	–	–	–	–	–	–	*A*	*G*	§	–	*T*	*T*
Hap5	*C*	*C*	*T*	*T*	–	–	*C*	*G*	*A*	*C*	*T*	*G*	*T*	*	–	*A*	–	–	*C*	*A*	*C*	*G*	–	–	–	–	–	–	*A*	*A*	§	–	T	T
Hap6	C	C	T	T	T	–	*C*	*G*	*A*	C	T	*G*	T	–	–	*A*	–	–	C	*A*	C	*G*	–	–	–	–	–	–	*A*	*A*	§	–	T	T
Hap7	C	C	T	T	–	–	*G*	*G*	*A*	C	T	*G*	T	–	–	*A*	–	–	C	*A*	C	*G*	–	–	–	–	–	‡	*A*	*G*	§	–	T	T

Table form Ref. [31]

*: CTGAAAAC; #: TTAGTAGTCTTTC; ¶: CTAGACTTATTTCTTTCCATTAAG; ‡: TTCCATTAAGAATAAATAAAG; ∮: AAAATT; £: GAAAAAAATA

4.5.4.1.2 Haplotype Network and Distribution

The 32 haplotypes of *S. baicalensis* and two haplotypes of *S. rehderiana* varied by at least ten mutations in the haplotype network (Fig. 4.11). The genealogical structure of 32 haplotypes of *S. baicalensis* showed a shallow gene tree with three visible centers: HapG, HapB, and HapC, which connected to the other 10, 6, and 4 haplotypes by just one mutation, respectively. Furthermore, both HapB and HapC were connected to HapG by one mutation. Remarkably, several haplotypes have only been detected in the wild populations, which are relatively distant from the three centers of the phylogenetic tree, such as A, H, M, T, W, X, and Y. The haplotypes of cultivated populations (symbolized by ovals and circles) consistently distributed across the haplotype network, which not showing any clusters.

The haplotype frequencies in each population and geographical distribution are presented in Fig. 4.12, Table 4.4, and Table 4.6. The haplotype distribution of wild and cultivated populations extremely differed based on Fig. 4.12. The distribution of

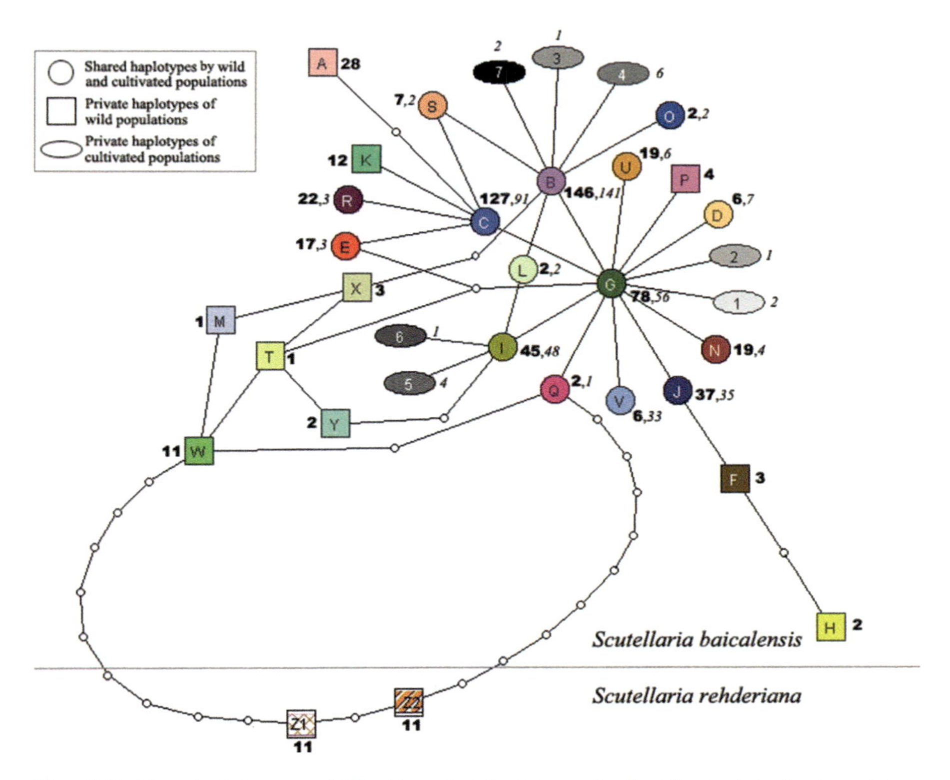

Fig. 4.11 Nested cladogram of 32 chloroplast haplotypes in *Scutellaria baicalensis* and two haplotypes in *Scutellaria rehderiana*. The bold and italic numbers besides haplotypes represented the number of wild and cultivated individuals with a certain haplotype, respectively. The small open circles represent the inferred interior nodes that were non-existent in the samples. Nevertheless, each branch demonstrated one mutation. (Figure form Ref. [31])

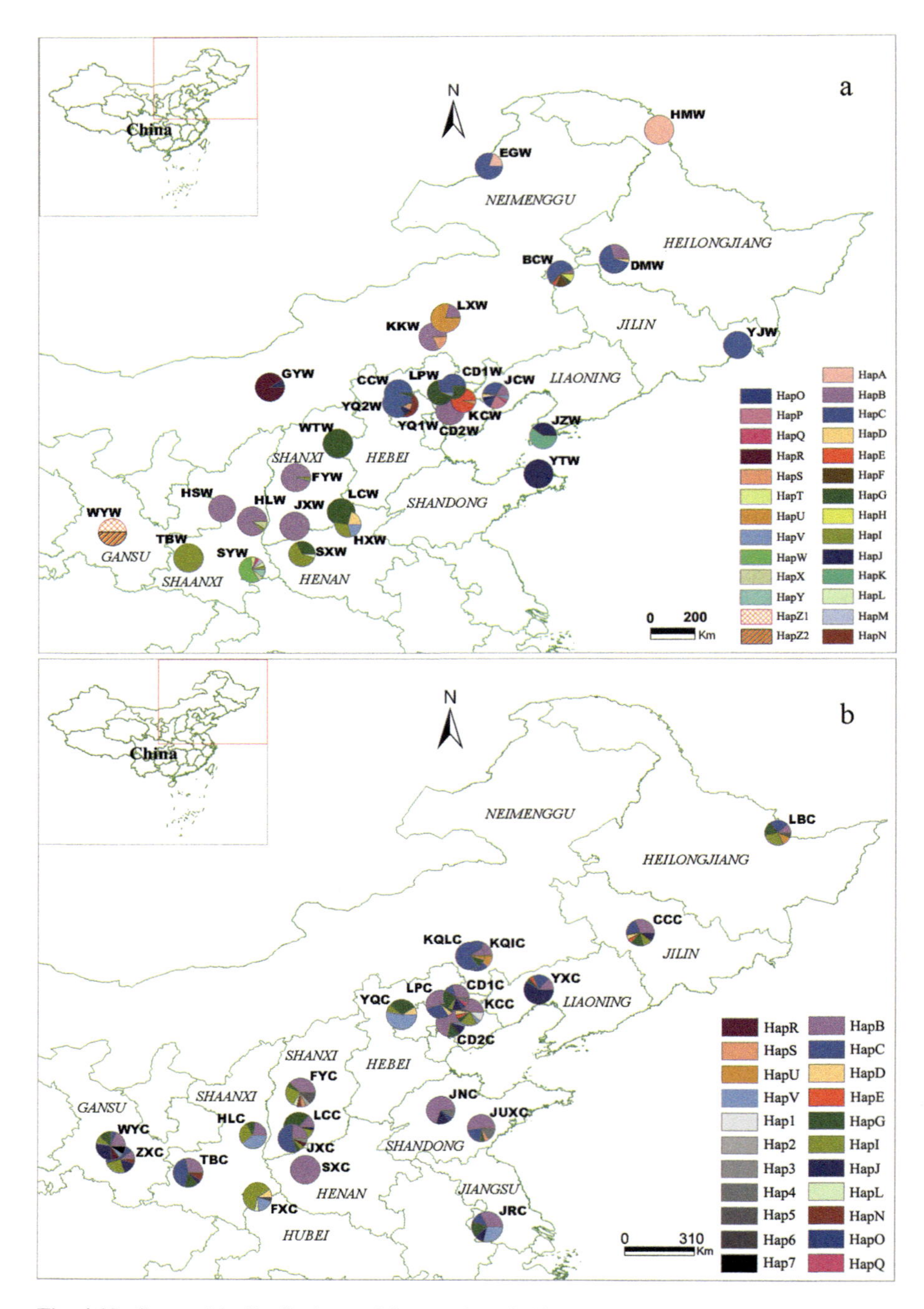

Fig. 4.12 Geographic distribution and frequencies of chloroplast haplotypes in wild *Scutellaria baicalensis* and *Scutellaria rehderiana* (WYW) (**a**) and cultivated *Scutellaria baicalensis* (**b**) populations. (Figure form Ref. [31])

Table 4.6 Chloroplast haplotype frequencies in 22 cultivated populations of *Scutellaria baicalensis*

		cpDNA haplotype																					
P	N	B	C	D	E	G	I	J	L	N	O	Q	R	S	U	V	1	2	3	4	5	6	7
KQLC	20	2	15			2									1								
KQIC	24	4	17												3								
LBC	17	2	4			3	5								2						1		
CCC	19	6	4	1	1	3	2	2															
YXC	22	3	4		1	1		13															
CD1C	17	5	2			4	1	2			1	1								1			
CD2C	21	15				3		2								1							
KCC	19	8	1	1	1	1	4									1	2						
LPC	20	11	6	1		2																	
YQC	23			2		7	2									12							
JNC	20	14						3			1									2			
JUXC	19	10	4			1								1						3			
JRC	24	8	3			4		2								7							
SXC	21	21																					
FXC	21			2			13		1							4						1	
FYC	21	8				1	5		1	1				1				1			3		
LCC	22	3				15	3	1															
JXC	20	5	11			1	1						1						1				
HLC	18	4				2	5									7							
TBC	22	6	10			3		1		2													
WYC	21	4	1			3	2	6					2			1							2
ZXC	20	2	9				5	3		1													
Total	451	141	91	7	3	56	48	35	2	4	2	1	3	2	6	33	2	1	1	6	4	1	2

Abbreviations: *P* population code, *N* number of sampled individuals

haplotypes in wild populations has a distinct geographic structure compared to that in cultivated populations. For example, HapG in wild populations (WTW, LCW, LPW, CD1W, SXW, and HXW) were mostly restricted to the central range of this species, but this haplotype occurred in 17 cultivated populations across the whole range. Another striking phenomenon is that most wild populations (10/28) have only one haplotype, while in the cultivated population, except for one population (SXC), which has only one haplotype, the rest are more than 2 haplotypes.

4.5.4.1.3 Genetic Diversity and Genetic Structure

Thirty-two haplotypes of *S. baicalensis* were detected, of which 25 (78% of the total number of haplotypes) were recovered in wild populations, and cultivated individuals carried 22 (69% of the total number of haplotypes) (Fig. 4.13, Tables 4.4 and 4.6). Fifteen haplotypes (47% of the total number of haplotypes) were combined with wild and cultivated populations (Fig. 4.13). Ten haplotypes were discovered in the wild populations but not in cultivated ones, and seven were found individually in cultivated populations (Fig. 4.13). The number of haplotypes and the relative abundance of each haplotype have a slight change under the anthropogenic influence

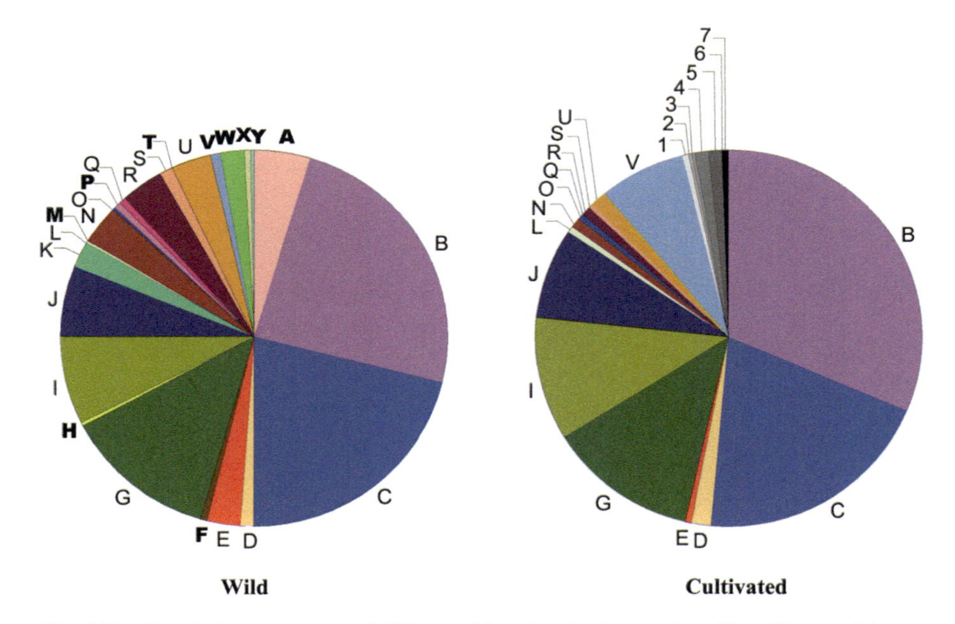

Fig. 4.13 The relative proportions of different chloroplast haplotypes found in wild and cultivated *Scutellaria baicalensis*. The color of haplotypes refers to Fig. 4.10. Twenty-five haplotypes were found in wild populations, and 22 haplotypes were recovered in cultivated populations. Fifteen haplotypes (thin letters) are shared by wild and cultivated groups. Ten (bold letters) and seven (numbers 1–7) haplotypes are specific to wild and cultivated groups, respectively. (Figure form Ref. [31])

Table 4.7 Comparisons of genetic diversity and genetic structure between wild and cultivated *Scutellaria baicalensis* populations

Parameter	Wild	Cultivated	*P*
Number of haplotype	25	22	0.733†
Total diversity, h_T	0.888 (0.0287)	0.832 (0.0234)	>0.05‡
Within-population diversity, h_S	0.265 (0.0526)	0.649 (0.0425)	**<0.001‡**
Population differentiation, G_{ST}	0.701 (0.0594)	0.220 (0.0449)	**<0.001‡**

Table form Ref. [31]
Parameters of population subdivision are followed by standard error in parentheses
Statistically significant comparisons are highlighted in bold type
†: x^2 test; ‡: Wilcoxon two-group test

Table 4.8 Heirarchical analysis of molecular variance for 50 populations of *Scutellaria baicalensis*. All populations are partitioned into cultivated and wild groups

HHierarchical level	Deg. of freedom	Sum of squares	Variance components	Percentage of variance	F-statistics	*P*-value*
Among groups	1	7.740	0.00055	0.09	$F_{CT} = 0.00091$	<0.001
Among populations within groups	48	356.792	0.34067	56.61	$F_{ST} = 0.56698$	<0.001
Within populations	1003	261.380	0.26060	43.30	$Fsc = 0.56659$	<0.001

Table form Ref. [31]
*Significance calculated using 1023 permutations

during the cultivation (Fig. 4.13, Tables 4.4 and 4.6). However, the results of the x^2 test (P = 0.733; Table 4.8) indicate that the difference in haplotype frequencies between wild and cultivated populations is not significant, as the cultivated population lacks only rare haplotypes (Fig. 4.13). The results of the Wilcoxon two-group test (Tables 4.7) are consistent with the results of the haplotype frequency distribution (Fig. 4.12, Tables 4.4 and 4.6). The total genetic diversity was not significantly different between cultivated and wild populations ($h_T = 0.832$ in cultivated and $h_T = 0.888$ in wild, P > 0.05), but the genetic diversity within the population of cultivated populations ($h_S = 0.649$) was significantly higher than the wild populations ($h_S = 0.265$, P < 0.001).

The population subdivision of wild populations ($G_{ST} = 0.701$) was significantly greater compared to that of the cultivated crops ($G_{ST} = 0.220$; $P < 0.001$, Wilcoxon two-group test; Table 4.7). Mantel test analyses indicated that genetic distance and geographical distance was significantly correlated ($r = 0.4346$, $P < 0.0010$; Fig. 4.14) in wild populations. However, this pattern was not improved in cultivated populations ($r = 0.0599$, $P = 0.2710$; Fig. 4.14). AMOVA analysis indicated that little genetic variation could occur between cultivated and wild populations (0.09%, $P < 0.001$). Most of the genetic variance was among populations (56.61%) and within populations (43.30%; Table 4.8). The genetic difference ($F_{ST} = 0.022$)

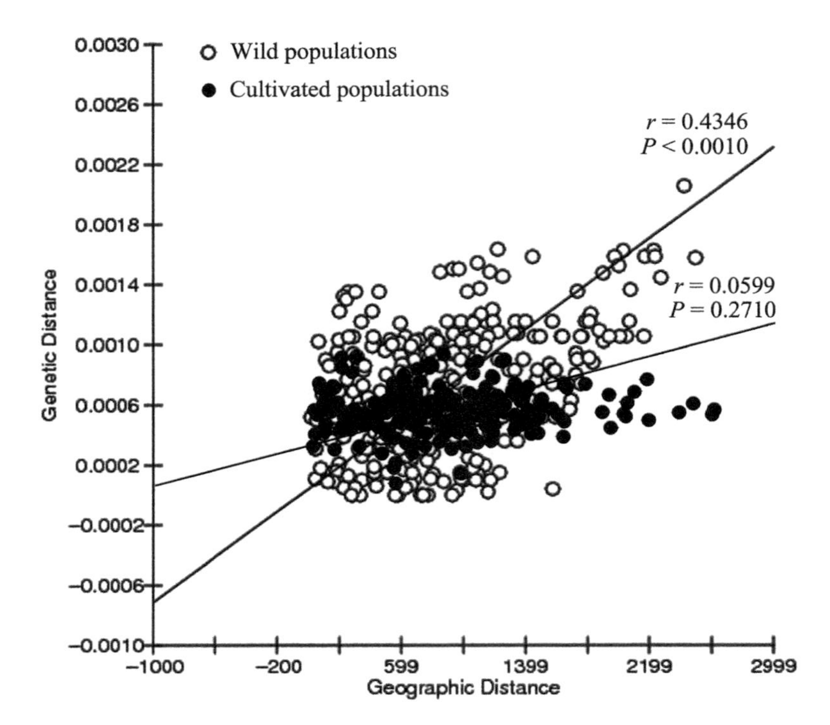

Fig. 4.14 Scatterplots of genetic distances (Kimura 2-parameter distance) against geographical distances (kilometers) separating each pairwise combination of populations within wild and cultivated *Scutellaria baicalensis*. (Figure form Ref. [31])

between the cultivated and wild groups was calculated by DNASP 4.00 and well consistent with the result of the AMOVA analysis.

4.5.4.2 Discussion

4.5.4.2.1 The Slight Reduction of Genetic Diversity and Little Alteration of Genetic Composition in Cultivated *S. baicalensis*

Cultivation of wild plants are always producing genetic bottlenecks and thus results in the loss of genetic diversity due to the founder impacts and unconscious or conscious choices [68]. For example, a total of 23% of chloroplast haplotypes had been detected in cultivated *Oryza rufipogon*, and cultivated *O. sativa* [69]. Nevertheless, the impact of cultivation bottlenecks are mostly focused on economically valuable plants with long domestication history and little impact of cultivated medicinal plants. In this study, a slight reduction of genetic diversity and a slight alteration of genetic composition during the cultivation of *S. baicalensis* were discovered (Fig. 4.12 and Table 4.7). The results are in sharp contrast to most crop species with long cultivation histories and might be clarified by the sizeable original population, size of cultivated *S. baicalensis*, the short cultivation history, and weak artificial selection pressure.

The extent of a genetic bottleneck during cultivation is controlled by two interacting factors: the size of the bottlenecked population and the bottleneck's period [68, 70]. Instinctively, cultivated plants of multiple origins (multiple places and multiple events) could hold a more significant population size and thus most possibly maintain greater genetic diversity [71], although artificial selection and founder-event-induced genetic drift throughout subsequent domestication can reduce genetic diversity in multiple-origin plants [28, 70]. A minority of empirical studies has supported this opinion (e.g., *Magnolia Officinalis* Var. *Biloba* [72], einkorn [73], *Spondias purpurea* [74]). In this study, the different lines of evidence indicated that cultivated *S. baicalensis* could have originated from different places multiple times, which results in a sizeable initial population size during cultivation. Initially, a large proportion of haplotypes (47% of the total number of haplotypes) were discovered among wild and cultivated populations. Additionally, these shared haplotypes were recovered from different wild populations (Figs. 4.12a and 4.13). This similar distribution pattern of shared alleles in cultivated and wild populations are strongly suggesting that the different geographic regions of genetically distinct individuals were taken for cultivation and subsequently distributed by humans [28, 74]. Additionally, no cluster of cultivated haplotypes was observed in the haplotype network (Fig. 4.11), indicating that cultivated *S. baicalensis* is of multiple origins due to crops of single origin often form a monophyletic clade in a phylogenetic tree that includes wild progenitors [72].

Huangqin was rarely cultivated untill the foundation of the People's Republic of China in 1949 consistent with the great literature of traditional Chinese medicine. Nevertheless, the wild resources of this species declined rapidly because of overexploitation over the past several decades. Large-scale cultivation began due to the growing demand for the root of this species since 1958. The combination with the large initial population size might be improbable to have a strong genetic bottleneck in such a short cultivation history, which resulting in a significant reduction of genetic diversity and change in the genetic composition of cultivated *S. baicalensis*. In addition, the other population genetic studies stated that the recently cultivated Chinese medicinal plants could have the high genetic diversity in cultivated populations due to their slight cultivation bottlenecks [72, 76, 77], which offers genetic raw materials for breeding new varieties of medicinal plants.

4.5.4.2.2 Extensive Seed Exchange in Cultivated *S. baicalensis* Populations

Cultivation influences not only the amount of genetic variation in cultivated populations but also the structure of this variation [74, 78, 79]. In this study, we discovered that the cultivated *S. baicalensis* populations showed a lower proportion of genetic variation in populations ($G_{ST} = 0.220$) than the wild populations ($G_{ST} = 0.701$). Furthermore, Mantel test analyses indicated that the genetic structure was significantly affected by the geographical isolation in the wild populations ($r = 0.4346$, $P < 0.0010$; Fig. 4.14) but not in cultivated populations ($r = 0.0599$, $P = 0.2710$; Fig. 4.14). These results proposed that cultivation has imposed a profound influence on the genetic structure of cultivated *S. baicalensis*. Nonetheless, these findings run contrary to the results of Hamrick and Godt, which stated that the mean value of genetic difference among crop species ($G_{ST} = 0.339$) is greater than that of the noncrop species ($G_{ST} = 0.212$).

There are two possible factors that might cause the unusual pattern revealed in *S. baicalensis*. Initially, there has been insufficient time for artificial selection and for breeding to have an effect on the genetic deviation of the cultiva of *S. baicalensis*. The crop species (i.e., *Zea mays*, *Hordeum vulgare*, etc.) synopsized in the review of Hamrick and Godt have long histories of cultivation. Moreover, long-term artificial selection and inbreeding within cultivars imposed by plant breeding could have resulted in the genetic variation occurring commonly among cultivars [78, 80, 81]. Conversely, *S. baicalensis* because of its short cultivation history, artificial selection and breeding cultivars are improbable to support the genetic differentiation of cultivated populations as suggested by a few *S. baicalensis* cultivars in China. Hence, this fact indicates that the standard industrial cultivation, a practice possibly imposing high artificial selection pressure on medicinal herbs, has not yet been established for *S. baicalensis*.

Subsequently, homogenization among cultivated populations, together with high within-population diversity, could have been simplified by extensive seed exchange among different geographic locations because of the highly developed transportation and commercial markets of modern times. A study of pepino (*Solanum muricatum*), a herbaceous Andean domesticated plant, indicated that human interchange among communities and countries could have weakened the geographic differentiation of genetic diversity of cultivated pepinos in the past 50 years [82]. Moreover, the bulking and mixing of seeds from different geographic locations is the common practice, which produced lower genetic differentiation among cultivated populations in Chinese cultivated medicinal plants, such as *Coptis Chinensis* and *Magnolia* [72, 77]. Nevertheless, the genetic markers (ISSR and AFLP) used in those studies are parentally transmitted, incapable of ruling out the effects of seed flow on genetic structure. In this study, to trace seed flow among populations by using maternally inherited chloroplast markers [69, 74], we discovered that geographical isolation has significant effect in wild populations ($r = 0.4346$, $P < 0.0010$; Fig. 4.14) but disappeared in the cultivated populations ($r = 0.0599$, $P = 0.2710$; Fig. 4.14). Furthermore, the haplotype distribution maps also picture that common haplotypes are much more widespread in cultivated *S. baicalensis* than in the wild populations. HapG in wild populations (WTW, LCW, LPW, CD1W, SXW, and HXW) is mostly limited to the central range of this species, whereas this haplotype occurred in 17 of 22 cultivated populations (Fig. 4.12). These patterns demonstrate that homogenization among cultivated populations and high within-population diversity should be mostly because of seed exchange facilitated by human activities under cultivation.

4.5.4.2.3 Conservation Implications of *S. baicalensis* Cultivation

There is always arguement about traditional agriculture can serve as an important reservoir of genetic variability [83, 84]. An example of jocote (*Spondias purpurea*), a small tree that bears fruit like little mongo in Mesoamerica, out of its natural, wild habitats and planting them in yards and other means of cultivation, farmers in the Mesoamerican region have helped to preserve the jocote's diversity [75]. The traditional methods of cultivation is also a successful way to preserve and conserve the gene pools of medicinal plants [72, 76]. Based on our investigation, traditional cultivation methods are also used for *S. baicalensis*. Generally, farmers pick up *S. baicalensis* seeds directly from the wild and then cultivate them in their fields. Occasionally, germplasm could be distributed to other places by their relatives or friends or by selling in the market. Moreover, a small reduction of genetic diversity and a slight change of genetic composition, as well as a few specific haplotypes of the cultivated group in cultivated *S. baicalensis*, confirmed the efficiency of local cultivation for saving plant diversity. This study further indicated that the introduction of wild medicinal plants into cultivation from different locations might help to expand the genetic background of cultivated populations, which will be valuable for the sustainable utilization of natural resources.

Cultivation is a successful strategy for conserving the genetic resources of *S. baicalensis*, but the wild resources need to be protected *in situ*. In this study, we did not find a significant decrease of genetic diversity and a visible change in the genetic composition of cultivated *S. bicalensis*, these patterns are mainly reflected the changes of common haplotypes rather than rare alleles. In addition, ten out of 25 haplotypes within wild populations have been lost during the cultivation of *S. bicalensis*. The results proved the theoretical expectation which is the bottlenecks of short duration might have a slight effect on heterozygosity (here represented by haplotype diversity, h_T) but the number of rare alleles will decrease severely [85]. Rare alleles are usually considered a minor element in genetic conservation programs and yet they could be hugely significant for long-term evolution or to encounter new breeding objectives such as resistance to introduced insects or diseases [85]. Therefore, it is necessary to preserve rare alleles of *S. baicalensis* through the *in situ* conservation of wild populations.

Seed exchange during the cultivation of *S. baicalensis* has a significant impact on the genetic architecture of cultivated *S. baicalensis*, and it might also have profound implications for the conservation and utility of *S. baicalensis* germplasm. Extensive seed exchange induced genetic mixture, which means that a fraction of cultivated populations may be sufficient to preserve the genetic variation of *S. baicalensis*. This will certainly facilitate the conservation of the genetic diversity of *S. baicalensis* through cultivation. Nevertheless, numerous theoretical and empirical studies have proposed that the genetic mixture of populations that are adapted to the different local conditions could result in outbreeding depression, and the decrease in fitness led to the breakdown of coadapted gene complexes [86]. Extensive seed exchange may increase the risk of maladaptation and reduced growth or fertility resulting from maladaptation could reduce the success of cultivation projects and endanger the long-term survival of wild populations [87]. The impact of extensive seed exchange on the evolution and actual production of *S. baicalensis* has remained unknown up to now. Thus, more researches are urgently needed to further monitor the possible negative impact of extensive seed exchange on the adaptation of wild populations and the growth of cultivated populations.

4.5.4.3 Conclusions

There were thirty-two haplotypes of *S. baicalensis* (HapA-Y and Hap1-7) been recognized when three chloroplast spacers were combined. These haplotypes established a shallow gene tree without apparent clusters for cultivated populations, which indicated the multiple origins of cultivated *S. baicalensis*. The cultivated populations ($h_T = 0.832$) kept comparable genetic variation with wild populations ($h_T = 0.888$), which demonstrated a slight genetic bottleneck due to the multiple origins of cultivation. Conversely, a considerable number of rare alleles (10 out of 25 haplotypes within wild populations) were lost during the cultivation of *S. bicalensis*. The genetic difference for the cultivated group ($G_{ST} = 0.220$) was significantly lower than that of the wild group ($G_{ST} = 0.701$). Isolation by distance analysis indicated that the influence of geographical isolation on genetic structure was significant among wild populations ($r = 0.4346$, $P < 0.0010$) but not in cultivated populations ($r = 0.0599$, $P = 0.2710$). These genetic distribution patterns demonstrated a transient cultivation history and an extensive seed change among different geographical areas during the cultivation of *S. bicalensis*. Although the total genetic diversity maintained in cultivated *S. baicalensis* is comparable to wild populations due to the large initial population size and the short cultivation history, substantial rare alleles have lost and extensive seed exchange has caused the homogenization of cultivated populations during the course of cultivation. This study indicates that conservation by cultivation is an efficient approach for protecting the genetic resources of *S. baicalensis*. Nevertheless, because the modern market and the development of transportation, the hybridization of germplasm resources is widespread, which is likely to cause problems such as impure variety and outcrossing; hence, this needs to be paid attention to during the cultivation process.

With the increased realization that some wild species are being over-exploited, cultivation of medicinal plants is not only a means for meeting current, but also a means of relieving harvest pressure on wild populations [88]. Nonetheless, this research indicates that the introduction and cultivation of a medicinal plant might have a profound impact on its genetic diversity patterns and even its evolutionary potential. Although, the cultivated *S. baicalensis* maintained the total genetic diversity as well as the wild populations because of its short cultivation history and large initial population size, but the substantial rare alleles have been lost, and extensive seed exchange has led to the genetic homogenization among the cultivated populations. This research not only provides baseline data for conservation genetic resource of *S. baicalensis* through conservation-by-cultivation approach, but also represents a paradigm for evaluating the genetic impacts of recent cultivation on medicinal plants, which may be instructive to future cultivation projects of traditional Chinese medicinal plants.

References

1. Arbogast BS: Phylogeography. The history and formation of species. Integr Comp Biol. 2001;41(1):134–5.
2. Moritz CDT, Brown WM. The mitochondrial DNA bridge between population genetics and systematics. Annu Rev Ecol Syst. 1987;18:269–92.
3. Bai WN, Zhang DY. Current status and future directions in plant phylogeography. Chin Bull Life Sci. 2014;26(2):125–37.
4. Avise JC, Arnold J, Ball RM, et al. Intraspecific phylogeography: the mitochondrial DNA bridge between population genetics and systematics. Annu Rev Ecol Syst. 1987;18:489–522.
5. Avise JC. Phylogeography: the history and formation of species. London: Harvard University Press; 2000.
6. Ge XJ, Chiang YC, Chou CH, et al. Nested clade analysis of Dunnia sinensis (Rubiaceae), a monotypic genus from China based on organelle DNA sequences. Conserv Genet. 2002;3 (4):351–62.
7. Ge XJ, Liu MH, Wang WK, et al. Population structure of wild bananas, Musa balbisiana, in China determined by SSR fingerprinting and cpDNA PCR-RFLP. Mol Ecol. 2005;14 (4):933–44.
8. Wang HW, Ge S. Phylogeography of the endangered *Cathaya argyrophylla* (Pinaceae) inferred from sequence variation of mitochondrial and nuclear DNA. Mol Ecol. 2006;15:4109–23.
9. Gao LM, Möller M, Zhang XM, et al. High variation and strong phylogeographic pattern among cpDNA haplotypes in Taxus wallichiana (Taxaceae) in China and North Vietnam. Mol Ecol. 2007;16(22):4684–98.
10. Yuan QJ, Zhang ZY, Peng H, et al. Chloroplast phylogeography of Dipentodon (Dipentodontaceae) in southwest China and northern Vietnam. Mol Ecol. 2008;17:1054–65.
11. Tian S, Li DR, Wang HW, et al. Mitochondrial DNA phylogeography of a montane pine (*Pinus kwangtungensis* Chun ex Tsiang) in south China: interglacial refugia and southward migration during cold stages. Ann Bot. 2008;102:69–78.
12. Xu XW, Ke WD, Yu XP, et al. A preliminary study on population genetic structure and phylogeography of the wild and cultivated *Zizania latifolia* (Poaceae) based on *Adh1a* sequences. Theor Appl Genet. 2008;116:835–43.
13. Rosenberg NA, Norborg M. Genealogical trees, coalescent theory and the analysis of genetic polymorphisms. Nat Rev Genet. 2002;2002(3):380–90.
14. Lei M, Wang Q, Wu ZJ, et al. Molecular phylogeography of *Fagus engleriana* (Fagaceae) in subtropical China: limited admixture among multiple refugia. Tree Genet Genomes. 2012;8:1203–12.
15. Templeton AR, Routman E, Philips CA. Separating population structure from population history: a cladistic analysis the geographical distribution of mitochondrial DNA haplotypes in the tiger salamander, *Ambystoma tigrinum*. Genetics. 1995;140:767–82.
16. Mayden RL. A hierarchy of species concepts: The denouement in the saga of the species problem. London: The units of biodiversity Chapman and Hall; 1997. p. 381–424.
17. Slade RW, Moritz C. Phylogeography of Bufo marinus from its natural and introduced ranges. R Soc. 1998;265:769–77.
18. Roca AL, Georgiadis N, Pecon-Slattery J, et al. Genetic evidence for two species of elephant in Africa. Science. 2001;293(5534):1473–7.

19. Gao J, Wang B, MAO JF, et al. Demography and speciation history of the homoploid hybrid pine Pinus densata on the Tibetan Plateau. Mol Ecol. 2012;21(19):4811–27.
20. Vandergast AG, Bohonak AJ, Hathaway SA, et al. Are hotspots of evolutionary potential adequately protected in southern California? Biol Conserv. 2008;141(6):1648–64.
21. Médail F, Diadema K. Glacial refugia influence plant diversity patterns in the Mediterranean Basin. J Biogeogr. 2009;36(7):1333–45.
22. Avise JC. Molecular population structure and the biogeographic history of a regional fauna, a case history with lessons for conservation biology. Oikos. 1992;63:62–76.
23. Moritz C, Faith D. Comparative phylogeography and the identification of genetically divergent areas for conservation. Mol Ecol. 1998;7:419–29.
24. Riddle BR, Hafner DJ, Alexander LF, et al. Cryptic vicariance in the historical assembly of a Baja California Peninsular Desert biota. Proc Natl Acad Sci. 2000;97(26):14438–43.
25. Zhang ZY, Wu R, Wang Q, et al. Comparative phylogeography of two sympatric beeches in subtropical China, Species-specific geographic mosaic of lineages. Ecol Evol. 2013;3 (13):4461–72.
26. Vilà C, Savolainen P, Maldonado JE, et al. Multiple and ancient origins of the domestic dog. Science. 1997;276(5319):1687–9.
27. Adam RB. The domestic dog: man's best friend in the genomic era. Genome Biol. 2011;12:216.
28. Olsen KM, Schaal BA. Evidence on the origin of cassava, Phylogeography of Manihot esculenta. Proc Natl Acad Sci USA. 1999;96:5586–91.
29. Molina J, Sikora M, Garud N, et al. Molecular evidence for a single evolutionary origin of domesticated rice. Proc Natl Acad Sci. 2011;108(20):8351–6.
30. Huang X, Zhao Y, Wei X, et al. Genome-wide association study of flowering time and grain yield traits in a worldwide collection of rice germplasm. Nat Genet. 2012;44(1):32–9.
31. Yuan QJ, Zhang ZY, Hu J, et al. Impacts of recent cultivation on genetic diversity pattern of a medicinal plant, *Scutellaria baicalensis* (Lamiaceae). BMC Genet. 2010;11:29.
32. Huang S, Jing SNBC. (The divine farmer's Materia Medica, translated by Yang SZ). Beijing: Traditional Chinese Medicine Ancient Books Press; 1982.
33. Joshee N, Mentreddy SR, Yadav AK. Mycorrhizal fungi and growth and development of micropropagated Scutellaria *integrifolia* plants. Ind Crop Prod. 2007;25:169–77.
34. Kumagai T, Müller CI, Desmond JC, et al. Scutellaria baicalensis, a herbal medicine, anti-proliferative and apoptotic activity against acute lymphocytic leukemia, lymphoma and myeloma cell lines. Leuk Res. 2007;31(4):523–30.
35. Murch SJ, Rupasinghe HPV, Goodenowe D, et al. A metabolomic analysis of medicinal diversity in Huang-qin (Scutellaria baicalensis Georgi) genotypes, discovery of novel compounds. Plant Cell Rep. 2004;23(6):419–25.
36. Doyle JJ, Doyle JL. A rapid DNA isolation procedure for small quantities of fresh leaf tissue. Phytochem Bull. 1987;19:11–5.
37. Hamilton MB. Four primer pairs for the amplification of chloroplast intergenic regions with intraspecific variation. Mol Ecol. 1999;8:521–2.
38. Sang T, Crawford DJ, Stuessy TF. Chloroplast DNA phylogeny, reticulate evolution, and biogeography of Paeonia (Paeoniaceae). Am J Bot. 1997;84:1120–36.
39. Thompson JD, Gibson TJ, Plewniak F, et al. The CLUSTAL_X windows interface, flexible strategies for multiple sequence alignment aided by quality analysis tools. Nucleic Acids Res. 1997;25(24):4876–82.
40. Caicedo AL, Schaal BA. Population structure and phylogeography of *Solanum pimpinellifolium* inferred from a nuclear gene. Mol Ecol. 2004;13:1871–82.
41. Nei M. *F*-statistics and analysis of gene diversity in subdivided populations. Ann Hum Genet. 1977;41:225–33.
42. Nei M. Molecular evolutionary genetics. New York: Columbia University Press; 1987.
43. Nei M. Analysis of gene diversity in subdivided populations. Proc Natl Acad Sci USA. 1973;70:3321–3.

44. Pons O, Petit RJ. Measuring and testing genetic differentiation with ordered versus unordered alleles. Genetics. 1996;144:1237–45.
45. Pons O, Petit RJ. Estimation, variance and optimal sampling of gene diversity. I. haploid locus. Theor Appl Genet. 1995;90:462–70.
46. Templeton AR, Crandall KA, Sing CF. A cladistic analysis of phenotypic associations with haplotypes inferred from restriction endonuclease mapping and DNA sequence data. III. Cladogram estimation. Genetics. 1992;132:619–33.
47. Clement M, Posada D, Crandall KA. TCS, a computer program to estimate gene genealogies. Mol Ecol. 2000;9:1657–9.
48. Jensen JL, Bohonak AJ, Kelley ST. Isolation by distance, web service. BMC Genetics. 2005; 6:13 v.13.11. http://ibdws.sdsu.edu/.
49. WenHsiung L. Molecular evolution. Sunderland: Sinauer Associates Incorporated; 1997.
50. Cavers S, Navarro C, Lowe AJ. A combination of molecular markers identifies evolutionarily significant units in *Cedrela odorata* L. (Meliaceae) in Costa Rica. Conserv Genet. 2003;4:571–80.
51. Zhang Q, CHIANG TY, George M, et al. Phylogeography of the Qinghai-Tibetan Plateau endemic Juniperus przewalskii (Cupressaceae) inferred from chloroplast DNA sequence variation. Mol Ecol. 2005;14(11):3513–24.
52. King RA, Ferris C. Chloroplast DNA phylogeography of *Alnus glutinosa* (L.) Gaertn. Mol Ecol. 1998;7:1151–61.
53. Dumolin-lapegue S, Pemonge MH, Petit RJ. Association between chloroplast and mitochondrial lineages in oaks. Mol Biol Evol. 1998;15:1321–31.
54. Fineschi S, Salvini D, Taurchini D, et al. Chloroplast DNA variation of Tilia cordata (Tiliaceae). Can J For Res. 2003;33(12):2503–8.
55. Huang H, Han X, Kang L, et al. Conserving native plants in China. Science (New York, NY). 2002;297(5583):935.
56. Dutech C, Maggia L, Joly HI. Chloroplast diversity in *Vouacapoua americana* (Caesalpinaceae), a neotropical forest tree. Mol Ecol. 2000;9:1427–32.
57. Huang S, Chiang YC, Schaal BA, et al. Organelle DNA phylogeography of Cycas taitungensis, a relict species in Taiwan. Mol Ecol. 2001;10(11):2669–81.
58. Lu SY, Peng CI, Cheng YP, et al. Chloroplast DNA phylogeography of *Cunninghamia konishii* (Cupressaceae), an endemic conifer of Taiwan. Genome. 2001;44:797–807.
59. Petit RJ, Duminil J, Fineschi S, et al. Invited review, comparative organization of chloroplast, mitochondrial and nuclear diversity in plant populations. Mol Ecol. 2005;14(3):689–701.
60. Posada D, Crandall KA. Intraspecific gene genealogies, trees grafting into networks. Trends Ecol Evol. 2001;16:37–45.
61. Avise JC. Phylogeography, the history and formation of species. Cambridge, MA: Harvard University Press; 2000.
62. Hewitt GM. The genetic legacy of the quaternary ice ages. Nature. 2000;405:907–13.
63. TABERLET P, FUMAGALLI L, WUST-SAUCY AG, et al. Comparative phylogeography and postglacial colonization routes in Europe. Mol Ecol. 1998;7(4):453–64.
64. Abbott RJ, Smith LC, Milne RI, et al. Molecular analysis of plant migration and refugia in the Arctic. Science. 2000;289(5483):1343–6.
65. Willis KJ, Whittaker RJ. The refugial debate. Science. 2000;287:1406–7.
66. Mengel RM. The probable history of species formation in some northern wood warbles (Parulidae). Living Bird. 1964;3:9–43.
67. Klicka J, Zink RM. The impotence of recent ice ages in speciation: a failed paradigm. Science. 1997;277:1666–9.
68. Doebley JF, Gaut BS, Smith BD. The molecular genetics of crop domestication. Cell. 2006;127:1309–21.
69. Londo JP, Chiang YC, Hung KH, et al. Phylogeography of Asian wild rice, Oryza rufipogon, reveals multiple independent domestications of cultivated rice, Oryza sativa. Proc Natl Acad Sci. 2006;103(25):9578–83.

70. Olsen KM, Gross BL. Detecting multiple origins of domesticated crops. Proc Natl Acad Sci USA. 2008;105:13701–2.
71. Allaby RG. The rise of plant domestication: Life in the slow lane. Biologist. 2008;55:94–9.
72. He J, Chen L, Si Y, et al. Population structure and genetic diversity distribution in wild and cultivated populations of the traditional Chinese medicinal plant Magnolia officinalis subsp. biloba (Magnoliaceae). Genetica. 2009;135(2):233–43.
73. Kilian B, Özkan H, Walther A, et al. Molecular diversity at 18 loci in 321 wild and 92 domesticate lines reveal no reduction of nucleotide diversity during Triticum monococcum (Einkorn) domestication, implications for the origin of agriculture. Mol Biol Evol. 2007;24(12):2657–68.
74. Miller A, Schaal B. Domestication of a Mesoamerican cultivated fruit tree, *Spondias purpurea*. Proc Natl Acad Sci USA. 2005;102:12801–6.
75. Miller AJ, Schaal BA. Domestication and the distribution of genetic variation in wild and cultivated populations of the Mesoamerican fruit tree *Spondias purpurea* L. (Anacardiaceae). Mol Ecol. 2006;15:1467–80.
76. Guo HB, Lu BR, Wu QH, et al. Abundant genetic diversity in cultivated Codonopsis pilosula populations revealed by RAPD polymorphisms. Genet Resour Crop Evol. 2007;54(5):917–24.
77. Shi W, Yang CF, Chen JM, et al. Genetic variation among wild and cultivated populations of the Chinese medicinal plant Coptis chinensis (Ranunculaceae). Plant Biol 2008. 1985;10 (4):485–91.
78. Ellstrand NC, Marshall DL. The impact of domestication on distribution of allozyme variation within and among cultivars of radish, *Raphanus sativus* L. Theor Appl Genet. 1985;69:393–8.
79. Hamrick JL, Godt MJW. Allozyme diversity in cultivated crops. Crop Sci. 1997;37:26–30.
80. Brown AHD. Isozymes, plant population genetic structure and genetic conservation. Theor Appl Genet. 1978;52:145–57.
81. Hyten DL, Song Q, Zhu Y, et al. Impacts of genetic bottlenecks on soybean genome diversity. Proc Natl Acad Sci. 2006;103(45):16666–71.
82. Blanca JM, Prohens J, Anderson GJ, et al. AFLP and DNA sequence variation in an Andean domesticate, pepino (Solanum muricatum, Solanaceae), implications for evolution and domestication. Am J Bot. 2007;94(7):1219–29.
83. Altieri MA, Merrick LC. In situ conservation of crop genetic resources through maintenance of traditional farming systems. Econ Bot. 1987;41:86–96.
84. Vargas-Ponce O, Zizumbo-Villarreal D, Martínez-Castillo J, et al. Diversity and structure of landraces of Agave grown for spirits under traditional agriculture: A comparison with wild populations of A. angustifolia (Agavaceae) and commercial plantations of A. tequilana. Am J Bot. 2009;96(2):448–57.
85. Allendorf FW, Luikart G. Conservation and the genetics of populations. Mammalia. 2007;2007:189–97.
86. Storfer A. Gene flow and endangered species translocations, a topic revisited. Biol Conserv. 1999;87:173–80.
87. Bower AD, Aitken SN. Ecological genetics and seed transfer guidelines for *Pinus albicaulis* (Pinaceae). Am J Bot. 2008;95:66–76.
88. Canter PH, Thomas H, Ernst E. Bringing medicinal plants into cultivation, opportunities and challenges for biotechnology. Trends Biotechnol. 2005;23:180–5.

Chapter 5
Salvation of Rare and Endangered Medicinal Plants

Xueyong Wang, Khabriev Ramil Usmanovich, Linglong Luo, Wen Juan Xu, and Jia Hui Wu

Abstract With the further development of Chinese medicine health services, the social demand for medicinal plant resources has further increased. Many wild animals and plants have become extinct or are in danger of extinction. It is imperative to protect wildlife resources. Reasonable development and utilization to achieve sustainable development of resources has become a concern and urgent problem for all sectors of society. So far, many medicinal plant resources are still mainly based on wild resources. After decades of long-term predatory development, many precious Chinese herbal medicines have been in short supply. Therefore, the protection of traditional medicinal plant resources is one of the key tasks of the current state in the field of traditional Chinese medicine research. This chapter conducts the theory, method, and practice of the protection of traditional Chinese medicine resources and introduces the definition of Chinese medicine resource protection, the reasons for the destruction of Chinese medicine resources and the significance of resource protection, the classification of endangered medicinal plants and animals, and the specific examples of protection of Chinese medicine resources in order to provide reference for researchers.

5.1 The Reasons for the Destruction of Medicinal Plant Resources

According to the "Red List of Endangered Species," published by the International Union for Conservation of Nature (IUCN) in November 2014, species are now extinct more than ten times faster than in the natural state. Many species of medicinal

X. Wang (✉) · L. Luo · W. J. Xu · J. H. Wu
International Center for New Resources of Chinese Materia Medica, Beijing University of Chinese Medicine, Beijing, China
e-mail: wxyph.d@163.comc; luolinglong1995@163.com; xuwenjuan0907@qq.com; wujiahui2017@126.com

K. R. Usmanovich
N.A. Semashko National Research Institute of Public Health, Moscow, Russia
e-mail: institute@nriph.ru

L.-q. Huang (ed.), *Molecular Pharmacognosy*,
https://doi.org/10.1007/978-981-32-9034-1_5

plants and animals in China tend to decline or endanger. The first volume of the *Chinese Plant Red Data Book: Rare and Endangered Plants* contains 354 species of plants, including 168 species of medicinal plants, some of which have been listed in the "National Key Protected Wild Medicine List" and the appendix of "Convention on International Trade in Endangered Species of Wild Fauna and Flora" (COTES) [1, 2].

The protection of medicinal plant resources refers to the protection of the renewable biological resources of medicinal animals and plants and the ecological environment and ecosystems closely related to them, in order to protect the diversity of organisms and save rare and endangered species of medicinal plants and animals [3]. The wild Chinese medicine resources which mainly protected in China are divided into three levels: Level 1, rare wild medicinal species that are on the verge of extinction; Level 2, wild medicinal species that are reduced in distribution area and in a state of exhaustion; and Level 3, serious resources that are mainly or commonly used wild Chinese herbal medicine species are reduced. The State Administration of TCM announced the first batch of 76 national key protected wild Chinese medicine varieties, including 58 species, 13 of which are protected by Level 1 and 2, including licorice, berberine, ginseng, eucommia, magnolia, cork, blood, and so on. There are 45 species listed in Level 3 protection, including *Fritillaria*, *Gentiana*, *Polygala*, *Asarum*, comfrey, *Schisandra*, vine, scorpion, *Dendrobium*, Awei, *Notopterygium*, *Acanthopanax*, radix *Scutellariae*, Tianmen Winter, pig scorpion, wind, Hu Huanglian, *Cistanche*, hawthorn, *Forsythia*, and so on. There are 18 species of animals, including 4 kinds of Level 1 protection, tiger, leopard, saiga, and sika deer, and 14 species of Level 2 protection, red deer, forest owl, original pheasant, black bear, brown bear, black snake, silver ring snake, five-step snake, pangolin, Chinese giant salamander, black frame dragonfly, forest frog, dragonfly, and so on. Among the more than 60 kinds of animal drugs listed in the *Pharmacopoeia of the People's Republic of China*, there are 9 species of national key protections. According to the national survey of Chinese medicine resources conducted in the 1980s, China has 12,807 species of Chinese medicine resources, including 11,146 medicinal plants and 1581 medicinal animals. Among the traditional Chinese medicine resources, there are more than 7000 species of folk herbs, accounting for 60% of the resources; and more than 4000 species of ethnic medicines that accounting for 30% of the total; and more than 1200 species of Chinese herbal medicines, accounting for 10% of medicinal resources. With the increase of population and the destruction of resources, the contradiction between supply and demand of Chinese medicine resources is more prominent. Whether Chinese medicine resources can be used sustainably is one of the most important issues in the survival and development of Chinese medicine. How to reduce human's demand for wild resources and establish a long-term sustainable use and protection mechanism for endangered medicinal plants and animals is an important problem to be solved.

There are many factors contributing to the reduction of resources of traditional Chinese medicine, which can be summarized as follows.

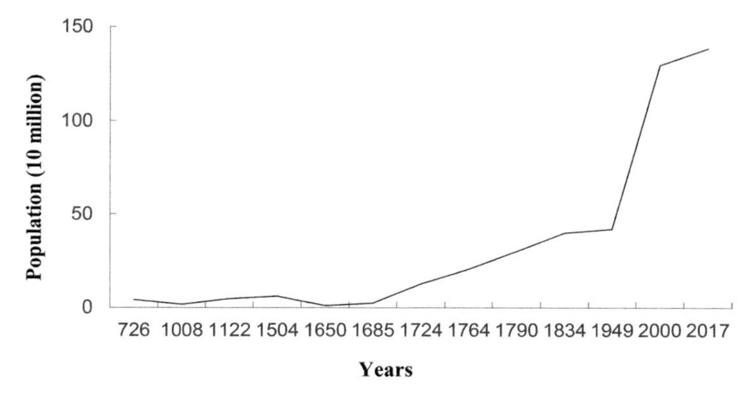

Fig. 5.1 China's population changes in different periods

5.1.1 *Population Increase*

Increasing population and increasing market demand for medicinal resources have led to overexploitation, resulting in a sharp decrease in wild medicinal resources. Despite our vast territory and vast resources, the population density in history is relatively low. Throughout the history of our country, the population has basically maintained at about 0.5 billion (only 1/2 of the current population of Henan Province). Before the Qing Dynasty, the population did not exceed 100 million at the most. After the Qing Dynasty, the population began to grow steadily and grew rapidly after the founding of the People's Republic [4], as shown in Fig. 5.1. In ancient times, the wild Chinese medicine resources produced by the broad land basically met the needs of people at that time. With the increase of population, the rapid development of Chinese medicine resources has led to an increase in the demand for Chinese herbal medicines. The contradiction between supply and demand in the market has increased the intensity of Chinese herbal medicines. Many kinds of medicines have exceeded the annual allowance and resources are becoming less and less.

5.1.2 *The Destruction of the Ecological Environment*

The destruction of the ecological environment has caused a sharp decline in the reserves of wild medicinal resources. The population increase requires more arable land, and the opening of farmland leads to the shrinking of the area of wild medicinal herbs. The improvement of people's living standards has promoted the rapid development of animal husbandry and the serious degradation of grassland. It is difficult for medicinal plants to obtain mature seeds, especially for pastoral activities such as grass cutting (mowing), which affects the regeneration of the population. Forestry mining also destroys the original ecological environment. Although afforestation has promoted the development of forestry, this artificially altered ecological environment plays an important role in the development of target forest species, but other species of the original ecology could not be effectively protected and developed.

5.1.3 Non-sustainable Harvesting Methods

Unsustainable harvesting methods undermine the natural renewal of species. Resources are becoming less and less. Non-sustainable and predatory medicinal material harvesting methods are another important factor leading to the destruction of wild Chinese medicine resources, such as cutting vines of *Schisandra chinensis* for fruits, peeling living cork tree, and grubbing out *Ephedra*. This is a kind of "fishing with exhaustion" or "killing chickens for eggs" harvesting method. According to the survey, the 1- and 2-year-old leeches listed in the specialized market of Chinese traditional medicine in 1999 reached 10 tons, and the output decreased to 60 tons in 2000. The leech resources were exploited in Huzhou District of Zhejiang Province in 1995. The peak output of leech reached 50 tons in 1998 and less than 20 tons in 2000.

5.1.4 Overreliance on Wild Herbs

Historically, Chinese herbal medicines are mainly derived from the wild. In the last century, especially after the 1970s, more and more varieties of Chinese herbal medicines appeared contradictory between supply and demand, and it became increasingly prominent at present. The development of modern agricultural technology has enabled many medicinal materials to be successfully cultivated, which can effectively alleviate or resolve this contradiction. However, due to changes in the ecological environment of cultivated plants, the appearance of the medicinal materials has undergone great changes and is generally considered to be inferior. For example, the price of ginseng grown is usually 100–200 yuan/kg, while wild ginseng (Santos) is usually several hundred thousand yuan/kg. For chemical components, the content of various saponins in wild and cultivated ginseng is close, and the content of some saponins is only 2–3 times different. Obviously, even if there is a difference in quality between the two, it will not be very large. The commodity price is not reflected by the actual value of the goods. The quality evaluation method from shape feature of the original medicinal materials is a summary by experiences and has certain objectivity, but these objectivities are derived from the original wild medicinal materials, and the quality evaluation of the cultivated needs to be further improved. The cultivated *radix Saposhnikovia* is only 1/4 of the wild price. However, from the analysis of various major active ingredients such as chromone and polysaccharide, the cultivated *radix Saposhnikovia* is high quality and has been proved in pharmacology.

5.2 The Significance of the Protection of Chinese Medicine Resources

Wildlife is not only an important part of nature but also a valuable resource endowed by nature. As long as it's properly utilized and properly harvested, wildlife resources can be continuously updated and become "inexhaustible and inexhaustible" natural asset. However, the renewal and recovery of wildlife resources has some limitations and self-regulation rules. Overutilization will lead to the decline of biological population and even extinction. The protection of traditional Chinese medicine resources is of great significance to the sustainable and healthy development of traditional Chinese medicine.

5.2.1 It Will Be Helpful for Biological Diversity

The main factors of extinction are habitat loss, invasion of alien species, overexploitation and use, and environmental pollution. These factors are actually caused by human activities and population growth. According to the inference, the rate of extinction of species is currently 0.5–25,000 times the rate of extinction of fossils, while the main objects of extinction are islands and species with smaller populations. However, for the mainland, the number of larger populations is also shrinking, and human activities often fragment large habitats, exacerbating the exchange of genetic information, leading to the "island" population. Smaller populations are more vulnerable to extinction than large populations.

Plant traits are determined by genes. It is generally believed that mutations can be caused by drastic changes in temperature, ultraviolet and chemical pollution, as well as some metabolic abnormalities in organisms or cells. Gene mutations are naturally and randomly formed and can occur at any time of individual development and in any cell of an organism. Mutations that occur in germ cells can be directly transmitted to offspring through fertilization. In a natural state, the frequency of genetic mutations is very low for an organism. It is estimated that in higher organisms, only about 100,000–100 million germ cells will have a germ cell mutation. The genetic mutation rates of different organisms are different. For example, microorganisms such as bacteria and phage have higher mutation rates than higher animals and plants. Different genes of the same organism have different mutation rates. Mutations occurring in somatic cells are generally not transmitted to offspring.

Mutation is a rare event which has a very weak evolutionary force in the short term. In addition to the formation of multiple varieties, the formation of species usually takes millions of years. In the short term, the choice is mainly due to the genetic diversity inherent in the species, not the mutation. Genetic diversity is determined by various alleles and genotypes in the population. There is usually a positive correlation between genetic diversity and population size. Larger genetic diversity is not susceptible to diseases, harsh environmental changes, pests, and other hazards, and it is easier to adapt to a changing environment.

Mutations in genes have no definite direction. A gene can mutate in different directions, producing more than one allele, which is the fundamental driving force for the evolution of species. Since any organism is a product of a long-term evolutionary process, they have been highly harmonized with environmental conditions. If a genetic mutation occurs, it is possible to destroy this coordination relationship. Therefore, most gene mutations are often harmful to the survival of organisms, a few are neutral, and very few mutations are beneficial.

In small populations of endangered species, loss of inbreeding and genetic diversity is inevitable. For harmful genes, inbreeding will show rare alleles in a homozygous form, making harmful genes easier to fix. For free genes, because the gametophyte selection of parents is random, some beneficial alleles, especially rare alleles, may disappear in the process of progeny transmission and reduce genetic diversity in small populations; in addition, due to environmental effects, the sex ratio of most animals and a few plants may be altered. Therefore, for small populations, they can reduce the reproductive and viability of the species in the short term; in the long run, they will weaken the evolutionary potential of the population to environmental changes and enter the "extinction vortex." The mutation rate of animal and plant germ cells is about $1/10^5$–$1/10^8$, and most of them are harmful or neutral mutations. It is difficult for extinct organisms to restore their natural competitiveness.

5.2.2 Providing Basic Materials for Breeding

This is a high-level historical process of utilizing biological resources after the development of biotechnology to a certain extent. According to genetics, each species has its own genetic characteristics, and different genetic characteristics should be considered as different germplasm. Because the mutation frequency of genes in nature is very low, each trait is formed in a long history. For example, *Acanthopanax* is in the transitional stage of bisexual flower to unisexual flower. The "male flower" still retains the ability to pollinate, but pistil's ability is extremely low; "female flower" still retains the stamen that can produce pollen, but the filament is very short, the pollen is almost aborted, but the process has been carried out for 29.9–455,000 years, and it is still continuing. Once each trait disappears, it is difficult to obtain it again. A large population often contains many excellent traits, and they may all be good materials for breeding. Therefore, collecting and studying these germplasm resources is very beneficial to human beings. Many research centers and institutions in the world have established a variety of "seed stocks" or "seed banks" for collecting germplasm resources, and even "gene pools" at the molecular level, using different germplasm for cross-breeding in order to meet different demands. In particular, the application of biogenetic engineering, which is marked by new technologies such as DNA cloning, hybridization, directed transplantation, and allogeneic expression, has shown great attractive prospects in the collection, research, and utilization of biological germplasm resources. However, due to the destruction of vegetation and the deterioration of the environment, the loss of germplasm in the world today is becoming more and more serious. The disappearance of germplasm cannot be recreated and needs to be preserved.

5.2.3 Ensuring the Sustainable Utilization of Chinese Herbal Medicines

Wild medicinal herbs are important part of current Chinese medicine resources. For renewable resources, the quantity should be basically consistent with the market demand. The amount of resources used must be compatible with the regeneration of resources, proliferation, replacement, and compensation capabilities to maintain a balance between supply and demand and sustainable development. Renewable resources have dual attributes. If they can be actively and effectively protected and rationally and orderly developed and utilized, limited Chinese medicine resources can maintain healthy development. On the contrary, if not effectively protected, indiscriminate taking will inevitably lead to the exhaustion of limited resources of traditional Chinese medicine, especially wild resources and nonrenewable resources, and even endanger many rare species, aggravating the contradiction between supply and demand of traditional Chinese medicine resources, resulting in negative effects that cannot even be recovered.

5.2.4 Benefiting the State's Strategic Planning of Traditional Chinese Medicine Resources

It is necessary to make top-level design, overall consideration, comprehensive utilization, and rational planning to include traditional Chinese medicine resources in the national strategic resources, build a new modern agricultural industry of traditional Chinese medicine, and protect and comprehensively develop and utilize traditional Chinese medicine resources. Among them, the protection of Chinese medicine resources is an important link. Since the National Traditional Chinese Medicine Administration launched the pilot survey of Chinese medicine resources in 2011, the census of Chinese medicine resources has been in full swing. The national dynamic monitoring and information service system for essential medicines for essential medicines has been basically established, and five seed breeding bases have been established. And two germplasm resource libraries strive to play a role in safeguarding the precious, bulky, and genuine medicinal plant germplasm resources of the country and promoting the sustainable use of traditional Chinese medicine raw materials to meet the requirements of the national essential medicines list demand.

5.3 Methods for Protecting Chinese Medicine Resources

Most of the traditional Chinese medicine resources are biological resources. As long as effective measures are taken to protect them, they can continue to self-renew and be used sustainably. The reasonable protection of traditional Chinese medicine resources should be based on the natural laws of biological growth and development. At the same time, the protection of traditional Chinese medicine resources is a kind of artificial intervention process, which requires corresponding laws and regulations to maintain it.

5.3.1 Technical Methods for the Protection of Chinese Medicine Resources

The protection of traditional Chinese medicine resources mainly includes in situ conservation (or in situ protection), ex situ conservation, and in vitro protection [5]. The basic measures include local protection through measures such as establishing nature reserves and wild tending to achieve in situ protection. Ex situ conservation can be achieved by introducing species from other places, like cultivating wild species, and constructing seed bank and germplasm resource nursery. In vitro conservation can be achieved through plant tissue culture and construction of gene germplasm bank.

5.3.1.1 In Situ Protection

In situ conservation protects threatened medicinal plant species by protecting natural ecosystems and the natural environment. This is the most effective form of protection. It not only protects the plant population that survives but also preserves the ecological environment on which plants depend. The most effective way to protect in situ is to establish various types of nature reserves [2]. The protection method can protect the medicinal plants and plants from human influence (including mining), and the population can be restored and developed in the adapted growth environment.

5.3.1.1.1 Nature Reserve

The establishment of nature reserves is also one of the most efficient and effective means of biological wild breeding. The Fourth National Symposium on the Classification of National Parks and Protected Areas in 1992 gave the following definitions for protected areas: "Protected areas" refer to the protection and maintenance of biodiversity, nature and related cultural resources, and adopted legislation or other effective means of managing land or sea. This definition is a generalized concept that encompasses nature reserves in the strict sense and protected areas in the general sense. In 1993, the State Environmental Protection Administration of China approved the "Classification of Types and Levels of Nature Reserves," which divided the nature reserves into nine categories: natural ecosystems, wildlife, and natural remains. The "Regulations on the Nature Reserve of the People's Republic of China" promulgated in 1994 stipulates that nature reserves refer to "protective objects such as representative natural ecosystems, natural concentrated distribution areas of rare and endangered wild animals and plants, and natural relics of special significance." In the land, land waters, or sea areas, a certain area is given according to the law to give special protection and management areas, which defines the functions and attributes of the nature reserves. The core issue of protected areas is to protect rare and endangered wildlife and corresponding ecosystems and to protect biodiversity through the construction and effective management of nature reserves. The construction of nature reserves has been widely promoted around the world. As early as 1993, 8619 nature reserves related to biodiversity conservation have been established around the world, covering an area of 79.226 million hm^2, accounting for about 6% of the Earth's area. China's nature reserves began in the Dinghushan Nature Reserve in Guangdong, established in 1956. After nearly several decades of efforts, 2697 nature reserves of various types and levels have been established nationwide, with a total area of 146.31 million hm^2. Among them, the area of terrestrial nature reserves has accounted for 14.77% [6] of the national land area, which plays a key role in protecting the medicinal and plant resources.

The nature reserves are mostly established in areas with rich biodiversity, good natural ecological environment, and relatively small impact on human activities. Among them, biodiversity is the most important ecological indicator for the selection of protected areas, and it is also one of the purposes of establishing protected areas. For example, there are 129 families, 481 genera, 1063 species, 130 varieties, and 33 varieties of wild medicinal plants in Jilin Changbai Mountain Nature Reserve, accounting for 47.60%, 22.71%, and 9.00% [7] of the medicinal plants in China, respectively. Protected areas should maintain the genetic diversity of natural populations as much as possible to better adapt to changes in the environment. The relationship between population size and genetic diversity can be expressed mathematically in a straight line, that is, as the population increases, its genetic diversity is also high. Under natural conditions, each species must have at least 500 individuals to maintain stable genetic variation and relatively long-term evolutionary growth [8].

The root cause of the destruction of traditional Chinese medicine resources is the human intervention in the environment. The protected areas exclude human disturbances through the protection of biological genetic diversity and rare and endangered species. In the natural state, the relationship between the number of species and the area is

$$S = C A^{Z}$$

Here S is the number of biological species, A is the area of the protected area, C is the distribution density of the species, and Z is the parameter related to the geographical coordinates. This formula indicates that in any protected area, the decay rate of a species increases as the area of the protected area decreases.

Nature reserves generally include three parts: core area, buffer zone, and experimental area. The core area is surrounded by a buffer zone. The area is mainly a native ecosystem or a concentrated distribution or breeding area of rare and endangered animals and plants. Any unit or individual is forbidden to enter; the buffer zone is replaced by some vegetation areas that may be restored to their original nature. The composition is only allowed to engage in scientific research; the experimental area is located at the outermost periphery, including secondary vegetation and barren hills and wasteland, etc., and can engage in scientific experiments, internships, tours, and other activities.

5.3.1.1.2 Wild Tending

Wild tending is a kind of protection measure combining wild resource protection and human positive intervention, and it is another important way for local medicine resources to be protected in situ. That is, according to the growth characteristics of animal and plant medicines and the requirements of ecological environment conditions, in their original or similar environment, the population is increased by artificial or natural means, and the community balance is maintained and utilized. The method of wild tending is mainly used for medicinal materials with harsh growth conditions, relatively high cost of planting (farming), or significant changes in the traits and quality of artificially cultivated medicinal materials. This method protectes the original ecological environment of medicinal plant and animal resources; it not only saves cultivated land resources and maintains the biodiversity of the original ecosystem but also restores the authenticity of Chinese herbal medicines and maximizes their quality. Wild tending plays an important role in the recovery and rejuvenation of rare and endangered species.

The basic ways of wild tending of traditional Chinese medicines are forest closure, artificial management, artificial replanting, imitation of wild cultivation, etc. [9].

Forest closure: For areas with severe resource damage but resilience, collection is prohibited. After the population is restored under natural conditions, it will be collected in moderation to maintain the sustainable development of resources.

Artificial management: On the basis of ban, artificially manage the medicinal material population and improve the growth environment of the bio-community, create favorable growth conditions, and promote the growth and reproduction of Chinese medicinal materials, such as improving the natural light conditions of *Schisandra*.

Artificial replanting: On the basis of ban, according to the characteristics of medicinal plants, seeding or planting seedlings in the original ecological environment, it can artificially increasing the population of medicinal materials.

Imitation of wild cultivation: In the native environment in which wild herbs are distributed or in similar environments, artificial planting is used to increase the population, but no field management is used.

In the process of wild tending, we should also pay attention to reasonable harvesting. Only in the right season, choosing a reasonable harvesting method and controlling a certain amount of harvesting can ensure the sustainability of the growth of the medicinal materials. "Maximum sustained production" is the maximum resource yield allowed under the premise of ensuring that the original resources remain stable. The annual allowable quantity refers to the quantity allowed to be harvested within 1 year, that is, the amount of harvest that does not affect its natural regeneration but guarantees sustainable use. It is a quantitative indicator to guide people to harvest wild herbs. In the case that the original ecological environment is not destroyed, if the actual wild excavation volume is less than the annual allowable amount per year, the amount of wild resources will be basically unchanged; if the actual wild excavation volume exceeds the annual allowable amount each year, the resources will cut back. There are still many problems in the artificial planting of the endangered medicinal material *Fritillaria*. The scholars such as Chen Shilin successfully established a large-scale wild nurturing system of *Fritillaria cirrhosa* in the Qinghai-Tibet Plateau about 4000 meters above sea level and explored a resource for the medicinal materials grown in the special environment of the plateau, which exploring ways to travel for the sustainable development of highland medicinal materials [10].

There are usually three ways to calculate the annual acceptance amount:

(A) The key to the calculation of the annual acceptance amount is its update period. Only by understanding the update cycle can the annual allowable volume be accurately calculated. НА Борисова proposed the following annual acceptance formula:

$$\text{Economic quantity} = \text{reserves} \times \frac{\text{standard and economical yields}}{\text{Reserve}}$$

$$\text{Annual allowance} = \text{economic volume} \times \frac{\text{Harvestable period}}{\text{Harvestable age} + \text{plant renewal cycle}}$$

For example, after investigating the economics of licorice in a certain area is 6 t, the harvest for root medicinal materials is usually 1 year, and the renewal period is 5 years, annual allowable amount is 1t.

Except for leaf, flower, fruit, seed, and other plant medicinal materials that can be harvested annually, most of the varieties are harvested once, and the harvestable life is usually 1 year.

(B) The update cycle is mainly obtained through an update survey. Since the renewal cycle for most wild medicinal plants has not been investigated, it can be calculated using the following formula:

$$\text{Annual allowance} = \text{economic volume} \times \text{ratio}.$$

Empirical data on ratio values: 0.3–0.4 for stem and leaf, 0.1 for roots and rhizomes, and 0.5 for flowers and fruits.

(C) The annual acceptance amount can also be calculated by the following formula:

$$\text{Annual allowance} = \text{medicinal material storage}/\text{harvesting cycle}.$$

In the above formula, the allowable amount can be visited by visiting the experienced drug farmers and pharmaceutical workers, especially the medicinal material purchasers in the survey area, to understand the purchase and sale of key medicinal materials, to analyze the historical data and data of medicinal materials, and to combine plant organisms. The characteristics of the study are analyzed and estimated, so the economic amount can be calculated by calculation. Through the actual field investigation, the relationship between economic volume and reserves can be obtained, so that the reserves of wild herbs can be obtained indirectly.

5.3.1.2 Ex Situ Conservation

Ex situ conservation, also known as moving out conservation, refers to the preservation and breeding of Chinese herbal medicine species resources outside the original growth environment of medicinal plants and animals. Rare and endangered medicinal species will be moved out of their natural habitats; preserved in protected areas, zoology and botanical gardens, and plantations; and researched on introduction and domestication. Through introduction and breeding, not only many rare and endangered species have been protected in zoos and botanical gardens, but also the provenance has been expanded. Ex situ conservation is more conducive to resource restoration, rejuvenation, and introduction [11] of protected species, which can protect protected species from vandalism and be replaced by other species [12].

5.3.1.2.1 Establishment of Plant (Moving) Garden

The origins of the botanical garden can be traced back to ancient China and the Mediterranean countries where first used as medicine and medicine teaching sites. In the modern China, since the 1920s, a few botanical gardens such as Zhongshan Botanical Garden, Taipei Botanical Garden, and Lushan Botanical Garden have been established. Since the 1950s, the number of botanical gardens has increased, and today there are about 160. In the botanical gardens of various places, special medicinal plant germplasm resources are basically set up. Among the 388 species (including varieties) of the first batch of China's rare and endangered plants, 332 species (including 86%) have been ex situ conserved in the botanical garden and 637 species of the second China rare and endangered plants list. There are 155 species (about 24%) in ex situ conservation. The protection of medicinal plant resources in China is still dominated by medicinal botanical gardens, and 5482 medicinal species have been introduced in the 4 major medicinal botanical gardens in Beijing, Guangxi, Yunnan, and Hainan. The Wuhan Institute of Botany introduced the rare and endangered plant species (including many medicinal plants) to be inundated in the Three Gorges Reservoir area in the Yangtze River to be protected in the germplasm resources near Yichang City and its plant sites, so that the Three Gorges reservoir area rare and endangered medicinal plant species are well preserved [13]. It should be noted that many medicinal botanical gardens generally only hold a single species or several strains or close relatives that are propagated by their seeds. Usually, the population is too low, which is not conducive to the protection and species of genetic diversity. As an important protection method for germplasm resources of medicinal plants, the maximization of genetic diversity of protected species by medicinal botanical gardens is also the primary consideration for preserving species [14, 15].

5.3.1.2.2 Artificial Cultivation and Breeding

Another important way for ex situ conservation is to transform the wild into domestic and intensive production, thereby expanding the source of drugs and protecting wild resources.

In recent years, China has made great achievements in artificial cultivation and breeding. In the research of medicinal plant cultivation, it has broken through the technical problems of large-scale introduction and cultivation of a batch of medicinal materials. There are many successful cases of trial planting, introduction, and wild house transfer in China. For example, in the 1960s, Yangchun Amomum was only produced in a few areas such as Yangchun County, Guangdong Province. The planting area and yield were limited, most of which depended on imports. After more than 10 years of research, it has break through the technical bottlenecks of pollination, management, disease prevention and control, then successfully introduced and cultivated in Xishuangbanna, which concluded a set of cultivation technology system of low quality, high efficiency, excellent quality, and easy to promote in Xishuangbanna mountainous area. The team of Wang Xueyong professor has successfully introduced *Salvia miltiorrhiza*, which are mainly produced in Shandong and Henan, to plant in Guizhou, realizing the planting and promotion of *Salvia miltiorrhiza*. There are nearly 300 species of cultivated medicinal materials in the country, most of which are reduced in resources, such as *Eucommia ulmoides*, *Phellodendron chinense*, *Magnolia*, *Gardenia jasminoides*, *Platycodon grandiflorus*, *Fritillaria cirrhosa*, *Cornus officinalis*, *Lonicera japonica*, etc., and are all wild in the 1950s–1960s or 1970s. Artificial cultivation is carried out under severely reduced resources and becomes a major source of commodities. In the 1950s, Yunnan Province began to cultivate 37 kinds of wild herbs, and most of them were successful. At the end of 1986, Huanglian, *Pinellia*, *Gentiana macrophylla*, *Artemisia scoparia*, *Vitex negundo*, areca nut, catechu, hematoxylin, millennium health, and Huhuanglian all had certain cultivation areas and replaced some wild herbs to a large extent. Once the introduction of wild herbs is successful, it will alleviate the contradiction between market supply and demand, reduce or get rid of dependence on wild resources, and have important positive significance for the protection of wild resources.

Some types of wild herbs are much better than cultivated herbs, which also exacerbates the predatory mining of wild resources. Therefore, strengthening the selection of excellent varieties and other measures to improve the quality of medicinal materials is also an important part of the protection of Chinese medicine resources.

5.3.1.2.3 Artificial Expansion and Naturalization of Population

The artificially expanded population pattern of introduction and breeding cannot completely replace the natural populations in a certain stage of natural evolution in wild habitats. Their population niche is obviously different. In the long-term cultivation state, the wild state will be lost. Many special genes are needed. Therefore, in order to scientifically protect the rare and endangered plant germplasm, it is necessary to transplant the artificially propagated populations into their original habitats, so that they can be naturalized, grow and multiply, and make more suitable wild plants under the action of natural selection. The state of genetic material is preserved. Therefore, another important strategy for the ex situ conservation of rare and endangered medicinal plants is to artificially grow and domesticate endangered species and then expand the scale and then carry out wild naturalization to achieve the purpose of protection.

In the process of artificial expansion and naturalization, it is realized in two steps: the first step can be based on the biological characteristics of the species, such as artificially breaking seed dormancy, artificial pollination, grafting, ramets, cuttings, tissue culture, etc. to solve bottleneck problems such as reproductive difficulties of these species. In the second step, the artificially cultivated species are wildly naturalized to achieve the purpose of wild natural reproduction. For example, China has successfully bred ex situ conservation [16] of more than 100 rare and endangered plants such as *Davidia involucrata*.

5.3.1.3 In Vitro Conservation

This is a method that makes full use of modern biotechnology to preserve some organs, tissues, cells, or protoplasts of medicinal plants and animals in an artificial environment (chiefly low-temperature environment) for a long period of time, so as to preserve the germplasm resources of medicinal animals and plants and their genetic integrity [13]. There are 20,000 germplasms in vitro preserved in Beijing, Guangxi, Yunnan, and Hainan botanical gardens in China. The number of germplasms of medicinal plants ranks first in the world. Among them are Guangxi botanical gardens (2863 species), Hainan botanical gardens (1598 species), Beijing botanical gardens (11806 species), and Yunnan botanical gardens (1122 species). In addition, 2300 species are preserved in Sichuan Medicinal Botanical Garden and 1100 species in Nanjing Medicinal Botanical Garden. At present, the methods of in vitro protection include seed bank, gene bank, and indirect protection.

5.3.1.3.1 Seed Bank

The seed contains complete genetic information of the species. It has many advantages, for example, the preservation of the seed is not restricted by the distribution area of species, small space demand, and low investment. It is the main means of ex situ conservation of plants at present and a direction for future development.

The seed bank is a germplasm conservation site specially established for the protection and utilization of germplasm resources (including seeds, strains, vegetative propagules, etc.) and is the most important way to preserve plant germplasm resources around the world. Some scholars have proposed the concept of "core collection" for seed bank. It refers to the selection of a certain number of germplasm resources (about 10%) from the existing seed bank, so that it can fully represent the genetic diversity of all germplasms. This concept has greatly improved the efficiency of resource management and development and utilization of seed bank and also provided scientific guidance for collection and conservation strategies of germplasm resources. In 1958, the world's first long-term seed bank has been built in the National Seed Storage Laboratory in Fort Collins, Colorado, USA. It is still the world's largest seed bank for preserving germplasm, with 230,000 seeds in storage. The number of seeds kept in seed bank has grown rapidly over the past few decades. According to the United Nations Food and Agriculture Organization (FAO), the number of seeds stored in the global seed bank is enormous, about 4.5 million. In 1983, China built the National Seed No. 1 warehouse. Ten years later, in 1994, the first national medicinal plant seed bank was built in Hangzhou. Currently, 200 kinds of 50,000 resources have been stored. The second medicinal plant seed bank is under construction at the Beijing Medicinal Botanical Garden. The third medicinal plant seed bank is under construction at the Chinese Medicine Resource Center of the China Academy of Chinese Medical Sciences.

The forms of germplasm preservation include seed cryopreservation bank, germplasm carp, and test-tube plantlet gene bank. The seed cryopreservation bank includes a long-term germplasm bank and a medium-term bank. In general, seed longevity can be doubled for every 5 °C reduction in seed storage temperature in the range of 0–50 °C; for every 1% reduction in water content in the range of 5–14% water content, seed longevity can be doubled, so low-temperature drying is an ideal condition for maintaining seed viability. The temperature of the long-term storage is generally maintained at −18 °C, the medium-term storage is 0–10 °C, and the water content of the seeds is controlled at 5%–8%. Long-term bank can generally hold seeds for more than 50 years and is mainly responsible for providing breeding seeds to the relevant medium-term or original seed-supplying units. The seed longevity in the medium-term bank is about 15 years. The function is to preserve germplasm resources in their area and distribute and exchange germplasm materials. It is mostly used in genetic breeding research.

5.3.1.3.2 Gene Bank

In vitro gene conservation can permanently preserve gene resources without living animals and plants, which is another important means of in vitro protection of traditional Chinese medicine resources. The core of in vitro gene protection is to establish germplasm gene bank of traditional Chinese medicine resources. Gene bank is a relatively new way to preserve germplasm resources. It mainly preserves somatic cells, germ cells, genomic DNA and DNA of animals and plants, etc. It can protects medicinal animals and plant genetic materials from natural or man-made destruction which usually cause gene loss. This method can not only retain valuable genetic resources such as genetic structure and genetic diversity of the population but also provide a material basis for research on functional genes related to key traits of the species.

Microorganisms become the most important carrier for in vitro gene protection because of the short growth cycle and rapid reproductive capacity. The most commonly used technical means for in vitro gene conservation is to construct transgenic strains, construct genomic library, and ultra-cryopreservation. As an important part of the strategy of in vitro conservation of traditional Chinese medicine resources, a group-based library of endangered wild Chinese herbal medicines and plant resources has been established, making it a “gene resource nature reserve” for the protection of endangered animals and plants. And it also provides essential genetic resources and platforms for further development and utilization of species and functional genes, which are associated with important economic traits and hereditary diseases.

5.3.1.4 Indirect Protection

The medicinal resources will be expanded through the research of tissue culture and substitutes. In this way, the original medicinal resources are also protected.

5.3.2 Administrative Means for the Protection of TCM Resources: Relevant Policies and Regulations

Legislation and construction of management mechanisms are indispensable administrative means to protect TCM resources. The state and local governments have also enacted relevant laws and regulations to protect TCM resources.

5.3.2.1 International Convention on the Protection of Biological Resources

5.3.2.1.1 Convention on International Trade in Endangered Species (CITES)

The Convention was signed in 1973 in Washington, USA, for a total of 25 articles, including three appendices. Three appendices specified the protected species of wild animals and plants when trading in the world, and the trade of more than 20,000 endangered species of wild animals and plants is restricted. Later, at the Twelfth Meeting of the States Parties (convened in San Diego, USA, in 2002), a number of amendments to Annexes I, II, and III of the Convention were adopted; at last, it went into effect on February 13, 2003. China officially joined the Convention on June 25, 1980 and became one of the member states.

According to the relevant provisions of the Convention, China has regulated the endangered species of medicinal animal and plant species under supervision of import and export, includes some of the medicinal values listed in the National Key Protected Wildlife, Plants, National Precious Species, Snakes.

5.3.2.1.2 Convention on Biological Diversity (CBD)

The United Nations adopted the world's first international convention on biodiversity at the United Nations Conference on Environment and Development in Rio de Janeiro, Brazil, in 1992, when more than 150 country leaders have signed the Convention, and it has now increased to 168 countries. In 1993, China officially joined the Convention.

5.3.2.1.3 The Cartagena Protocol on Biosafety (CPB)

It is adopted in Montreal, Canada, in February 2000. It is an important international legislative activity following the CBD and is also a concrete implementation of the CBD framework system. On May 19, 2005, the State Environmental Protection Administration officially announced China's accession to the Cartagena Protocol on Biosafety. This is another important commitment of China in biosafety management.

5.3.2.1.4 The Convention on the Protection of New Varieties of Plants (UPOV)

In 1961, 12 European countries discussed and signed the relevant draft in Paris which took effect in 1968. Currently there are two texts that are in force, 1978 and 1991, respectively. They are mainly aimed to protect new varieties of medicinal plant resources. The text in 1991 is more protective, versatile, and effective for medicinal organisms. China officially joined UPOV's 1978 text in 1999 and became the 38th member state.

5.3.2.2 Major Laws and Regulations Concerning the Protection of TCM Resources

5.3.2.2.1 Regulations on the Protection and Management of Wild Medicinal Herb Resources

In order to protect and rationally utilize wild medicinal materials resources and meet the needs of people's medical care, the State Council promulgated the Regulation on October 30, 1987, and formally implemented it on December 1, 1987. This is China's first professional law to protect the resources of traditional Chinese medicine in the form of law. In *Regulations on the Protection and Management of Wild Medicinal Herb Resources*, China has stipulated the specific license system for the collection of medicinal materials and the export licensing system for wild medicinal materials. The *Regulations* stipulate that it is forbidden to hunt the first-class protected wild medicinal materials. Article 9 also stipulates that hunting of species at the second and third levels must be approved by the medical management departments at or above the county level and must submit an application to the wildlife authorities at the same level for approval by the competent medical authorities at the next higher level. Drug acquisition activities can only be carried out after obtaining a license for drug acquisition. Any unit or individual that violates the provisions of the Regulations shall be punished according to the seriousness of the circumstances. The formal implementation of the Regulations makes the protection and management of Chinese medicine resources legally applicable and perfects and enriches the content of resource protection. It is of great significance for maintaining the balance of ecosystems and protecting and rationally developing and utilizing Chinese medicine resources.

5.3.2.2.2 List of Species of Wild Medicinal Materials Under State Key Protection

According to the abovementioned ***Regulations on the Protection and Management of Wild Medicinal Herb Resources***, the State Administration of Medicine and the State Council's Wildlife Management Department and relevant experts have jointly formulated the first batch of "*List of Species of Wild Medicinal Materials Under State Key Protection*" which contains 76 species of wild medicinal materials, including 18 medicinal animals and 58 medicinal plants.

5.3.2.2.3 Notice of the State Council on Prohibiting the Trade of Rhinoceros Horn and Tiger Bone

It was promulgated and implemented by the State Council on May 29, 1993. The notice pointed out that the medicinal standards for rhinoceros horn and tiger bone were cancelled. Pharmaceutical use of rhinoceros horn and tiger bone is prohibited. The proprietary Chinese medicine preparations containing rhino horn and tiger bone shall be sealed up within 6 months from the date of notification and are forbidden to sell.

5.3.2.2.4 Notice on the Prohibition of Collecting and Selling of Nostoc flagelliforme, Stopping the Problems Related to Over-excavation of Licorice and Ephedra, Protecting Liquorice and Chinese Ephedra Resources, and Implementation of the Monopoly Franchise and License Control System

The notice was issued and implemented by the State Council in June 2000. It said that collecting *Nostoc flagelliforme* without restriction and digging licorice and ephedra indiscriminately are prohibited. We should prevent damage to the ecological environment and prevent grassland degradation and desertification.

5.3.2.3 Establishing a Management System for the Protection of Local Chinese Medicine Resources

For example, *Regulations on the Protection of Wild Plant Resources in Hunan Province*, *Interim Regulations on the Protection of Wild Plants in Jilin Province*, *Interim Provisions on the Protection and Management of Licorice Resources in*

Xinjiang Uygur Autonomous Region, and *Provisions on the Protection of Licorice Resources by the People's Government of Ningxia Hui Autonomous Region*.

The formulation of these laws and regulations has played a positive role in the protection of wild Chinese medicine resources in China. However, with the development of China's social economy and the rapid growth of market demand, the phenomenon of wild Chinese medicine resources being indiscriminately excavated in China has not been effectively curbed. The main reason is that the present laws and regulations are still not perfect and are not strictly enforced.

5.4 Classification of Endangered Medicinal Plants and Animals

In order to better protect rare and endangered species of animals and plants, we have classified them internationally and domestically and determined their species and levels.

5.4.1 *Species of Wild Medicinal Materials Under State Key Protection*

The Convention on International Trade in Endangered Species was signed on March 3, 1973. The Convention protects endangered species by including them in Appendices I, II, and III to the Convention. The Convention regulates endangered species involved in international trade by issuing permits and convenes a conference of contracting states every 2–3 years to discuss which species need to be listed for protection and to change the level of inclusion in the appendix according to the facts of the situation. China also specializes in grading medicinal endangered wildlife. In 1987. The State Council issued *Regulations on the Protection and Management of Wild Medicinal Materials Resources*, which proposed the list and grade standards of wild medicinal resources species. According to this standard, the biological species of wild medicinal materials protected in China were divided into three levels.

Level 1: Rare and precious wild medicinal species that are on the verge of extinction
Level 2: Important wild medicinal species that are facing problems such as shrinking distribution areas and exhaustion of resource
Level 3: Wild medicinal species feature severe reduction in resources

The catalogue contains 76 species of wild medicinal materials, including 18 medicinal animals. It has four species under first-class protection including tigers, leopards, saigas, and sika deer. They are strictly forbidden to hunt, 14 species belong to second-class protected wild animals, including scorpion, horse, black bear, black snake, silver snake, brown bear, pangolin, Chinese giant salamander, black-framed stork, Chinese forest frog, five-step snake, dragonfly, etc. There are 58 species of medicinal plants, 13 of which belong to second-class protection and 45 of which belong to the third class.

5.4.2 *The Classification of Endangered Medicinal Plants and Animals*

The endangered status and grade of endangered medicinal plants can be calculated and evaluated by the following method: calculating the endangerment coefficient, an assessment system of species endangerment can be established to determine the degree of species endangerment.

According to the calculation method of endangerment coefficient of medicinal plants [9], we can calculate the endangerment coefficient of each species by the following formula:

$$\text{Cendangered} = \sum_{i=1}^{4} Xi / \sum_{i=1}^{4} X_{\text{maxi}}$$

$C_{\text{endangered}}$ is the endangered coefficient, i is the evaluation index, X_i is the score of the i-th evaluation index of the species, and X_{maxi} is the highest score of the i-th evaluation index.

According to the calculation results of plant endangerment coefficient, referring to IUCN (International Standard for Threatened Species), the endangered status of plants is divided into five grades: critically endangered species, endangered species, vulnerable species, near-risk species, and safe species. See Table 5.1.

Table 5.1 Endangered statistical tables of medicinal plant species

Endangered level	Number	Endangerment factor	Representative medicinal plants
Extremely dangerous species	6	>=0.90	*Schisandra chinensis*
			Phellodendron amurense
			Acanthopanax sessiliflorus
Endangered species	18	0.7–0.9	Painted tree (*Toxicodendron vernicifluum*)
			Acanthopanax senticosus
			Aralia chinensis
Vulnerable species	17	0.60–0.69	*Scutellaria baicalensis*
			Tartary buckwheat (*Fagopyrum tataricum*)
			Laportea macrostachya
Near-risk species	43	0.50–0.59	*Physaliastrum sinicum*
			Delphinium grandiflorum
			Wild linen (*Linum stelleroides*)

5.4.2.1 Critically Endangered (CR)

Critically endangered species have characteristics such as narrow distribution range and special growing environment and have high economic value and scientific research value. Besides, their natural propagation is difficult. The number of mature individuals is estimated to be less than 1000, and the number of species is still decreasing year by year under the influence of human activities. A taxon is critically endangered when the best available evidence indicates that it meets any of the criteria A to E for critically endangered (shown as follows) and it is therefore considered to be facing an extremely high risk of extinction in the wild.

(A) A reduction in population size based on any of the following:

1. After observing, estimating, inferring, or speculating on any of the following data, for the next 10 years or within three generations (for a longer period of time), the cause of the decrease is clearly reversible and can be understood as have ceased. Population size decreased by at least 90%:

   (a) Direct observation
   (b) A richness index suitable for the taxon
   (c) Occupying area, shrinking distribution areas, and/or degradation of habitat quality
   (d) Actual or potential level of development
   (e) Adverse effects due to biological invasion, hybridization, disease, pollution, competitors, or parasites

2. According to the data of any aspect of (a)–(e) below A1, it suggests, estimates, speculates, or guesses that the factors for reduction may not have stopped, or we can see the cause as reversible in the next 10 years or within three generations (for a longer period of time); the taxon will be reduced by at least 80%.

(B) Geographic range in the form of either B1 (extent of occurrence) or B2 (area of occupancy) or both:

1. Extent of occurrence estimated to be less than 100 km^2 and estimates indicating at least two of a–c:

   (a) Severely fragmented or known to exist at only a single location
   (b) Continuing decline, observed, inferred, or projected, in any of the following: (i) extent of occurrence; (ii) area of occupancy; (iii) area, extent, and/or quality of habitat; iv number of locations or subpopulations; vnumber of mature individuals
   (c) Extreme fluctuations in any of the following: (i) extent of occurrence; (ii) area of occupancy; (iii) number of locations or subpopulations; (iv) number of mature individuals

2. Area of occupancy estimated to be less than 10 km^2 and estimates indicating at least two of a–c:

(a) Severely fragmented or known to exist at only a single location
(b) Continuing decline, observed, inferred, or projected, in any of the following: (i) extent of occurrence; (ii) area of occupancy; (iii) area, extent, and/or quality of habitat; (iv) number of locations or subpopulations; vnumber of mature individuals
(c) Extreme fluctuations in any of the following: (i) extent of occurrence; (ii) area of occupancy; (iii) number of locations or subpopulations; (iv) number of mature individuals

(C) Population size estimated to number fewer than 250 mature individuals and either:

An estimated continuing decline of at least 25% within 3 years or one generation, whichever is longer (up to a maximum of 100 years in the future), OR a continuing decline, observed, projected, or inferred, in numbers of mature individuals AND at least one of the following (a–b):

(a) Population structure in the form of one of the following:

(i) No subpopulation estimated to contain more than 50 mature individuals
(ii) At least 90% of mature individuals in one subpopulation

(b) Extreme fluctuations in number of mature individuals

(D) Population size estimated to number fewer than 50 mature individuals.
(E) Quantitative analysis showing the probability of extinction in the wild is at least 50% within 10 years or three generations, whichever is longer (up to a maximum of 100 years).

5.4.2.2 Endangered (EN)

The distribution range is narrow and the growth environment is special. The number of mature individuals of trees is estimated to be less than 5000, the distribution is scattered, and natural reproduction is difficult. The species has high economic value, and the population size tends to decline due to external influence. A taxon is endangered when the best available evidence indicates that it meets any of the criteria A to E for endangered (shown as follows), and it is therefore considered to be facing a very high risk of extinction in the wild.

(A) Reduction in population size based on any of the following:

1. The causes of a population size reduction of $\geq$70% over the last 10 years or three generations, whichever is longer, are clearly reversible AND understood AND ceased.

(a) Direct observation
(b) A richness index suitable for the taxon
(c) Occupying area, shrinking distribution areas, and/or degradation of habitat quality

(d) Actual or potential level of development
(e) Adverse effects due to the introduction of alien organisms, hybrids, diseases, pollution, competitors, or parasites

2. The causes of population size reduction of ≥50% over the last 10 years or three generations, whichever is longer, may not have ceased OR may not be understood OR may not be reversible.
3. A population size reduction of ≥50% may be projected or suspected to be met within the next 10 years or three generations, whichever is longer (up to a maximum of 100 years).
4. An population size reduction over any 10-year or three-generation period, whichever is longer (up to a maximum of 100 years in the future), where the time period must include both the past and the future, is projected or suspected to be ≥50%. The reduction and its causes may not have ceased OR may not be understood OR may not be reversible.

(B) Geographic range in the form of either B① (extent of occurrence) OR B② (area of occupancy) OR both:

1. Extent of occurrence estimated to be less than 5000 km^2 and estimates indicating at least two of a–c:

(a) Severely fragmented or known to exist at no more than five locations
(b) Continuing decline, observed, inferred or projected, in any of the following: (i) extent of occurrence; (ii) area of occupancy; (iii) area, extent, and/or quality of habitat; (iv)number of locations or subpopulations; (v) number of mature individuals
(c) Extreme fluctuations in any of the following: (i) extent of occurrence; (ii) area of occupancy; (iii) number of locations or subpopulations; (iv)number of mature individuals

2. Area of occupancy estimated to be less than 500 km^2 and estimates indicating at least two of a–c:

(a) Severely fragmented or known to exist at no more than five locations
(b) Continuing decline, observed, inferred, or projected, in any of the following: (i) extent of occurrence; (ii) area of occupancy; (iii) area, extent and/or quality of habitat; (iv) number of locations or subpopulations; (v) number of mature individuals
(c) Extreme fluctuations in any of the following: (i) extent of occurrence; (ii) area of occupancy; (iii) number of locations or subpopulations; (iv) number of mature individuals

(C) Population size estimated to number fewer than 2500 mature individuals and either:

An estimated continuing decline of at least 20% within 5 years or two generations, whichever is longer (up to a maximum of 100 years in the future), OR a continuing decline, observed, projected, or inferred, in numbers of mature individuals AND at least one of the following (a–b):

(a) Population structure in the form of one of the following:

(i) No subpopulation estimated to contain more than 250 mature individuals
(ii) At least 95% of mature individuals in one subpopulation

(b) Extreme fluctuations in number of mature individuals

(D) Population size estimated to number fewer than 250 mature individuals.
(E) Quantitative analysis showing the probability of extinction in the wild is at least 20% within 20 years or three generations, whichever is longer (up to a maximum of 100 years).

5.4.2.3 Vulnerable (VU)

The distribution range is narrow, the growth environment is special, the number of mature individuals is more than 5000, natural reproduction is easier, and it has certain economic value. The number of individuals is decreasing due to human influence. A taxon is vulnerable when the best available evidence indicates that it meets any of the criteria A to E for vulnerable (shown as follows), and it is therefore considered to be facing a high risk of extinction in the wild.

(A) Reduction in population size based on any of the following:

1. The causes of a population size reduction of ≥50% over the last 10 years or three generations, whichever is longer, are clearly reversible AND understood AND ceased.

   (a) Direct observation
   (b) A richness index suitable for the taxon
   (c) Occupying area, shrinking distribution areas, and/or degradation of habitat quality
   (d) Actual or potential level of development
   (e) Adverse effects due to the introduction of alien organisms, hybrids, diseases, pollution, competitors, or parasites

2. The causes of population size reduction of ≥30% over the last 10 years or three generations, whichever is longer, may not have ceased OR may not be understood OR may not be reversible.
3. A population size reduction of ≥30% may be projected or suspected to be met within the next 10 years or three generations, whichever is longer (up to a maximum of 100 years).
4. An population size reduction over any 10-year or three-generation period, whichever is longer (up to a maximum of 100 years in the future), where the time period must include both the past and the future, is projected or suspected to be ≥30%.

The reduction and its causes may not have ceased OR may not be understood OR may not be reversible.

(B) Geographic range in the form of either B1 (extent of occurrence) OR B2 (area of occupancy) OR both:

1. Extent of occurrence estimated to be less than 20,000 km^2 and estimates indicating at least two of a–c:
   (a) Severely fragmented or known to exist at no more than ten locations
   (b) Continuing decline, observed, inferred, or projected, in any of the following: extent of occurrence; area of occupancy; area, extent, and/or quality of habitat; number of locations or subpopulations; number of mature individuals
   (c) Extreme fluctuations in any of the following: extent of occurrence; area of occupancy; number of locations or subpopulations; number of mature individuals
2. Area of occupancy estimated to be less than 2000 km^2 and estimates indicating at least two of a–c:
   (a) Severely fragmented or known to exist at no more than ten locations
   (b) Continuing decline, observed, inferred, or projected, in any of the following: extent of occurrence; area of occupancy; area, extent, and/or quality of habitat; number of locations or subpopulations; number of mature individuals
   (c) Extreme fluctuations in any of the following: extent of occurrence; area of occupancy; number of locations or subpopulations; number of mature individuals

(C) Population size estimated to number fewer than 10,000 mature individuals and either:

An estimated continuing decline of at least 10% within 10 years or three generations, whichever is longer (up to a maximum of 100 years in the future), OR a continuing decline, observed, projected, or inferred, in numbers of mature individuals AND at least one of the following (a–b):

(a) Population structure in the form of one of the following:

(i) No subpopulation estimated to contain more than 1000 mature individuals
(ii) All mature individuals in one subpopulation

(b) Extreme fluctuations in number of mature individuals

(D) Population very small or restricted in the form of either of the following:

1. Population size estimated to number fewer than 1000 mature individuals
2. Population with a very restricted area of occupancy (typically less than 20 km^2) or number of locations (typically five or fewer) such that it is prone to the effects of human activities or stochastic events within a very short time period in an uncertain future and is thus capable of becoming critically endangered or even extinct in a very short time period

(E) Quantitative analysis showing the probability of extinction in the wild is at least 10% within 100 years.

5.4.2.4 Near Threatened (NT)

A taxon is near threatened when it has been evaluated against the criteria but does not qualify for critically endangered, endangered, or vulnerable now, but is close to qualifying for, or is likely to qualify for, a threatened category in the near future.

5.4.2.5 Least Concern (LC)

A taxon is least concern when it has been evaluated against the criteria and does not qualify for critically endangered, endangered, vulnerable, or near threatened. Widespread and abundant taxa are included in this category.

5.4.3 Assessment of the Level of Medicinal Plant Protection

The genetic loss coefficient is a quantitative evaluation of the degree of genetic loss that may occur to a given species after the extinction or the potential genetic value of the threatened plant species [17]. The genetic loss coefficient was calculated based on four indicators including the value of genetic breeding, the ancient plague, and the unique situation. The quantitative grading of each evaluation index is as follows:

Species: Scoring according to the number of species belonging to the threatened plant species and the species contained in the family: only 1 genera and 1 species of the species or their genus are only 1 species, 10 points; 6 points; the species belongs to more than 5 species, 1 point.

Germplasm resources and genetic breeding value: Scoring according to the value of the national key protected plants and the wild relatives of important cultivated plants: wild plant species protected by the state (all plants in the Chinese Plant Red Book and the National Key Protected Wild Plants List), 10 points; wild species of the same genus of important cultivated plants, 6 points; and 1 point for other species.

Ancient plaques: Relict plants (also known as living fossil plants) refer to those that have long-standing origins, widely distributed in the Cenozoic Tertiary or earlier. And most of them have been extinct due to geological and climatic changes; only a small range of plants survived. The shapes of these plants are basically the same as those found in fossils, retaining the original shape of their ancient ancestors, and their close relatives are mostly extinct, so they evolve slowly and are relatively isolated. The grading was divided into the ancient relics of Tertiary and pre-Tertiary, 10 points; the relics of Quaternary, 6 points; and the others, 1 point.

Unique situation: Scoring and dividing according to the size of distribution area, for example, we can score 10 points to Beijing endemic species; 6 points to North China endemic; and 1 point to other species.

According to the scores of the indicators above each species, the following formula can be used to calculate the correlation genetic loss coefficient:

$$\text{Cendangered} = \sum_{i=1}^{4} Xi / \sum_{i=1}^{4} X_{\text{maxi}}$$

where Cendangered is the coefficient of genetic loss, X_i is the score in the *i*-th bid of a species, and X_{maxi} is the highest score of the *i*-th indicator.

5.5 Examples of Research on Medicinal Plant Resources

5.5.1 Case 1: Status and Protection of Wild Resources of Coptis omeiensis [17]

Coptis omeiensis is a *Coptis* genus of the family Ranunculaceae. It is a perennial evergreen herb, also known as Emei Yelian, which is called Yanlian Yelian and Fengweilian. *Compendium of Materia Medica* records: "*Coptis chinensis* is good only for those who are fat and strong in Shujun, but for those who are in Lizhou in Tang Dynasty, Wu and Shu today, but good only for those in Yazhou and Meizhou." Historically, Emei and its surrounding areas are the real estate areas of *Coptis chinensis*. Wild *Coptis chinensis* have been used as Tugong medicinal materials and local specialties in the past dynasties. Long-term overexploitation has made the resources of wild *Coptis chinensis* very rare. The rhizome of *Coptis omeiensis* has important differences from other medicinal materials. Its rhizome has a single branch and is slightly curved, and the surface is dark brown. The golden yellow section is deeper than the *Coptis deltoidea* color, and the taste is bitter and the quality is better. *Coptis omeiensis* was listed as a national second-class endangered plant in 1984, and its products originated from the wild. *Coptis omeiensis* has low natural fertility and small population size. If it does not take effective protective measures, it will face an endangered situation. This article combines "Chinese Flora," "Flora of China," "Sichuan Flora," "Sichuan Chinese Medicine" related literature, local county records and many other historical records, as well as recent literature reports, a more comprehensive investigation of the wild *Coptis omeiensis* Distribution, in order to systematically assess the status of *Coptis omeiensis* resources, habitats, wild populations, development and utilization and conservation.

Table 5.2 Overview of wild distribution sites of *Coptis omeiensis*

Numbering	Location	Altitude/m	Longitude	Latitude
1	Hongya County Gaomiao	1210	29°31.050′	103°17.758′
2	Huangwan, Emeishan City	1251	29°33.772′	103°15.841′
3	Huangwan, Emeishan City	1160	29°31.178′	103°16.461′
4	Wawushan Town, Hongya County	1470	29°32.389′	102°59.894′
5	Wawushan Town, Hongya County	1337	29°34.867′	102°59.243′
6	Wawushan Town, Hongya County	1306	29°34.062′	103°08.383′
7	Heilin Town, Hongya County	1346	29°35.295′	103°09.309′
8	Huangwan, Emeishan City	1116	29°35.423′	103°21.317′
9	Huangwan, Emeishan City	1130	29°35.391′	103°21.312′
10	Huangwan, Emeishan City	1208	29°35.282′	103°21.374′
11	Huangwan Township, Emeishan City	1652	29°34.480′	103°20.399′
12	Huangwan Township, Emeishan City	1666	29°34.483′	103°20.379′
13	Huangwan Township, Emeishan City	1124	29°35.759′	103°18.692′
14	Huangwan Township, Emeishan City	1201	29°35.782′	103°18.703′
15	Huangwan Township, Emeishan City	1231	29°35.173′	103°18.554′
16	Huangwan Township, Emeishan City	1101	29°35.294′	103°21.496′
17	Huangwan Township, Emeishan City	1041	29°35.589′	103°21.430′
18	Huangwan Township, Emeishan City	1201	29°35.208′	103°21.231′
19	Huangwan Township, Emeishan City	1287	29°35.245′	103°21.127′
20	Huangwan Township, Emeishan City	1171	29°35.373′	103°21.175′

5.5.1.1 Investigation on the Wild Distribution of *Coptis omeiensis*

The investigation of *Coptis omeiensis* medicinal resources involves the relevant districts and counties such as Sichuan Emei, Hongya, and Ebian areas and neighboring Ya'an City. Field surveys mainly collect and record geographic coordinates, soil, slope direction, plant community and resource status of *Coptis omeiensis*, and related information such as mining, utilization, and acquisition. By processing the above data, 36 counties of Emeishan City, Hongya County, Embian County, Xingjing County, Yucheng District of Ya'an City, and Jinkouhe District of Leshan City are determined and 5 investigation points after field investigation identified. Among the 41 survey sites, 20 survey sites found the distribution of wild plants of *Coptis omeiensis*. The results are shown in Table 5.2 (to protect wild plants, the detailed names are omitted from the table).

5.5.1.2 The Characteristics of the Wild Ecological Environment of *Coptis omeiensis*

According to the survey results, the habitat requirements for the wild *Coptis omeiensis* are very harsh, the distribution area is extremely narrow; suitable for cold weather, can withstand low temperature below 0 °C, in the winter and early spring under the frost and snow cover leaves can maintain the evergreen; Avoid direct sunlight, like steep slopes with wet water seepage or wet rock walls with relative humidity above 85%. The soil of the wild habitat of *Coptis omeiensis* is mountain yellow soil or yellow brown soil with the pH of 5.15–6.15. At present, the distribution points of wild plants in *Coptis omeiensis* are between 1000 and 1900 m, with dolomite as the base rock on the damp and steep slope (more than 50°) or the shaded damp suspension in the low mountain evergreen broad-leaved forest. Although the distribution of wild *Coptis omeiensis* has a large altitude, it has the common points of shading, moistness, steepness, and high soil organic matter and is most suitable for growth in neutral humus soil. In addition, from the perspective of water demand, *Coptis omeiensis* needs shade and moist soil environment, but there is no distribution in the large slope, good drainage, and stagnant water and stagnant soil.

5.5.1.3 Community Characteristics

In the field investigation, it was found that the wild *Coptis omeiensis* was mostly aggregated and distributed little by individual, which may be closely related to the suitability of its ecological environment. The vertical distribution range of *Coptis omeiensis* almost runs through the evergreen broad-leaved forest belt and the evergreen deciduous broad-leaved forest belt in the Emeishan and Wawushan areas and reaches the lower limit of coniferous and broad-leaved mixed forest. Due to the particularity of the rock wall and steep-slope ecological environment, the associated plants have obvious shady and lithologic characteristics. *Coptis omeiensis* is also closely related to certain associated species. In the Wawushan area, *Coptis omeiensis* is mainly distributed in the wet sub-forests and mixed forests of the wet and damp rock walls and the steep slopes. The main companion plants are *Cyrtomium fortunei*, *Sinopteris grevilleoides*, *Neolepisorus ovatus*, *Clematis florida*, *Anisochilus carnosus Wall*, *Elatostema involucratum*, *Ainsliaea glabra Hemsl*, etc. The shrubs are *Piper nigrum*, *Begonia grandis*, *Stachyurus yunnanensis*, *Rosa rubus*, *Phoebe zhennan* and so on. In the Emei Mountain area, *Coptis omeiensis* is mainly distributed in the shaded and damp steep rock wall in the evergreen broad-leaved forest. Commonly associated plants include *Anisochilus carnosus Wall*, *Coniogramme japonica*, *Rhododendron molle*, *Paris*, *Pilea notata*, *Emei*, *Mazus japonicus*, etc. The shrubs have wax plum flowers, *Deutzia*, *Rubus*, *Berchemia*, *Begonia limprichtii*, etc. The trees have *Bothrocaryum controversum*, *Helwingia japonica*, and so on.

5.5.1.4 *Coptis omeiensis* Wild Resource Reserves

5.5.1.4.1 Sample Survey of Resource Reserves

Determine the number of plants of different sizes (different ages) per unit area, and then select 5–15 strains from each sample (depending on the growth conditions, it is generally appropriate to select about 10 strains. Too little is not representative. Sexuality, and the weighing, is not accurate. Too many of them are sometimes damaged by the collection of wild species in the sample). The plants are excavated and cut, and the fresh stems and air-dried weights are determined for the rhizomes, stems, and leaves, and the plants are determined. The average weight of the part is then counted by the number of squares. The statistical results of the resource reserves of *Coptis omeiensis* in different places are shown in Table 5.3.

Table 5.3 Statistical table of resource reserves of *Coptis omeiensis* in different samples

Sample number	Sample area product (m^2)	Dry weight per unit area (g/m^2)	Sample medicinal material (g)
20101021001	2×2	5.57	22.26
20101021002	2×2	8.72	34.89
20101026-1	2×2	38.4	153.6
20101026-2	2×2	16.5	66
20101026-3	2×2	6.15	24.6
20101026-4	2×2	1.29	5.18
20101027-1	2×2	0.38	1.51
20101027-2	2×2	6.44	25.76
20101103001	2×2	20.06	80.24
20101105001	2×2	9.8	39.2
20101103002	2×2	20.81	83.23
20101105002	2×2	6.75	27
20101105004	2×2	5.44	21.75
20101203001	2×2	18	72
20101203002	2×2	4.5	18
20101203003	2×2	5.625	22.5
20101204001	2×2	23.6	94.5
20101211001	2×2	45	180
20101211002	2×2	6.75	27
20101211003	2×2	9	36
20101212001	2×2	7.9	31.5
Total	-	-	1066.72

5.5.1.4.2 Wild Resource Reserves of *Coptis omeiensis*

Through the on-the-spot investigation of the distribution area of *Coptis omeiensis* and the survey of surrounding farmers, the distribution area of *Coptis omeiensis* in each survey area was determined, and based on the sample survey, the average yield of the plot was multiplied by the actual distribution area, then estimate the wild reserves in each survey area. At the same time, because the growth environment of *Coptis omeiensis* is weak illumination, the generally medicinal plants need to be more than 10 years, so the annual allowance is calculated according to 10% of the reserves of the producing area, and the annual demand is based on the actual harvest of farmers in the past three years (2007–2010) based on the survey of the distribution area. Statistics on the actual harvested amount of farmers. The statistical results are shown in Table 5.4.

The survey results show that in the Wawu Mountain area of Emei Mountain, the actual distribution area of the wild resource population of *Coptis omeiensis* is about 364 ha, the potential reserves are about 2000 kg (dry weight), and the maximum allowable harvest is about 200 kg (dry weight)/year/year. According to the survey, the actual spontaneous excavation volume of the drug farmers in the past 3 years is about 115 kg (dry weight) per year/year.

Table 5.4 Statistical table of wild resources of *Coptis omeiensis*

	Actual area of regional distribution	Average yield per plant	Regional reserves	Annual allowance	Average annual harvest
Survey area	/ha	/g dry weight	/kg dry weight	/kg dry weight	/kg dry weight
20101021001	20	8.55	171	17.1	3.2
20101021002	18	8.81	159	15.9	4.5
20101026-1-4	40	9.20	372	37.2	25.0
20101027-1-2	35	3.91	132	13.2	6.3
20101103-5001	20	7.13	143	14.3	7.8
20101103002	20	4.25	85	8.5	5.8
20101105002	20	6.02	120	12.0	6.6
20101105004	20	5.33	107	10.7	5.3
20101203001	30	2.25	68	6.8	7.2
20101203002	25	4.00	100	10	8.7
20101203003	20	3.25	65	6.5	2.2
20101204001	22	4.23	93	9.3	4.5
20101211001	15	3.19	47	4.8	4.1
20101211002	18	5.62	101	10.1	8.8
20101211003	16	3.16	51	5.1	4.5
20101212001	25	6.11	153	15.3	10.2
Total	364	-	1968	196.8	114.7

5.5.1.5 The Problem Faced by the Wild Resources of *Coptis omeiensis*

Coptis omeiensis mainly faces the following problems.

First, the distribution range is shrinking and biodiversity is decreasing. The collection of *Coptis omeiensis* basically relies on wild resources. The high-intensity collection in the past millennium has led to the narrow distribution of the original narrow range to the core area. Currently, only Emei and Hongya have distribution. Our large-scale survey shows that the actual distribution area of wild *Coptis omeiensis* is about 364 ha and the actual reserves of wild medicinal materials are not more than 2000 kg. In addition, with the expansion of human activities, the scope and intensity of various types of development have increased. The ecological environment damage of *Coptis omeiensis* has become increasingly serious, and the distribution area has been greatly reduced and fragmented. This situation may lead to a decrease in biodiversity of *Coptis omeiensis*.

The second is the difficulty of artificial cultivation. According to the investigation of local drug farmers, the wild seed storage time of *Coptis omeiensis* is short, and the germination rate after the traditional method treatment is only about 20%. It is difficult to germinate seeds more than half a year. According to the survey, *Coptis omeiensis* is different from *Coptis chinensis* or *Coptis deltoidea*. It is separated from the damp and steep cliffs. According to the ordinary *Coptis chinensis* or *Coptis deltoidea*'s open-cover shading cultivation technology, the survival rate is extremely low, and it is difficult to cultivate successfully, which makes the cultivation scope of *Coptis omeiensis* suffer great constraints. At present, the yield of *Coptis omeiensis* is low. Under the wild condition, the soil thickness of *Coptis omeiensis* generally grows below 10 cm, which showing acidity. The light conditions of this environment are poor and the plant growth is slow. We rarely found in the investigation a medicinal material that meets the specifications of commercial medicinal materials. Therefore, the yield per plant of wild perennial plants is much lower than that of *Coptis deltoidea* and *Coptis chinensis*, which are the same age as cultivated. In addition, the wild resources of *Coptis omeiensis* are extremely small. Although it have been used as an local precious medicinal material in history, it does not entered the *Pharmacopoeia* catalogue at present. Therefore, the economic value of cultivation is difficult to be guaranteed, which also to a certain extent limits the introduction and domestication of large-scale cultivation and related biotechnology and theory research.

Third, basic research work is weak and lacks long-term protection and management measures. Given the fact that the recent editions of the *Pharmacopoeia* did not contain *Coptis omeiensis* objectively avoiding the destruction of the wild resources of *Coptis omeiensis*, but also neglected all aspects of the research on *Coptis omeiensis*, especially the introduction of domestication, reproduction, conservation of germplasm resources and so on. As a national secondary protected plant, *Coptis omeiensis* has not caused special protection in the local area, nor has it established a special conservation area for *Coptis omeiensis* germplasm resources, nor has it caused the attention of relevant authorities or introduced corresponding policies to prevent humans and animals from ecologically interference and destruction of the environment and resources.

5.5.1.6 Protection and Development Strategy of Wild Resources in *Coptis omeiensis*

The wild population of *Coptis omeiensis* is small and scattered, and it is easily destroyed by devastating destruction. Once it disappears, it will cause irreparable damage. The investigation found that the wild resources of *Coptis omeiensis* have reached the edge of extinction, and it is imperative to increase the protection of the wild resources of *Coptis omeiensis*. At present, although Emei Mountain and Wawu Mountain have already established protected areas, human disturbance, ecological environment damage, and disorderly collection of *Coptis omeiensis* are difficult to eliminate. It is recommended to establish wild *Coptis omeiensis* resources in the concentrated distribution areas of Emei Mountain and Wawu Mountain, which clarifies the protection of *Coptis omeiensis*. In recent years, research on genetics, germplasm resources, biological characteristics of pollination, improvement of seed germination rate, and some important secondary metabolites have been carried out in *Coptis omeiensis*. It is recommended to continue to collect the germplasm resources of wild populations of *Coptis omeiensis* on this basis. The germplasm resources of the population are collected and preserved through the establishment of germplasm resources nursery, tissue culture techniques, and cryopreservation of seeds.

5.5.2 Case 2: Priority Management Ranking of Forest Plant Community in Puwa [18]

The priority of conservation of endangered species of medicinal plants is also an important part of in situ conservation. This chapter uses the Beijing Forestry University's research as a typical example to illustrate the priority ranking of the forest plant community in Puwa.

The establishment of nature reserves is recognized as the most effective means of protecting endangered species. The ultimate focus of protecting endangered species is to protect the environment in which endangered species live. There are many reasons for the endangerment of species, but habitat destruction is the most important factor. Once the living environment of endangered species is disturbed, endangered species may be rapidly extinct. Therefore, determining the survival of endangered species in one or more geographic units is the first and most important step in effective conservation. An important basis for the functional zoning of protected areas is to distinguish the importance of different geographic units and further quantify the differences and reflect them numerically. An important concept

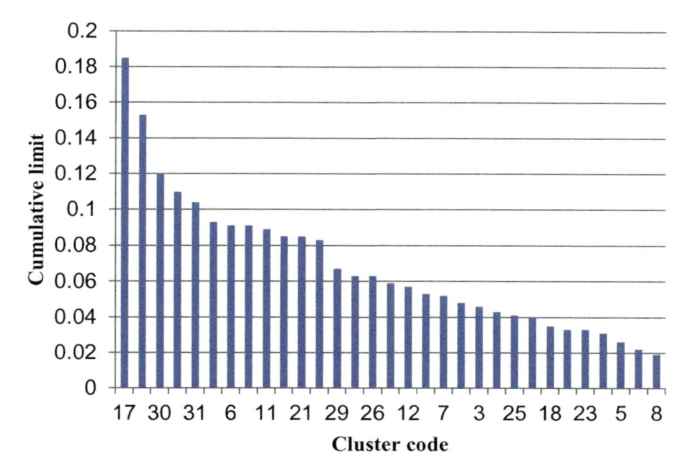

Fig. 5.2 Cluster cumulative limit ordering

in ecology, "the exact limit," helps to deepen our understanding of the above issues. The exact limit refers to the extent to which a species is confined to a certain type of plant community. Braun-Blanquet can calculate the exact limit of a certain target species in a certain cluster type according to the exact limit of the plant type. It provided us the information: What types of habitats are of concern for protected species, which species are more environmentally demanding in endangered plants, and which types of endangered plants may be included in a type of habitat. By summing the exact limits of different targets in the same cluster or constructing a calculation formula, it can reflect the difference in the conservation value of different habitats. Since these differences are numerically reflected, sorting them gives the basis for priority protection. On the vegetation map (Fig. 5.2), these clusters with different conservation values are classified, and a reasonable basis for the functional area division of the protected area is obtained.

5.5.2.1 Research Methods

Select the cluster types that participate in the sorting, number them separately (see Table 5.5), and enter the computer to create an Excel database. Consider that the composition of the meadow and the general forest vegetation is quite different. *Lespedeza davurica*, *Ass. Carex lanceolata*, *Phalaris arundinacea* shrub, and *Spiraea pubescens*, and *Ass. Carex lanceolata* shrub were excluded from the ranking. In most protected areas, meadow vegetation is often distributed at higher elevations, and the division into core areas is generally not controversial. In addition, in vegetation classification, *Ass. Carex lanceolata*, *Atractylodes lancea*, *Fraxinus rhynchophylla*, *Quercus wutaishansea Mary* forest, *Ass. Carex lanceolata*, *Leptopus chinensis*, and *Fraxinus rhynchophylla* were merged into *Ass. Carex lanceolata*,

Table 5.5 Group cluster classification tables participating in priority protection sorting

Vegetation type	Group	Cluster type	Code
Warm coniferous forest	*Pinus tabuliformis* forest	*Dendranthema chanetii – Spiraea pubescens – Pinus tabuliformis* forest	1
		Dendranthema chanetii – Vitex negundo – Pinus tabuliformis forest	2
	Platycladus orientalis forest	*Rabdosia rubescens – Spiraea pubescens – Platycladus orientalis* forest	3
Cold-temperate coniferous forest	*Larix principis-rupprechtii* forest	*Ass. Carex lanceolata – Spiraea pubescens – Larix principis-rupprechtii* forest	4
		Aster ageratoides Turcz – *Spiraea pubescens – Larix principis-rupprechtii* forest	5
		Ass. Carex lanceolata, Lespedeza davurica (Laxm.) Schindl – Larix principis-rupprechtii forest	6
Deciduous broad-leaved forest	*Syringa reticulata* var. *amurensis* forest	*Ass. Carex lanceolata – Spiraea pubescens – Syringa reticulata* var. *amurensis* forest	7
	Syringa pekinensis forest	*Ass. Carex lanceolata – Spiraea pubescens – Syringa pekinensis* forest	8
	Artificial *Robinia pseudoacacia* forest	*Deyeuxia arundinacea* – artificial *Robinia pseudoacacia* forest	9
	Carpinus turczaninowii forest	*Ass. Carex lanceolata – Deutzia parviflora – Carpinus turczaninowii* forest	10
		Ass. Carex lanceolata – Myripnois dioica – Carpinus turczaninowii forest	11
		Aconitum carmichaelii – Deutzia grandiflora – Carpinus turczaninowii forest	12
		Ass. Carex lanceolata – Spiraea pubescens – Carpinus turczaninowii forest	13
	Juglans mandshurica forest	*Amphicarpaea trisperma Baker – Spiraea pubescens – Juglans mandshurica* forest	14
		Ass. Carex lanceolata – Spiraea pubescens, Juglans mandshurica forest	15
		Viola acuminata – Deutzia parviflora – Juglans mandshurica forest	16
	Quercus wutaishanica forest	*Ass. Carex lanceolata – Spiraea pubescens – Quercus wutaishanica* forest	17
		Dendranthema chanetii – Spiraea pubescens – Quercus wutaishanica forest	18
		Ass. Carex lanceolata – Carpinus turczaninowii – Quercus wutaishanica forest	19
		Ass. Carex lanceolata – Leptopus chinensis, Fraxinus rhynchophylla – Quercus wutaishanica forest	20
		Ass. Carex lanceolata – Atractylodes lancea, Fraxinus rhynchophylla – Quercus wutaishanica forest	21
	Pteroceltis tatarinowii forest	*Rabdosia rubescens – Vitex negundo – Pteroceltis tatarinowii* forest	22

(continued)

Table 5.5 (continued)

Vegetation type	Group	Cluster type	Code
	Populus davidiana forest	*Ass. Carex lanceolata – Leptopus chinensis – Populus davidiana* forest	23
		Ass. Carex lanceolata – Spiraea pubescens – Populus davidiana forest	24
		Phlomis umbrosa – Spiraea pubescens – Populus davidiana forest	25
	Acer mono Maxim forest	*Phlomis umbrosa – Acanthopanax senticosus – Acer mono Maxim* forest	26
Deciduous broad-leaved shrub	*Vitex negundo* shrub	*Ass. Carex lanceolata – Vitex negundo* shrub	27
	Carpinus turczaninowii shrub	*Ass. Carex lanceolata – Carpinus turczaninowii* shrub	28
	Elsholtzia stauntonii bush	*Ass. Carex lanceolata – Elsholtzia stauntonii* bush	29
	Spiraea pubescens shrub	*Ass. Carex lanceolata – Spiraea pubescens* shrub	30
Evergreen broad-leaved shrub	*Rhododendron micranthum* shrub	*Saussurea nivea – Rhododendron micranthum* shrub	31

Fraxinus rhynchophylla, and *Quercus wutaishansea Mary* forest clusters. In order to compare and verify the scientific nature of this partitioning method, they are separated at the time of sorting and treated as two different clusters.

The total plant inventory generated by plot data is compared to the list of plants in each cluster to list the frequency of occurrence of a single species in all clusters. This frequency is actually the exact limit of this species in the Puwa Nature Reserve (the limit value is greater than or equal to 1 is less than or equal to the total number of cluster types. The smaller the value, the greater the possibility that the plant will be destroyed by habitat destruction). Sort all species by exact limit, select the number of species which has an exact limit value equal to 1–10, and record the number of species with an exact limit of 1 as n_1 and the number of species with an exact limit of 2 to 3 is n_2; the number of species with a limit value of 4–10 is n_3 (only species with an exact limit of less than 10 are selected for statistical purposes to highlight the status of endangered species in the cluster, while avoiding the next addition some clusters are not consistent with the actual situation due to the inclusion of non-endangered species).

Count the number of species with an exact limit of 1 in a single cluster, denoted as q_1. Similarly, the number of species with an exact limit of 2–3 and 4–10 is q_2 and q_3.

Calculate the cumulative exact limit of the individual clusters separately, and formulate the cumulative exact limit formula as follows:

$$\text{Accumulated exact limit}\, Q = (q_1 \times 5 + q_2 \times 3 + q_3 \times 1)/(n_1 \times 5 + n_2 \times 3 + n_3 \times 1)$$

The parameters 5, 3, and 1 in the formula are determined by experts, and different research areas can be assigned different values according to actual conditions.

The cumulative order of the same clusters is given the priority order of protection for different cluster types in the entire Puwa Nature Reserve.

5.5.2.2 The Sorting Results

5.5.2.2.1 Priority Protection Sorting

According to statistics, among the 31 cluster types, 293 species of plants were recorded. Among them, there are 100 species with an exact limit value of 1 and 81 species with an exact limit of 2 to 3. There are 66 species with an exact limit of 4 to 10 (see Table 5.6 for details); then $n_1 = 100$, $n_2 = 81$, and $n_3 = 66$. The number of species that are only limited to one cluster is 100, indicating that the composition of forest plants in the Puwa Nature Reserve presents certain complexity and uniqueness.

Calculate the cumulative exact limit of each cluster according to the cluster cumulative exact limit formula. The results are shown in Table 5.7.

Sort the 31 clusters from high to low according to the cumulative exact limit, and draw a column chart as follows:

The results show that different clusters belonging to the same group may have great differences in conservation value, such as *Ass. Carex lanceolata*, *Spiraea pubescens*, and *Quercus liaotungensis*, with *Ass. Carex lanceolata*, *Leptopus chinensis*, *Fraxinus rhynchophylla*, and *Quercus liaotungensis*, which belong to the *Quercus liaotungensis* group, but in the whole sort there is a big gap in the position.

The sorting result has an important practical significance in the planning and management of the protected area. Taking the above classification results as an example, if the two values of 0.05 and 0.1 are critical points, the forests of the entire

Table 5.6 Statistical table of forest plant species exact limits in Puwa Nature Reserve

Exact limit	1	2	3	4	5	6	7	8	9	10	10–31
Number of species	100	49	32	15	15	10	6	10	5	5	46

Table 5.7 Statistics on the cumulative exact limits of the population of the Puwa Nature Reserve

Cluster type	Number of species	q_1	q_2	q_3	Cumulative exact limit
1	41	3	4	8	0.043
2	40	3	1	14	0.040
3	25	5	1	9	0.046
4	45	5	1	11	0.048
5	44	1	1	13	0.026
6	51	10	2	18	0.091
7	40	5	3	8	0.052
8	21	1	2	4	0.019
9	36	6	1	10	0.053
10	73	16	8	20	0.153
11	50	8	8	8	0.089
12	35	5	5	6	0.057
13	65	11	2	13	0.091
14	52	10	3	8	0.083
15	71	8	2	23	0.085
16	50	11	7	13	0.110
17	103	18	9	33	0.185
18	45	2	2	12	0.035
19	65	8	6	17	0.093
20	39	3	2	6	0.033
21	57	6	6	21	0.085
22	22	7	4	4	0.063
23	45	2	1	14	0.033
24	60	6	1	15	0.059
25	41	3	3	9	0.041
26	29	7	3	7	0.063
27	42	2	1	12	0.031
28	33	2	0	8	0.022
29	39	5	5	14	0.067
30	77	11	6	24	0.120
31	49	10	4	22	0.104

Puwa Nature Reserve can be divided into three different levels according to their importance. The five-population cluster with cumulative exact limit greater than 0.1 is the most important, and the region belonging to this vegetation type should be considered as the core region; the regions with cumulative exact limit values below 0.1, above 0.5, and below 0.5 can be considered separately the buffer and experimental area.

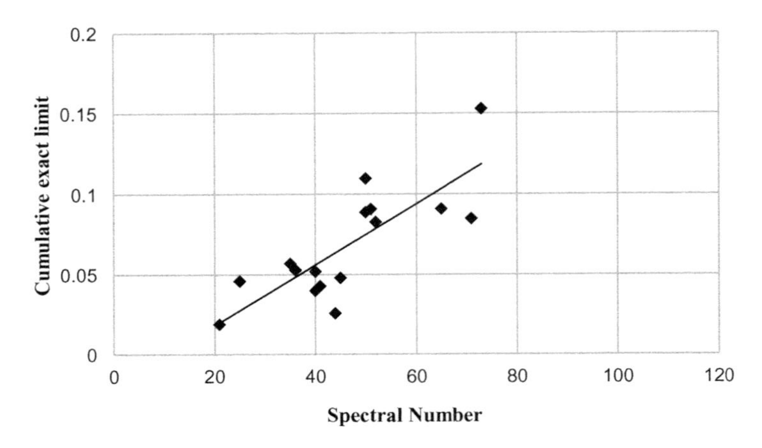

Fig. 5.3 Spectral number – cumulative exact limit scatter plot

5.5.2.2.2 The Comparison Between the Exact Limits of the Clusters and the Number of Species in the Cluster

There may be some clusters, the composition of which is very simple, and these simple components have high exact limit values. Finally, it is found that the clusters with higher cumulative exact limit values are actually too single. The phenomenon is that the result is too different from the expected goal. Based on this question, the number of clusters is taken as the abscissa, and the cumulative exact limit of the cluster is plotted as the ordinate as a scatter plot as follows (Fig. 5.3).

Further, the two groups of data from 31 clusters were input into the SPSS statistical software for correlation analysis. The results showed that the number of cluster species was significantly correlated with the accumulation exact limit of 0.01 at the correlation level, and the correlation coefficient was 0.812. This result indicates that the number of cluster species and the accumulation exact limit are positively correlated; usually, the larger the number of species, the higher the cumulative exact limit value; in the case of exact limited human and material resources, the populations of each cluster are roughly counted. The number of species can measure the status of a cluster within the overall conservation goal.

References

1. Li DZ, Ji YG. Protection and sustainable reuse of traditional Chinese medicine resources. Tradit Chin Med. 2016;16(18):23–4.
2. Di Z, Wang ZY, Yu XF, et al. Analysis on the present situation of grassland medicinal plant resources and the comprehensive utilization of 3 main medicinal plants in the west of Heilongjiang. Chin Med Resour Cultiv, Chin Pharm. 2015;26(4):570–3.

3. Liu JH, Wei JW, Liu GY. Current situation and future of sustainable development of traditional Chinese medicine resources. Dig Latest Med Inform World. 2019;19(09):32–3.
4. Zhong LF. Statistics of Chinese household registration, field and field assignments. Beijing: Zhonghua Book Company, edition; 2008.
5. Wang LX, Yin CM. A brief discussion on the new viewpoint of Chinese medicine resources protection. 2010 China Pharmaceutical Conference and the 10th China Pharmacist Week; 2010. p. 1–6
6. Zi LG, Wang QC, Cui GF. Situation and prospect of functional zoning of nature reserves in China. World For Res. 2016;29(05):59–64.
7. Dai YH, Jin H, Zhao Y, et al. Present situation and protection countermeasures of medicinal plant resources in Changbai Mountains. Jilin Agric. 2017;19:83.
8. Franklin IA. Evolutionary change in small population. In: Soule ME, Wison BA, editors. Conservation biology: an evolutionary ecological perspective. Sunderland: Sinauer Associates; 1980. p. 135–50.
9. Jiao PN. Study on the isolation of endophytic fungi from wild tending ginseng and their effects. Master thesis; 2012.
10. Chen SL, Xiao SY, Wei JH, et al. Sustainable development model of chinese medicinal materials based pharmaceutical products: case studies from Bulbus Fritillariae Cirrhosae. Asia-Pac Tradit Med. 2006;2:72–5.
11. Que L, Chi XL, Zang CX, et al. Species diversity of ex-situ cultivated Chinese medicinal plants. Chin J Tradit Chin Med. 2018;43(05):1071–6.
12. Si SS. Introduction and ex-situ conservation of plants in China. Agric Technol. 2017;37(08):28.
13. Yan XB. Botanical Garden. Pop Sci Res. 2009;4(22):74–9.
14. Yu YH, Liao JP, et al. Global botanic gardens and plant diversity conservation. Chin Bot Garden. 2008:14–24.
15. Xiao PG, Chen SL, Zhang BG, et al. Ex-situ conservation and utilization of Chinese medicinal plant germplasm resources. Mod Chin Med. 2010;12(6):3–6.
16. Li HZ, Cai XZ, Zhang HL. Endangering mechanism and strategy for conservation of endangered plant. J Hunan Inst Human Technol. 2006;15(6):43–6.
17. Ji HQ, Jiang SY, Feng CQ, et al. Study of present situation of mild coptis omeiensis resources and protection approach. J Chengdu Univ (Natural Science Edition). 2011;30(4):298~304.
18. Chuan CW, Bo Z, Xing SH, et al. Sequence of preferential protection of associations of forest vegetation in Puwa Nature Reserve, Beijing. J Northeast For Univ. 2006;5:62–4.

Chapter 6
Gene Modification of Medicinal Plant Germplasm Resources

Rong-min Yu, Jian-hua Zhu, and Chun-lei Li

Abstract Germplasm resources refer to the basic materials for selecting and cultivating new species, which include propagating materials of cultivated species and wild species from different plants and artificially creating genetic materials of various plants based on the propagating materials. Chinese herbal medicine covers about 11,146 species, possessing rich medicinal plant germplasm resources. Medicinal plant germplasm resources are mainly composed of all of the medicinal genetic resources from medicinal plants, including the seeds, the roots, stems, embryos, cells, protoplasts, etc., and are a cornerstone for introducing the cultivation and sustainable use of resources as well as the bases for ensuring the quality of medicine. Hence, collection and evaluation of medicinal plant germplasm resources plays a vital role in studying and utilising medicinal materials and bioactive metabolites. The investigation of medicinal plant germplasm resources mainly includes photographs and records of their morphological characteristics, biology, medicinal organs, medicinal values, other economic values, etc. The purpose preserving germplasm resources is for the sustainable use of genetic resources. The stored germplasm resources should maintain the original viability, degrees of genetic variation, and amounts of samples. The evaluation of germplasm resources includes the species identification of germplasm, genetic diversity analysis, and the determination of genetic relationships. The genetic manipulation of medicinal plant germplasm resources that includes conventional selective breeding, cross-breeding, and plant genetic engineering is a process in which the genome of a medicinal plant species is modified in order to produce desired traits. Through genetic modification, functional genes in medicinal plant germplasm resources are reformed to ensure the efficacy and quality of traditional medicines and their products. This chapter introduces the medicinal plant germplasm resources, investigation methods and gene modification

R.-m. Yu (✉) · C.-l. Li
Biotechnological Institute of Chinese Materia Medica, Jinan University, Guangzhou, China
e-mail: tyrm@jnu.edu.cn; lcl992@126.com

J.-h. Zhu
Institute of Molecular Plant Sciences, The University of Edinburgh, Edinburgh, Scotland, UK
e-mail: v1jzhu14@exseed.ed.ac.uk

L.-q. Huang (ed.), *Molecular Pharmacognosy*,
https://doi.org/10.1007/978-981-32-9034-1_6

techniques of germplasm resources, and reviews gene modification applications in medicinal plant germplasm resources.

6.1 Concept and Significance of Medicinal Plant Germplasm Resources

6.1.1 Concept of Medicinal Plant Germplasm Resources

Germplasms are the genetic materials of organisms, which can be inherited to the individual or colony progeny by propagation. In terms of plant, it not only includes the seeds but also the roots, stems, embryos, cells, protoplasts, and even DNA fragments. The word “germplasm”, brought forward by a German geneticist called Weismann in 1892, originates from “germplasm theory” [1]. It claimed that germplasms were the genetic materials descended from the parental generation to offspring, producing habits that were not influenced by the environment.

Germplasm resources, also called “genetic resources”, are the basic materials of selecting and cultivating new species. They include propagating materials of cultivated species and wild species from different plants and various plant genetic materials artificially created by the propagating materials above. In a narrow sense, germplasm resources usually refer to a specific species, in which all available genetic vehicles are involved, including cultivated species, wild species, relative species, and special inheritable materials (wildly or artificially mutated materials like polyploidy and haploid number).

Medicinal plant germplasm resources are composed of the medicinal genetic resources from algae, fungi, lichens, bryophytes, pteridophytes, gymnosperms, angiosperms, etc. Narrowly speaking, medicinal plant germplasm resources refer to a specific species, like “the germplasm resources of *Panax*”, “the germplasm resources of *Caladium*”, and “the germplasm resources of *Salvia miltiorrhiza*”.

6.1.2 Purpose and Significance of Research on Medicinal Plant Germplasm Resources

China possesses rich biotic resources and has the longest history worldwide of medicinal plant application [2]. Chinese herbal medicine mainly comes from medicinal plants and covers about 11,146 species (including those mentioned below). Recently, with the improvement of life quality and the rise of “going back to nature”, the exploration of medicinal plant resources is becoming more popular globally. Besides being used as herbal materials and pharmaceutical materials, medicinal plants are wildly applied to many novel areas, such as health food, beverages, flavouring agents, perfumes, cosmetics, plant pesticides, and drugs for birds and domestic animals. Under such a promising situation, pharmacognosy research has to meet a stricter requirement; medicinal plant resources should be protected and used effectively. However, we are still in the very first step of researching medicinal plant germplasm resources, and there are many problems during production. Most of the cultivated herbal drugs are introduced blindly and acclimated in a short time without screening evaluations. As a consequence, cultivated plant germplasm resources can be mixed up. Even if the species obtained from farmers go through primary screening, they are only usually appraised according to their appearance and production and are in need of further screening.

6.1.2.1 Medicinal Plant Germplasm Resources Are a Cornerstone for Introducing the Cultivation and Sustainable Use of Resources

Medicinal plant germplasm resources serve as an essential factor for the quality and production of Chinese herbal medicine and are also the material bases for maintaining stable drug quality. Nowadays, facing the resource shortages of Chinese herbal medicine, we should deal with the cultivation of Chinese herbal medicine and the protection of core production areas in order to realise the sustainable use of medicinal plants. Economic indices should be considered before introduction, which means that the most useful genetic resources would be screened out, so as to obtain more products while consuming the same amount of materials. Take *Taxus* as an example; research on germplasm resources has found that this genus contains 11 kinds (four kinds in China with one mutated variety) [3]. Below the genus level, there are many geographic races and cultivated races; *Taxus baccata* L. has no less than 40 geographic races, whose active component (taxinol) content is barely 0.01%. In China, the main cultivation type of *T. baccata* L. is the hybridised *Taxus* × *media* from North America, which is selected according to the investigation of germplasm resources and contains 0.014% taxinol. It has a lot of merits, such as strong germination, fast production, easy sticking and propagation, and strong adaptability to the environment.

6.1.2.2 Medicinal Plant Germplasm Resources Are the Material Bases for Cultivation

Modern cultivation is characterised by the wide collection, storage, and deep study of germplasm resources. Germplasm resources with different genetic properties are the material bases for cultivation. The more plentiful the germplasm resources are, the more the cultivation could be anticipated. Therefore, it will be more possible for some new and potential races to be obtained. Medicinal plant germplasm resources play basic roles in cultivation. Whether cultivation has significant achievement depends on the mutation, discovery, and use of these key characters.

In order to avoid poor genetic resources of species, it is necessary to make use of more germplasm resources [4]. Due to the promotion of good medicinal species, and the selection of some germplasm resources with the same genetic backgrounds during cultivation by some institutions or scientists, it may lead to the following bad results: first, the quality of herbal drugs could decline; second, medicinal plants may have poor resistance and low productivity; third, the favourable restructuring could be limited by the narrow genetic base, which may slow down the progress of breeding, even leading to stagnation. Therefore, to avoid poor genetic bases of medicinal species, it is necessary to collect more germplasm resources and establish germplasm resource nurseries of young plants or gene pools.

Medicinal plant germplasm resources are one kind of important natural resource, which are formed by long-term natural evolution and artificial creation. They have great influence on the production of excellent herbal medicines. To some extent, the formation of many effective authentic ingredients is attributable to the effect of regional varieties. The cultivation of medicinal plants pertains to the utilisation, processing, and transformation of various germplasm resources by different methods according to demand. Thus, medicinal plant germplasm resources are the main origins for people to select novel varieties through plant genetic mutation. In addition, they also serve as the material basis for medicinal plant germplasm cultivation. Whether the cultivation of medicinal plants is successful or not mainly depends on the quantity and quality of species resources, which we have explored and mastered. Their research extension and depth play key roles as well.

The cultivation of medicinal plants in China and even worldwide has already established a primary basis [5]. Many Chinese herbal medicines, such as ginseng, *Chrysanthemum*, mint, saffron, medlar, *Rehmannia*, *Fritillaria*, yam, *Polygonatum*, Campanulaceae, woad, hemp, *Ginkgo biloba*, *Coix*, *Dendrobium*, motherwort, honeysuckle, *Eucommia*, *Codonopsis*, sage root, *Radix Angelica*, etc., have become good local “varieties”. However, because a great number of “varieties” lack the involvement of breeders, their genetic purities are still unknown, and they could not be counted as true species. At present, we are facing the ambitious task of breeding a growing number of medicines, so it is necessary to speed up the introduction and application of new techniques and methods for breeding.

6.1.2.3 Medicinal Plant Germplasm Resources Are the Bases for Ensuring the Quality of Medicine

Germplasm resources are a source of production of Chinese herbal medicine, and their quality exerts critical effects on the quantity and quality of medicine. With the development of production and scientific technology, human beings will continue exploring more novel drug varieties from wild medicinal plant resources, so as to fulfil the growing needs of production and living. The more germplasm resources we have, the more advantages of innovative varieties we can use to occupy the international market. Therefore, many countries have spared no expense to collecting large numbers of germplasm resources through multiple channels. Recently, among the 800 kinds of commonly used Chinese herbal medicine, most of them come from wild resources, and only about 300 kinds have been introduced and cultivated, with the other 200 kinds being planted in large areas [6]. For example, coptis root, cultivated extensively in Sichuan and Chongqing, was found during resource surveys and later promoted in the 1960s. Moreover, according to the research on the origin of long-term cultivated Chinese herbal medicines like Feng Dan, *Chrysanthemum*, Chinese yam etc., it has been revealed that Feng Dan originated from the authentic medicine *Paeonia ostii*, which formed after long-term cultivation in the Tongling area. Thus, its wild species or related species should be analysed extensively.

At the same time, the main cultivating varieties at present are from the domestication of wild plants in different historical periods. Besides, the major cultivating varieties of Chinese herbal medicine are interwoven seriously without new species cultivation and purification.

6.2 Types of Medicinal Plant Germplasm Resources

Since there is a wide range of germplasm resources, for convenience of research and use, the methods of classifications by different disciplines according to their use are not exactly the same. Usually, the classification methods are based on kinships, ecological types, sources, and value. In this book, on the basis of characters of medicinal plant germplasm resources, there are two ways to classify them.

6.2.1 Categorisation According to Sources

According to sources, germplasm resources are commonly divided into local, foreign, wild, and artificial germplasm resources.

6.2.1.1 Local Germplasm Resources

Local germplasm resources refer to old regional varieties and improved varieties being currently promoted. They own several characters. First, local germplasm resources are highly adaptive to local natural conditions and ecological characteristics. Second, they reflect the needs of production and livelihoods of local people. Third, they have special properties. Finally, these old local varieties are fertiliser intolerant and have low yield and poor disease resistance.

Take ginseng as an example. It can be classified into Da-Ma-Ya (big horse teeth in Chinese)-like rhizoma, Er-Ma-Ya (small horse teeth in Chinese)-like rhizoma, Chang-Bo (long neck in Chinese)-like rhizoma, and other types or local varieties, according to the morphology of its roots and rhizomes [7]. In addition, based on commodity-type characteristics, it can be divided into general ginseng, Biantiao ginseng, and Shizhu ginseng. Moreover, regarding the morphology above-ground, classifications include yellow fruit ginseng, red fruit ginseng, orange fruit ginseng, purple stems ginseng, green stems ginseng, tight panicle ginseng, loose panicle ginseng, etc.

6.2.1.2 Foreign Germplasm Resources

Foreign germplasm resources stand for the medicinal species and plant types introduced from other countries or areas, whose characters are as follows: first, it can reflect the natural conditions and production properties of original areas; second, most of the foreign germplasm resources are poorly adaptive to the local conditions. These can include resources such as safflower, *Gardenia*, pomegranate, walnut, and garlic, which were introduced in ancient times, and American ginseng, Japan chuanxiong, and Japanese angelica, which were introduced in modern times.

6.2.1.3 Wild Germplasm Resources

Wild germplasm resources refer to the related species of various medicinal types and valuable species of wild plants. Since they formed through long-term natural selection under specific conditions, they usually have some essential traits that cultivated Chinese medicinal plants generally lack, such as high resistance to pest and pathogens, unique quality, and male sterility. At present, most medicinal plant germplasm resources still belong to wild germplasm types, such as bidentata and yellow and red sage root.

6.2.1.4 Artificial Germplasm Resources

Artificial germplasm resources are the hybrid, mutant, and intermediate materials produced by different methods (hybridisation, physical and chemical mutagenesis, etc.). For example, when buds of Mongolian *Astragalus* seedlings are treated with an improved agar smear, tetraploid plants will be obtained, which serve as a foundation for further cultivating new varieties of *Astragalus* with polyploidy. The generation of ginseng × American ginseng hybrids is performed in a similar way.

6.2.2 *Categorisation According to Relationship*

According to relationship, the gene pool is divided into three types:

(i) Primary gene pool (Gp-1): includes materials within the same species, equivalent to the traditional biological species concept. They can mutually hybridise with fertility and transfer genes between relatively simple materials.

(ii) Secondary gene pool (Gp-2): contains closely related wild species and related genera of plants, including some species and related species that can be hybridised with.
(iii) Tertiary gene pool (Gp-3): composed of materials that can cause severe cross-infertility and cross-sterility when hybridised with distant species.

6.2.3 *Categorisation According to Research Range*

Germplasm resources are divided into local germplasm resources and single germplasm resources, based on research range.

(i) Local germplasm resources refer to all medicinal germplasm materials in one district or area.
(ii) Single germplasm resources refer to all the medicinal germplasm materials of some specific medicinal species.

6.3 Characteristics of Medicinal Plant Germplasm Resources

China has vast territories of both land and sea, spanning latitude above 49° from north to south and longitude 60° from east to west. Its topography and climate are complicated, going from frigid, temperate, and tropical zones from north to south, respectively, with diverse ecological environments. Besides, China is considered to have one of the most abundant germplasm resources in the world. The core indicator for germplasm resource evaluation is clinical effect, which consequently determines the following characteristics of medicinal plant germplasm resources, compared to farm crops.

6.3.1 *Unique Division and Selection Criteria*

Crops focus on yield, nutrition, and flavour. However, medicinal plant germplasm resources are in pursuit of drug quality [8]. The most effective components in medicinal plants are secondary metabolic products and abnormal secondary metabolic products, such as alkaloids, saponins, flavonoids, terpenoids, coumarins, and tannins. Their contents in plant are very low, only accounting for several percentage points or several parts per 10,000 that can be hardly observed directly; thus, we have to use physical and chemical analysis to obtain them. Meanwhile, these elements are susceptible to the environment in plants. As a result, it is difficult to determine the genetic and environmental effects. Molecular biology methods to directly analyse the DNA of plant material serve as an important way to study medicinal plant germplasm resources.

6.3.2 Obvious Regional Characters of Medicinal Plant Germplasm Resources

In the traditional production of Chinese herbal medicine, the authenticity of medicine was stressed, which shows that medicinal plant germplasm resources place more emphasis on region than any other field crops. In terms of an authentic medicinal plant or a farmed type, similar to ecological factors, genetic factors also play a significant role. Genuine medicines are often produced in a particular area by long-term natural and artificial selection, such as Ludang in Shanxi Changzhi, Shizhu ginseng in Jiangsu Zhenjiang, and Gongju in Anhui Shexian. These species can best adapt to the climatic and environmental conditions to accumulate most bioactive metabolites, as well as cropping systems. Increasing evidence has demonstrated that there are some local excellent varieties of traditional authentic medicinal ingredients that have been selected for many years, but are not being purified. They serve as a valuable resource for us to study medicinal plant germplasm resources.

6.4 Collection of Medicinal Plant Germplasm Resources

6.4.1 Investigation of Medicinal Plant Germplasm Resources

Medicinal plant germplasm resources are extremely plentiful; their investigation and collection are the most basic work and also the most important one.

6.4.1.1 Preparations Before Investigation

To investigate the germplasm resources of medicinal plants, especially in large-scale investigations, careful plans should be drafted beforehand and adequate preparations should be made, which are the preconditions to obtain satisfactory outcomes. Before investigation, the following works should be completed:

Making an Investigation Plan

Before making a plan, it is necessary to read the relevant references to determine the location, route, time, expedition, equipment, and materials of the investigation.

(a) Determination of the location and route of investigation

Investigations of germplasm resources of medicinal plants are mainly on local germplasm resources and single species. Investigations of local germplasm resources should be focused on three regions:

(a) Primary and secondary origin centres of medicinal plants and regions with the highest plant diversities.

(b) Regions where medicinal plant germplasm resources have suffered the most losses and threats. The location and route of investigations on single species of germplasm resources should be designed according to the main production areas, genuine production areas, and areas with rich diversity.

After the location and route of the investigation have been determined, the staff participating in the study should refer to the data for the area investigated and understand the following information, landform, vegetation, terrain, hydrology, and climate, and background data, including social structure, ethnic distribution, living habits, economic status, social change situation, arable land, crop species, cultivation methods, and the occurrence of major pests and diseases, to finish the preparation for the investigation.

(b) Duration arrangement

Due to the specificity of Chinese herbal medicine, the investigation duration should be determined according to the growth characteristics and medicinal harvesting period. Sufficient time is needed to finish the investigation, but it should not be too long for the consideration of cost and labour process. The specific duration will be determined in accordance with the investigation scope and purpose.

(c) Preparation of commodities and equipment

Commodities and equipment needed in field investigations of germplasm resources include transportation vehicles; sample collection tools; record forms for surveying; instruments, tools, and containers used for investigation; and articles for personal use (including medicine).

(d) Expedition and team

The number of expeditions should be determined according to the investigation tasks, objectives, and area. Team members should receive appropriate skill, safety, and survey method training.

6.4.1.2 Contents of Medicinal Plant Germplasm Resource Investigation

This is the most critical step for the investigation of medicinal plant germplasm resources. The investigation contents of regional germplasm resources mainly involve species of wild medicinal plants (including varieties), species of cultivated medicinal plants, formed species or strains, and so on. We should take photos and records of their morphology, biological characteristics, medicinal organs, medical value, and other economic values.

Character (Video) Records

The characteristics of various types of medicinal plants have great differences. During the investigation, the record contents and the number of descriptors are different, but the following information should be inspected and recorded for all kinds of medicinal plant germplasm resources (Table 6.1).

Accession is the basic unit in the registration of germplasm resources. A so-called single accession of germplasm should have a similar genetic unit (such as

Table 6.1 Catalogue of medicinal plant germplasm resources – table of accession

1. Passport information								
1.	Number of germplasm	Plant classifications and codes in China (GB/T14467-93) + XXXXX (five natural order numbers)						
2.	Chinese name of germplasm							
3.	Foreign name of germplasm							
4.	Germplasm types	Population	Individual	Cultivated variety				
		Population relationship	Somaclone	Strain				
		Source		Genetic material				
		Family (Full-sib)	Famous tree	Others				
		Family (Half-sib)	Local variety					
		Other (Specified):						
5.	Germplasm source	County (District/City): Province (Autonomous region/Municipality): Country:						
6.	Storage institute							
7.	Name of family (Chinese name)							
8.	Name of genus (Chinese name)							
9.	Name of species (Chinese name)							
10.	Name of species (Latin name)							
2. Extended information								
11.	Internal number in the storage institute							
12.	Germplasm characteristics							
		Picture 1, 2, 3, 4 (Optional)						
13.	Germplasm application							
14.	Germplasm genealogy							
15.	Bred time							
16.	Germplasm habitat	Climate zone: Longitude: ° ′ ″ Latitude: ° ′ ″ Altitude: m Soil type:						
		Average annual temperature: Average annual rainfall: Extreme factors:						
17.	Stored material	Plantlet	Root	Bud	Pollen	Cell	Gene	
		Stem	Flagellum	Seed	Tissue	DNA		
		Others (Specified):						
18.	Storage method	*In situ* storage	*Ex situ* storage	Equipment conservation				
19.	Storage address	County (District/City): Province (Autonomous region/Municipality): Country:						
20.	Reproductive mode	Seed	Cutting	Grafting	Layering	Tissue culture		
		Other (Specified):						
21.	Exchangeable material	Seedling	Seed	Cutting	Scion	Pollen	Seminal root	Bud
		Other (Specified):						

(continued)

Table 6.1 (continued)

3. Communication information					
22.	Name of institute				
23.	Address				
24.	Name				
25.	Email				
26.	Tel.				
27.	Fax				
28.	Remark:				

population/provenance, family, somaclone, etc.) and cannot be repeated with the germplasm heredity that has been registered. For example, a germplasm in the same family collected at a different time should not be counted as two accessions, because they have the same genetic composition. Thus, they are one replicated accession (i.e. genetic homogeneity is one accession, but genetic heterogeneities are multiple accessions). Sometimes, the genetic homogeneity has been recorded as two accessions in actual operations. This is a gradual process, which needs genetic detection at different levels (population relationship, phenotype, physiology, ecology, molecule etc.), building up the "quasi-core" and "core" germplasm resources gradually.

A. Information of germplasm habitats

(a) Geographic location
This includes the natural and administrative regions of investigation spots (specify the names of province, county, country, village, and place), latitude and longitude, altitude, direction, and distance to the nearest town, topography, slope direction, slope degree, etc.

(b) Conditions of water and soil
These involve the soil type, soil depth, hydrological conditions, etc.

(c) Climate conditions
These include the average annual temperature, average monthly temperature, highest temperature, lowest temperature, non-frost period, solar radiation, rainfall, etc.

(d) Ecological conditions
These include the vegetation, accompanying plants (wild resources), rotation crops (intercropping, relay cropping), etc.

B. Information on germplasm resources

(a) Germplasm types
Germplasm types include wild type (species, varieties, variants), cultivation type (local varieties, breeding materials, bred varieties), etc.

(b) Germplasm source
This includes field, farmland, research institutes, etc.

(c) Germplasm name
Such as Chinese name, Latin name, alias, local name, etc.
(d) Germplasm number
The number involved in the investigation.
(e) Utilisation values
These values mean the real utilisation values, such as the yield, quality, usage, storage, and transport characteristics, and the potential utilisation values, such as the utilisation status of wild resources by local residents, utilisable part of plant, and some specific usage of cultivated plants.
(f) Utilisation status
This includes the proportion, cultivation history, and distribution of local economic crops.
(g) Botanical characters
Type and distribution of roots; type and size of stems; plant height; branching situation; type, morphology, and adherent patterns of leaves; morphological characteristics of flowers; type and morphology of fruit; reproduction methods; seeds morphology; etc.
(h) Biological characters
Biological characters include the growth habits, florescence, fruiting periods, reproductive cycles, and main stress responses (such as pests, diseases, and other various stresses).

C. Other related information

(a) Investigators and their home institutions
(b) Video information
Photos and video should be taken and recorded in detail. For the samples that are dehydrated and deformed easily after collection, video should be recorded as soon as possible.

Real Samples

During the collection of germplasm resource samples, the location, sampling technique, and sampling quantity should be considered.

A. Selection of location
The locations for collection of germplasm resource samples are mainly the genuine production areas and main production areas, including farmlands of cultivated species and the natural habitats of wild species.
B. Sampling technique
As the differentiation incidences in germplasm are imbalanced, the germplasm samples should be collected selectively. If adopting a random method, many valuable materials are often omitted. The number of and distance between sampling sites should be determined according to the following factors: the variety of local plants, distribution methods, the density of the target germplasm, differentiation situation of individuals, etc. The purpose is to obtain the largest variability in the smallest sample number. The optimal sampling number should be determined by the degree of diversity. Local species, wild species, and wild

relatives are all farraginous populations, and the collection numbers of various germplasm materials should be as large as possible. Generally speaking, collections from different sites are better than collections all from the same site in the investigation area.

C. Plant specimen and medicinal parts should be collected during sampling

   As the main basis of plant classification is plant characters, the specimens should be collected with flowers (fruit); the medicinal parts are important evidence for evaluating the germplasm resources of medicinal plants; thus, the collected medicinal parts should be the crude drug samples in harvest.

6.4.1.3 Sorting of Specimens and Data

Serial Number of Samples

The specimens collected should be labelled and numbered at all times. Various numbers of one kind of specimen should be consistent. One serial number is for one material and should be clear and not repeated in every situation. For example, the number of specimens collected from Mount Emei, Sichuan, is coded as "schemsh" (the Pinyin initials of Mount Emei, Sichuan), and numbers of the same kind of materials can be set sequentially as schemsh-1, schemsh-2, etc. Fill in the original record card during specimen collection.

Preparation of Specimen

Paper should be changed regularly during plant specimen collection in the investigation. After they are dry, they should be prepared into exsiccated specimens, which are beneficial for storage. The medicinal specimens should be dried rapidly to prevent mildewing. It is necessary to prevent moisture and pests from entering during storage. If possible, it is better to store them in specimen museums with the necessary facilities.

Original Record Card

The specimen collected could possibly not completely represent all characteristics, production areas, ecological environments, etc. Therefore, careful observation is needed, and the results should be filled in on the original record cards one by one during the field investigation. The original record cards of medicinal plants should prominently reflect all of the featured contents.

Video Information

Photos and videos of the ecological environment of main germplasm materials and their collection spots should be taken, especially for samples that are easily dehydrated and deformed after collection. These video data could help us to identify and classify the germplasm resources. In addition, for medicinal plants with high trunk (stems), only certain parts of the plant can be collected during the investigation; thus, video information must be used to present the whole appearance of plants.

6.4.1.4 Investigation Summary

During the whole investigative process, we should summarise frequently. If shortages are found, check back to the investigation spot. After investigation, it is necessary to review the investigation work comprehensively and write the summary reports. This is a very important process, which can integrate, systematise, and theorise the obtained data and materials during the investigation. As the valuable results of investigation work, the summary should be as detailed as possible, so as to provide references for the current and future use of germplasm resources.

The contents of the report should include the following items:

1) Basis and purpose of investigation
2) Location, objects, and time of investigation
3) Natural and ecological environments of the investigated germplasm resources
4) Local production situations
5) Local planting area, altitude distribution, species, and their changing histories
6) Plant communities and associated plants of wild species and wild relative plants; their value, geography, and altitude distribution ranges; and changing situations
7) Obtained samples and their features and characters, their positions in plant classification and plant origin evolution, their values in scientific studies on genetics and breeding, etc.
8) Comprehensive evaluation and recommendation on local agricultural production and developing the utilisation of medicinal plant germplasm resources
9) Experience and lessons from the investigation

6.4.2 Collection of Germplasm Resources of Medicinal Plants

Collection of germplasm resources is the gathering of germplasm resources for a special purpose, including general surveying, collection of special classes, domestic collections, international exchanges, etc.

6.4.2.1 Collection Objects of Germplasm Resources

The collection objects of germplasm resources include the following:

1) Medicinal wild species and cultivated species, especially those in danger of extinction and local rare high-quality species
2) Species that were cultivated in the past but have been abandoned
3) Wild relative plants of cultivated medicinal plants
4) Special germplasm resources, such as mutated cultivated species, homozygous types, and intermediate hybridisation types
5) Wild plants with potential value for humans

6.4.2.2 Collection Principles of Germplasm Resources

Comprehensive Property
During the collection of germplasm resources, all varieties should be found, as the germplasm resources of medicinal plants are usually mixed communities. It is necessary to observe carefully and collect the germplasm resources of all existing variants of medicinal plants. Only in this way can the integrity of germplasm resources be represented, especially those valuable germplasm resources, which could be omitted from collection. The collection quantities of seeds or propagating materials for each accession of germplasm resources should be based on the actual situation. Generally, an accession of seeds carries in the range of 200–2500.

Integrity
The collected specimen of germplasm resources is required to be whole, especially the flower and fruit of medicinal plants, which are important bases for plant classification. When collecting specimens of woody medicinal plants, owing to their large size, complete plants often cannot be collected. Therefore, it is acceptable to only obtain branches with flowers and fruits. For dioecious plants, the female and male plants should be collected individually. Specimens of plants whose leaves grow after blooming are collected twice.

Representatives
Each collected specimens, seeds, or asexual propagation material should come from community plants. Only in this case can they represent the features and characters of germplasm resources. Each collected material should be made sure to express the genetic variation of this germplasm resource completely. At the same time, the interference of non-genetic variation factors must be avoided. The collected seeds or asexual propagation material must be normally developed and fully mature.

6.4.2.3 Germplasm Resource Collection Methods

Collection of Germplasm Resources
The collection of medicinal plant germplasm resources is organised by local staff and delivered to administration units, after receiving the documents or papers sent by the national administration department or medicinal/agricultural research departments. However, this method is used rarely.

Investigation and Collection of Germplasm Resources
Investigation and collection by field expedition is currently one of the main methods to collect and accumulate medicinal plant germplasm resources. Three large-scale general surveys were conducted in the twentieth century in China, and basic resource coordination work was conducted for important medicinal plants in the 7th, 8th, and 9th Five-Year Plan, respectively. At present, basic knowledge on medicinal plant germplasm resources in China has been obtained, but the collected and stored germplasm resources of medicinal plants are relatively few; there is a lot of in-depth work to still be carried out.

6.4.2.4 Establishment of Germplasm Resources Files

Database
The database includes the collection number, species, material names, collectors, collection time, collection channels, source, origin, collection quality, and main characteristics. Plants are classified according to the origin of the name as the first-level classification order and the material names as the second-level classification order. Then, the database will be established based on the order of family, genus, species, subspecies, variety, and cultivar. Image data and evaluation results over the years should be digitised and added to the database. When all collected germplasm resources have been assessed and main characters have been identified, the germplasm collection directory should be summarised and printed.

Real Specimens
Real specimens include wax leaf specimens, seeds, fruit, and other propagation material, as well as medicines and other samples.

6.5 Molecular Evaluation of Medicinal Plant Germplasm Resources

There are many methods to identify germplasm resources, including morphological observation, group determination methods, population identification mode, phenotypic determination, origin analysis, and molecular measuring methods. Different means and standards could distinguish germplasm resources at different levels, such as subspecies, variety, species, population, family, origin, clone, and strain.

The ultimate criterion to evaluate the quality of germplasm resources is the clinical effect, for which the material foundation is the active components. Therefore, the ratio of the active ingredients is a feasible standard to determine whether medicinal plant germplasm resources are high quality or not. Since most active ingredients of medicinal plants are secondary metabolites with low contents, which are controlled by environment and genetics, it is hard to evaluate the quality of medicinal plant germplasm resources based on the active components. As such, a key focus of research is how to eliminate the effects of environmental conditions and evaluate the germplasm. Molecular marker technology, which is not affected by the environment, might reflect genetic differences in germplasm resources [9]. This section mainly introduces the molecular evaluation of medicinal plant germplasm resources.

6.5.1 Species Identification of Germplasm Resources

Generally, when collecting medicinal plant germplasm resources, we choose to collect seeds. Thus, it is necessary to identify the seed collected, especially for those with long growth periods and many relative species, to ensure its authenticity. Given that DNA could essentially reflect the features of medicinal plants, analysis at the DNA level could not only be used in the species identification of germplasm materials but also to resolve disputed genetic relationships and provide evidence for the discovery of new medicinal resources. For example, the genetic diversity and relationships among 10 kinds of *Dendrobium* resources were analysed by RAPD technology [10]. The results showed that the 10 kinds of Dendrobium resources had three categories, which revealed the genetic background and relationships among them at the molecular level. It is obvious that DNA molecular marker technology, as an effective method, can be used to compare the genetic relationships among different medicinal plant species, genera, families, and other taxonomic groups, or within each individual one. It brings new vitality to the species identification of medicinal plant resources and the discovery of new drug sources.

6.5.2 Purity Test of Medicinal Plant Germplasm Resources

In germplasm resource research, analysis of genetic diversity is an important concept. At present, the genetic diversity of medicinal plants can be studied in the following ways.

6.5.2.1 Identification of Species and Purity

With the rapid development of DNA analysis, species and their purity can be identified more accurately and reliably. There are many traditional ways to identify the purity of species, such as using seedling, plant, and seed morphologies. Compared to traditional field morphological identification, the application of DNA molecular markers provides an objective, accurate, and rapid channel for the identification of crop varieties. The test object is the DNA fragments (genes) of seeds without organ specificity, which can be detected in various tissues at all developmental stages. They are not influenced by the environment, having nothing to do with expression. Moreover, these DNA fragments are innumerable throughout the entire genome, with rich polymorphism. This process also has high accuracy and good repeatability. DNA samples isolated under appropriate conditions can be preserved for a long time, which is very beneficial to retrospective identification or arbitration.

6.5.2.2 Establishment of Germplasm Conservation Strategies

The genetic diversity analysis of medicinal plants provides a theoretical basis for the proper development of germplasm collection, conservation, and utilisation programmes, mainly for the analysis of collected germplasm resources, narrowing the scope of preservation, to save financial and material resources.

6.5.3 Determination of Genetic Relationships

Wild relatives of cultivated medicinal plants are the ancestors of cultivated plants or wild species with close genetic relationships and serve as important carriers of disease resistance, insect resistance, and stress resistance genes, due to their long-term survival in natural adversity. There are countless examples of the cultivation of high-yield varieties from genes of wild relatives at home and abroad. For example, during the 1970s, China supported three lines of rice to cultivate hybrids, substantially increasing rice yield [11]. The key to this success is the discovery and use of male sterile wild rice. In addition, after long-term cultivation and human breeding activities, the biological characteristics of some medicinal plants have degraded, and they become clones or wild species are unclear, as seen in Chuanxiong and ginger. The protection and use of wild relatives of medicinal plants play an essential role in improving the genetic diversity of cultivated medicinal plants.

If we plan to protect and use wild relatives of cultured medicinal plants, the primary task is to determine what belongs to wild relative species. At present, among numerous identification technologies, DNA molecule appraisal technology, with its unique advantages, serves as an important means of carrying out this work. For example, the study of the genetic relationship among 26 classifications of Dendranthemas and the phylogenic relationship among seven kinds of wild *Chrysanthemum*, verified at the molecular level that modern cultured *Chrysanthemum* is a culture–hybrid complex [12]. It is the result of the hybridisation of natural hybrids between Maohua *Chrysanthemum* and wild *Chrysanthemum*, and *Purple Chrysanthemum* and *Chrysanthemum nankingense* in later hybrids, which is produced after artificial selection. Moreover, using RAPD technology to explore the relationship between the genetic differentiation of wild and cultivated *Paeonia lactiflora* Pall and the authentic formation of red peony root and white peony root, evidence was found for the authentic formation of red peony root and white peony root at the molecular level [13]. Moreover, research on the relationship between four types of cultivated *Angelica dahurica* and three wild relative species (*A. dahurica*, *A. dahurica* var. *formosana*, and *A.porphyrocaulis*), using RAPD and ITS sequencing, showed that the contemporary available wild species of Chinese medicine *Angelica dahurica* come from Taiwan *Angelica dahurica*, which is distributed throughout south-eastern areas of China only, mainly Taiwan [14]. Thus, the evidence from molecular biology has an important application value for the confirmation of medicinal plant wild relatives.

6.6 Genetic Modification

6.6.1 Concept of the Genetic Modification of Medicinal Plant Germplasm Resources

The genetic manipulation of medicinal plant germplasm resources is a process in which the genome of a medicinal plant species is modified in order to produce desired traits. In the past, this was achieved through selective breeding. Selective breeding can be described as the following: a medicinal plant would be born with a desired trait, and a farmer would breed this plant to produce more organisms with that trait. These desired character changes, along with natural evolutionary changes, have led to different biological species that are now genetically various from their ancestors. Selective breeding is the reason why we have cabbage, cauliflower, broccoli, kale, and Brussels sprouts [15]. All of these plants are variants or cultivars of wild mustard, *Brassica oleracea*. Similarly, Jin Yin Hua, Flos Lonicerae, is the dry flower buds collected in early summer before the buds bloom and is used as an antiphlogistic and antibacterial agent to treat catarrh, cold, fever, abscesses, laryngeal problems, dysentery, and erysipelas. The sources of this medicinal plant comprise *Lonicera japonica L.*, *L. confusa* DC., *L. hypoglauca* Miq., and *L. dasystyla* Rehd. Each cultivar was bred for thousands of years for specific features, and the features seen in each cultivar are the result of naturally occurring genetic differences found in the genome.

Over time, conventional plant breeding techniques, such as selective breeding, cross-breeding, and mutation breeding, have become successful approaches to developing new varieties of crops and medicinal plants. More stress-tolerant and high-yield crops, as well as high-quality medicinal plants, were picked and widely cultivated. Hence, the thriving development of modern society should be attributed to conventional plant breeding in some ways. However, breeding programmes applied with the traditional breeding methods have relied on natural and mutant-induced genetic variations to select favourable genetic combinations, and several drawbacks have emerged. The conventional methods such as intergeneric crosses, translocation breeding, and mutagenesis are non-specific. Moreover, sometimes, large parts of the genome are transferred or shifted instead of a single gene, or sometimes hundreds of nucleotide bases are altered instead of a single nucleobase. Therefore, after decades-long studies on molecular biology and genetics, transgenic breeding techniques appeared at the end of the twentieth century as a brand-new method of manipulating recombinant DNA to create plant variants with new characteristics [16]. Transgenic breeding techniques were firstly used in crop breeding. Multiple important agronomic species of plants were transgenically modified, including tobacco, corn, tomato, potato, banana, alfalfa, and canola. Furthermore, more than 10 kinds of medicinal plants have been involved in transgenic modification worldwide, such as *Houttuynia cordata*, *Morinda officinalis*, patchouli (*Pogostemon cablin*), *Fructus aurantii*, *Atropa belladonna*, coastal glehnia root (*Glehnia littoralis*), dandelion, mint, indigo, lily, alfalfa, liquorice, and *Echinacea purpurea*. Some transgenic modifications are aimed at improving disease and insect resistance; for example, the antimicrobial peptide gene and herbicide-resistant glufosinate acetyltransferase gene were introduced into *Houttuynia cordata* and ginseng plants, respectively [17, 18]. Some transgenic modification aimed at increasing the content of active ingredients, such as the transformation of the 4S-limonene synthase gene into mint to increase the plant monoterpene and menthone contents and the transformation of the hyoscyamine-6β-hydroxylase gene to enhance the production of the anticholinergic scopolamine in *Atropa belladonna* [19, 20].

However, transgenic techniques have their defects; they are unable to achieve the precise knockout of target genes, they show uncontrollable expression of exogenous genes, and they also have security problems. The expression of these genes is not controllable in the new genetic background, and whether there is a potential threat to human and animal health is still inconclusive.

With the booming development of genome sequencing and modern genetic manipulation techniques, the twenty-first century is regarded as the genomic era, and genomics are rapidly improving in terms of their importance to molecular plant breeding programmes. Medicinal plants coupled with genetic editing tools have opened new doors in genome-based breeding programmes and germplasm resource development [21]. In recent years, genome editing has been continuously developed and improved in the field of life sciences and has become a technological breakthrough with the same influence as life science technologies such as polymerase chain reaction (PCR). Gene editing technology can generate sequence-specific nucleases (SSNs) to generate double-stranded breaks (DSBs) at specific sites in the genome, enabling gene knockout, gene site-directed insertion or substitution, and chromosome recombination. So far, mature gene editing technologies include the first-generation zinc finger nuclease (ZFN) gene editing system, the second-generation gene editing system transcription activator-like effector nucleases (TALENs), and the third-generation gene editing system clustered regularly interspaced short palindromic repeats (CRISPR)/CRISPR-associated 9 (CRISPR/Cas9) [22]. CRISPR/Cas9 technology has a simple experimental design, efficient operation, and low cost [23]. It has quickly replaced the first two generations of gene editing technology and played an important role in promoting bio-gene transformation and the gene modification of medicinal plant germplasm resources.

6.6.2 Genetic Modification Strategies

6.6.2.1 Conventional Techniques

Simple Selective Breeding

The easiest method of plant genetic modification is selective breeding, which was first discovered by our ancestors and continues to be used today. A simple selection procedure is used to select plants with superior traits, such as plants with good palatability and high yield, by continuous propagation and examining genetically heterogeneous plant populations. Other plants that do not have good traits are eaten or discarded. Seeds from selected plants are sown to produce a new generation of plants in which all or most of the plants will carry and express the desired trait. Over the years, these plants or their seeds have been preserved and replanted, which has increased the population of selected plants and altered the genetic population to be dominated by superior genotypes. Today, this ancient breeding method has been enhanced by modern technology. One application of modern selection methods is marker-assisted selection, which uses molecular analysis to detect plants that may express the desired characteristics, such as resistance to one or more specific pathogens in the population [24]. Applying marker-assisted selection enables the faster and more efficient identification of candidate individuals who may have "good features".

Cross-Breeding

Cross-breeding methods are often applied to plant breeding. Hybrids are produced when a plant breeder removes pollen from a plant and brushes it onto the pistil of a sexually compatible plant, creating a hybrid that carries the gene from the parent. Hybrid progenies can also be used as parents when they reach flowering maturity. Useful features of both plants are often desired for combination. For example, by crossing a disease-resistant but low-yielding gene from one plant to another, the disease resistance gene could be retained while the poor genetic characteristics of low yield could be eliminated or attenuated. Because of the random nature of recombinant genes and traits in hybrid plants, researchers often have to do a lot of work to create hundreds or thousands of hybrid progenies to create and identify plants with good traits.

Intraspecific crossing could also help improve plant traits. Genes from one species can naturally integrate into distantly distant genomes under certain conditions. Some plants carry genes that originate from different species and are transferred through natural and human intervention. For example, common wheat varieties carry genes from rye, and the common potato *Solanum tuberosum* can be crossed with relatives of other species, such as *S. acaule* or *S. chacoense*. Chromosome engineering involving chromosomal deletions, inversions, or translocations with defined endpoints is the genetic manipulation of recombinant DNA, and species with different phylogenetic relationships can be recombined by chromosome translocation. This process has proven to be valuable for transferring properties that are not available, such as pest or disease resistance for crop species. However, the utility of this technique is limited because the shifting of large segments of chromosomes also transfers many neutral or harmful genes. Chromosome engineering is now applied to the transformation of a variety of crops, such as corn, soybeans, rice, barley, and potatoes.

Somaclonal Variation

Somaclonal variation is the name of a spontaneous mutation that occurs when a plant cell grows in vitro. Plants regenerated from years of cultured cells sometimes have novel features. It was not until the 1980s that this method was considered to provide a new source of genetic variation, and that some mutant plants might confer valuable attributes to plant breeders. Since the 1980s, plant researchers have grown plants in vitro and regenerated potentially valuable varieties of a range of different crops [25]. New varieties of several crops, such as flax, have been commercialised. At that time, government regulators did not require molecular analysis of these new varieties, nor did the developers perform molecular analysis to identify potential genetic changes. Some plant breeders still use somatic clonal variation, especially in developing countries, but this non-genetic engineering technology is largely replaced by more predictable genetic engineering techniques.

Mutation Breeding: Induced Chemical and X-Ray Mutagenesis

Mutation breeding involves exposing a plant or seed to a mutagen (such as ionising radiation) or a chemical mutagen (such as ethyl methanesulfonate) to induce random changes in the DNA sequence. By adjusting the dose of the mutagen, it is sufficient to cause some mutations, but not enough to kill. A large number of plants or seeds are usually mutagenised and grown to reproductive maturity to assess the phenotypic expression of potentially valuable new traits in offspring. There is no way to control the action of a mutagen or to target a particular gene or trait, except by changing the dose. Mutagenic effects appear to be random throughout the genome, and harmful mutations can occur even if useful mutations occur in specific plants. Once a useful mutation has been identified, the study of reducing deleterious mutations or other undesirable characteristics in the mutant plant has to be performed. However, crops from mutant breeding are still likely to carry DNA changes beyond the specific mutations that provide superior properties.

6.6.2.2 Genetic Engineering

The genetic manipulation of medicinal plant germplasm resources is a process based on the variation and recombination of plant genetic material with the aim of producing high-yield, high-quality, disease-resistant, and insect-resistant plant varieties. Plant genetic engineering technology is a biotechnology that has emerged with the development of DNA recombination technology, genetic transformation technology, and tissue culture technology. Over the past few decades, medicinal genetic engineering has come a long way and has gained significant attention. Genetic modification technology has developed rapidly and been applied to several crops, such as *Arabidopsis thaliana*, *Oryza sativa*, *Nicotiana tabacum*, *Sorghum bicolor*, *Triticum aestivum*, and *Hordeum vulgare* [26].

Zinc Finger Nucleases (ZFNs)

In 1983, the zinc finger protein (ZFP) was first discovered in *Xenopus laevis* transcription factor IIIA [27]. The first modified endonucleases, zinc finger nucleases (ZFNs), were first researched and applied in the late 1990s and had epoch-making significance. ZFNs consist of the zinc finger protein (ZFP) and the nuclease domain of the FokI endonuclease, the former responsible for recognition and the latter for the cleavage of DNA. ZFP is a naturally occurring protein structure composed of a zinc finger (ZF). The ZF recognises a specific three consecutive base pairs, so the recognition specificity of ZFNs can be adjusted by the number of ZFs in series. FokI is connected to the ZFP through the *N*-terminus. Since FokI acts in the form of a dimer, the ZFN needs to be paired in use. As a new gene editing tool, ZFNs have been used for the gene editing of different species, including plants, since 2001. Several studies have shown successful gene editing through the ZFN approach, including the high-frequency modification of tobacco genes, targeted inactivation of *Arabidopsis* genes, and addition of a herbicide resistance gene as well as disruption of a target locus in maize [28]. In addition, ZFN was also applied for trait stacking in maize [29].

Transcription Activator-Like Effector Nucleases (TALENs)
ZFN technology has led to the era of genetic editing that no longer relies solely on naturally occurring DSBs; however, it has significant limitations, such as its high cost and difficulty in multi-target editing. The discovery of the transcription activator-like effector (TALE) motif led to the second generation of gene editing technology, TALE nucleases (TALENs). The structure of TALENs is similar to that of ZFN. The TALE motif is cascaded into a DNA recognition module that determines the targeting and is linked to the FokI domain. Unlike the ZF motif, a TALE motif recognises a base pair, so the tandem TALE motif has a one-to-one correspondence with the identified base pairs. The study found that TALENs have the same cutting efficiency as ZFNs for the same target, but the toxicity is usually lower than that of ZFNs, and the construction is also easier than ZFNs. However, TALENs are much larger in size than ZFNs, and there are more repeats, and the coding genes are more difficult to assemble in *Escherichia coli*. Because of the convenience of site-directed gene manipulation through the TALEN approach, this gene modification system has been successfully used in plant species, such as rice, wheat, *Arabidopsis*, potato, and tomato [30].

Clustered Regularly Interspaced Short Palindromic Repeats (CRISPR)
The clustered regularly interspaced short palindromic repeats (CRISPR) system was originally an adaptive immune system evolved by bacteria and archaea to protect against foreign viruses and plasmid DNA. The Type II CRISPR/CRISPR-associated (Cas) system relies on the integration of foreign DNA fragments in CRISPRs, which are transcribed and cleaved to produce short CRISPR RNAs, crRNAs, and trans-activating crRNAs (tracrRNA) anneals, and then directs CRISPR-associated protein 9 (Cas9)-mediated sequence-specific exogenous DNA degradation. Scientists found that the crRNA and tracrRNA that Cas9 relies on for targeted cleavage can be integrated into sgRNA. Subsequently, several research groups have reported that the CRISPR/Cas9 system can be used for targeted gene editing in human cells. Compared to ZFN and TALEN technology, CRISPR/Cas9 is designed to be much simpler and less expensive, and for the same target, CRISPR/Cas9 has comparable or even better targeting efficiency [31]. Thus, due to its efficiency, simplicity, and wide capabilities, the CRISPR/Cas9 system has already become a popular system in a short time and has been used in the gene modification of various crops and medicinal plants.

CRISPR/Cas9 Derivative Technology
With the deepening of research on the CRISPR/Cas system, the catalytic mechanism of Cas9 nuclease has also been revealed. Through mutation of amino acids at specific sites, Cas9 nickase (Cas9n) with single-strand-targeted cleavage function and dead Cas9 (dCas9) were obtained. These different Cas9 nucleases have spawned a wider range of gene editing systems.

CRISPR/Cas9-Nickase Technique

The crRNA used in the Cas9 system could withstand a certain degree of mismatch resulting in off-target effects, limiting the application of the CRISPR/Cas9 system in high-precision editing. In order to improve the accuracy of the Cas9 editing system, scientists used the $Cas9^{D10A}$ mutant to obtain the function of nickase activity and designed the strategy of "one site, double sgRNA targeting", named the CRISPR/Cas9-nickase gene editing technique [32]. The principle is similar to that of ZFN and TALEN. Two Cas9n/sgRNA complexes simultaneously target one site and cut one of the DNA strands to achieve a double-stranded break and induce nonhomologous end joining (NHEJ) or homologous recombination (HR) repair. Using this strategy, the off-target efficiency of gene editing in cell lines can be reduced by up to four orders of magnitude.

CRISPR/dCas9-FoKI Technique

For addressing off-target issues with CRISPR/Cas9 technology, Guilinger adopted a dCas9-based strategy. In theory, dCas9/sgRNA only plays a simple targeting role and cannot induce DNA breaks, similar to the DNA binding domain in ZFN or TALEN. In order to achieve DNA cleavage, the cleavage domain of the introduced FoKI endonuclease was ligated with dCas9 to form the fusion protein fCas9, which is similar to the design strategy of ZFN and TALEN. In the genetic editing of human cells, fCas9 is 140-fold more specific than wild-type Cas9 [33]. At a highly similar off-target site, fCas9 is at least four times more specific than Cas9n. The application of fCas9 will further enrich the Cas9 toolbox and provide a more complete genetic editing tool.

Single-Base Editing Technique Based on CRISPR/dCas9

The genetic modification of medicinal plants often requires genetic point mutation techniques, which may lead to plants with new traits if repaired by precise means. In the case of providing a homologous recombination template, site-directed mutagenesis can be achieved by CRISPR/Cas9, but the random insertion or deletion of bases that may be caused by the induced NHEJ repair is a potential risk factor. The Cas9 mutant dCas9 does not cleave double-stranded DNA, but can serve as a target-searching and location tool. It is possible to construct a CRISPR/Cas9n/dCas9-oriented single-base editing technique with consideration of the CRISPR/dCas9-FoKI system design, if a protein/domain that catalyses a specific base switch is available. Scientists fused cytosine deaminase (APOBEC1) derived from rats to dCas9 and found that it can convert C to U at a fixed point and then realised a C:G base pair to T:A transformation under subsequent DNA replication or repair [34]. Subsequently, the Kondo group of Kobe University in Japan and the Changxing Group of Shanghai Jiao Tong University in China also reported similar research results. That is, the free conversion between C to T and G to A bases has been achieved, and it will be possible to perform the arbitrary conversion of four bases in the future.

Gene Expression Regulation Technique Based on CRISPR/dCas9

In addition to directly editing DNA, the CRISPR/Cas system also plays a role in gene expression regulation. For example, the use of dCas9 without endonuclease activity but still binding to DNA can directly impede the binding of its DNA to other factors and affect gene expression. If a transcriptional repressor or activator is fused to dCas9, inhibition and activation of the target gene could be achieved, providing a flexible means of manipulation for gene function studies.

CRISPR/Cas12a Gene Editing Technique

CRISPR/Cas9 technology is limited by the G-base-rich protospacer adjacent motif (PAM) sequence and does not target any other sequence. Actually, the Cas9 protein is too large in molecular weight and in some cases inconvenient to use. Almost all of the archaea and numerous bacteria use the CRISPR/Cas mechanism for immune defence, including a variety of CRISPR/Cas systems. The class II CRISPR/Cas system that has been characterised belongs to the type II CRISPR/Cas system using the Cas9 family nuclease as an effector, while the other is present in the genera *Pristus* and *Francis* (Prevotella and Francisella 1). The class C CRISPR/Cas system is classified as a type V CRISPR/Cas system. In 2015, Zhang's team reported that CRISPR from Prevotella and Francisella 1 (Cpf1, also called Cas12a) in the type V system is a functional bacterial immune mechanism and can mediate effective gene editing in human cells [35]. CRISPR/Cas12a has advantages that CRISPR/Cas9 does not appear to have. One of them is that Cas12a requires a T-base-rich PAM sequence to enable its use in genome-rich A/T base-rich species. Another research team fused the DNA-free cleavage dCas12a with rat-derived cytosine deaminase (APOBEC1) and found that it is similar to a Cas9-based base editor and could effectively catalyse C to T base conversion in human cells. Due to the recognition of T-base-rich PAM sequences, the Cas12a-based base editing system complements the Cas9-based base editing system, providing more comprehensive technical conditions for basic research and future applications in germplasm resources.

6.6.3 Current Gene Modification Techniques in Medicinal Plant Germplasm Resources

As the new direction of genome editing technology, CRISPR has been widely used in plant gene modification. Multisite simultaneous mutation or single-site directed mutation of the polyploid wheat gene has been performed using the CRISPR/Cas9 system. Cas9 and the synthetic sgRNA were transferred into an orange by *Agrobacterium*-mediated transformation, and the *Citrus* × *sinensis* phytoene desaturase (*CsPDS*) gene was successfully edited to mutate [36]. In addition, the endogenous phytoene dehydrogenase gene in bananas was also successfully knocked out to obtain mutant strains [37]. A multi-target modification was performed on the PDS gene of *Populus tomentosa*, and an albino mutant was obtained in the T_0 generation transformant. Researchers designed four different gRNAs to edit the poplar PDS gene (*PtoPDS*) and adjacent genes of different genomic loci in *P. tomentosa*. After *Agrobacterium*-mediated transformation, transgenic poplars showed obvious albino phenotype. The CRISPR technique was used to delete the polyphenol oxidase (PPO) gene family that encodes browning in *Agaricus bisporus*, and the activity of the PPO enzyme was reduced by 30%, which induced *A. bisporus* to delay browning or have browning resistance [38]. The CRISPR/Cas9 system was successfully applied to edit the soybean acetyl-lactone synthesis gene acetylated lactic acid synthesis (ALS11ALS1) and obtain the chlorsulfuron resistance gene. Through the CRISPR system, the regulation of the tomato fruit ripening gene *ripening inhibitor* (*RIN*) was edited to extend the shelf life. Further, the *Solanum lycopersicum* 1-aminocyclopropane-1-carboxylic acid synthase (*SlACS2*) gene was modified in tomato. *SlACS2* is part of the tomato system II of ethylene synthesis. The rate-limiting enzyme, which regulates the overexpression of ethylene in system II, prevents the tomato from overripening. CRISPR/Cas9 technology was used to knock out the *SlA-GO7* gene, which controls the morphology of tomato leaves. The leaves of the mutant plants were narrower or even needle-like compared to those of the wild-type, thus demonstrating the function of the tomato *SlA-GO7* gene. In addition, the CRISPR/Cas9 system has also been widely used in plants, such as sorghum, potato, watermelon, and cucumber.

The genetic editing of medicinal plants has been performed. Keasling used CRISPR/Cas9 technology to engineer *S. cerevisiae* to increase the mevalonate content of the mutant strain by increasing the metabolic flux of the mevalonate (MVA) pathway, reducing the metabolic efficiency of sterols, and cutting off the synthesis of diterpenoids [39]. The mevalonate content of the mutant strain is 41-fold higher than that of the wild-type strain. In addition, the use of CRISPR/Cas9 technology was also applied to regulate the content of many important secondary metabolites, such as knocking out key genes on the tanshinone biosynthesis pathway and transferring the metabolic flux of geranylgeranyl diphosphate (GGPP) to the synthesis pathway of tanshinones. At present, the Chinese herbal medicines used in clinical practice are mainly obtained by artificial planting and field excavation. The medicinal plant pests and diseases have always been a serious problem for the farmers, such as wilt and leaf spot of *Salvia miltiorrhiza*, powdery mildew of *Astragalus* (*Angelica sinensis*) and *Coptis*, and the rust of ginseng and American ginseng. It is possible to learn about the genetic control of certain medicinal plant diseases and insect pests through CRISPR/Cas9 technology field crop research. On the other hand, some medicinal plants produce toxic metabolites to the human body during their growth, which limits their clinical application, such as aristolochic acid from Guan Mu Tong of the Aristolochia family, which has renal toxicity [40]. Many patients have developed renal injury from taking Long Dan Xie Gan pills consisting of Guan Mu Tong and other herbal medicines for a long time. If the metabolic pathways of related harmful components can be blocked by CRISPR/Cas9 technology, this will also be a new research strategy for the improvement of medicinal plant varieties.

6.7 Searching for Essential Functional Genes of Medicinal Plant Germplasm Resources and Molecular Assisted Selection

6.7.1 Searching for Essential Functional Genes of Medicinal Plant Germplasm Resources

Functional gene studies are one of the most active and core research areas on medicinal plant germplasm genomes in the post-genomic era. They emphasise the development and application of experimental methods as a whole (at the genomic level or systemic level) to analyse genome sequences and clarify their function. The basic strategy has changed from a single gene or protein in the past to all genes or proteins at a systemic angle. In the study of functional genes in medicinal plant germplasm resources, the cloning and characterisation of biosynthesis-related genes from DNA genomes is the most active research area, which is the most related to active ingredient formation. Due to the increasing practical demand for natural medicines in medicine, and the continuous development of gene cloning and expression technology, there has been a trend of rapid growth in functional gene cloning and characterisation of medicinal plants in recent years.

6.7.1.1 Functional Gene Cloning from Medicinal Plant Genomes

Herbal drugs, which are relatively safer than conventional drugs, are widely used by 80% of the world's population. As of 2019, the World Health Organization has listed 21,000 medicinal plants; over 10,000 medicinal plant species are listed in China, accounting for ~87% of the total Chinese medical materials [41]. Conventional drugs are expensive and have serious side effects, so researchers have been focusing on herbal medicines or drugs derived from botanical sources for the past few decades. The main ingredients of herbal drugs are secondary metabolites (alkaloids, terpenoids, and phenolic compounds) that are metabolised by plants. So far, only a small percentage of plant metabolites have been explored. The development of new herbal genomics research, advanced sequencing techniques, and the construction of a medicinal plant genomics consortium may help the speedy discovery of large scales of previously unknown metabolites through the identification of functional genes and potential biosynthetic pathways. As of January 2019, several medicinal plants' whole-genome sequencing projects have been completed. Two well-known Chinese medicinal plants that are in heavy demand, *Panax ginseng* [42, 43] and *Artemisia annua* [44], have had their whole genomes sequenced. In addition, representative plants whose pharmaceutical components are relatively unambiguous and that have typical secondary metabolism pathways, such as *Salvia* medicinal plants [45, 46], *Glycyrrhiza uralensis* (Chinese liquorice, Fabaceae) [47, 48], and *Lycium chinense* (Chinese boxthorn, Solanaceae) [49], have also been the subjects of whole-genome sequencing.

Cloning of Flavonoid-Related Genes and P_{450} Cytochrome Oxidase Gene
Flavonoids possess a variety of pharmacological activities, which are closely related to the flower colours. Thus, the study of the flavonoid biosynthesis pathway began in the 1980s and has made great progress in its biosynthesis steps, enzymes during each step, and genes. A statistical analysis of flavonoid synthesis genes in 26 plants, including *Arabidopsis*, *Antirrhinum*, pine, peach, petunia, and grape, showed that at least 103 flavonoid biosynthesis genes were cloned, such as genes of chalcone synthase, the enzymes catalysing the hydroxylation of flavonoid skeletons, and glycosidase.

The other genes studied often were cytochrome P450 genes, which belong to a class of oxidases containing haem proteins with a variety of catalytic functions. P450 in plants exerts an important role; it potentially catalyses a number of primary and secondary metabolic reactions. Due to their essential functions in secondary metabolism and plant resistance (anti-herbicide), and potential application in biological decontamination, studies on P450 attract particular concern. Currently, more than 600 P450 genes have been cloned, more than 100 of which were expressed successfully in bacteria, yeast, and baculoviruses along with function identification. Besides the model plant *Arabidopsis thaliana*, the P450 enzymes in *Catharanthus* have been studied widely.

Cloning of Taxol Biosynthesis Genes
The cloning of taxol biosynthesis genes is the most typical example of the cloning of plant medicine functional genes, which represents the research and application trends of natural product chemistry in the postgenomic era. Croteau laboratory in Washington State University has cloned and expressed 11 genes related to the biosynthesis of the anticancer drug taxol from different species of yew, using homologous primer PCR, differential display combined with *Taxus* cell cultures, and reverse genetics methods [50]. These genes include taxane skeleton formation genes, three hydroxylase genes catalysing oxygen replacement in the taxane skeleton, and five acetyl transferase genes responsible for side chain formation. They are all expressed in *E. coli,* yeast, or insect expression systems, whose expression products are also functionally identified. On this basis, it is determined that taxol formation from the common precursor of a diterpenoid goes through at least 20 steps. Besides, the biosynthetic pathway of taxol, with eight oxygen substituents, has been elucidated. Moreover, the specificity of recombinant enzymes and the order of various substituents in the biosynthesis have been determined with different substrates.

6.7.1.2 Conclusion

The "Chinese Medicine Genome Project" combines molecular biology, bioinformatics, and other modern biotechnology with modern scientific technology to research and develop traditional Chinese medicine. In addition, the United States has put forward the "Human Genome Project" (HGP), "Plant Genome Project", and "Microbial Genome Project" and has achieved breakthroughs. Proposed and implemented in China, the "Chinese Medicine Genome Project" is the significant strategic initiative in the modernisation of Chinese medicine research and development, among which functional genomics is one of the most important parts. Through study on functional genomics and proteomics, the technical system of important functional genomics research and the discovery of important functional genes from Chinese medicinal plants should be established and improved. The profound meaning of Chinese herbal medicine should be reflected. Furthermore, genes with Chinese-independent intellectual property rights, clear function, and potential application should be acquired as well. After understanding the genetic expression and regulation rules in Chinese medicine, we can up-regulate the genes responsible for effective components so as to promote their production. Additionally, through the control of genetic information, the active ingredients of Chinese medicine can be analysed and synthesised, and their mechanism will be clarified, which lays the foundation for the modernisation of Chinese medicine and its acceptance into the international market. Last but not least, it can promote the development of the related pharmaceutical industry, making a significant contribution to human health.

In functional gene studies of plant medicine, compared to genome sequencing and expressed sequence tags, genes related to active ingredient biosynthesis attract more attention. The growing number of their cloned genes suggests the beginning of the postgenomic era of natural products and possible changes in the production methods for some natural products. Therefore, our hopes for biotechnological methods may be realised through new technology with a combination of genetic engineering and chemistry.

On the basis of having studied *Dendrobium* for many years, our laboratory carried out some research on its functional genes. Using chemical mutagen EMS (ethylmethane sulfonic acid) to directly induce the mutation of protocorm in *Dendrobium* tissue cultures, we obtained a stable mutant without *Dendrobium* alkaloids. Moreover, using differential display, antisense RNA technology, and transgenic techniques, genes concerned with *Dendrobium* alkaloid biosynthesis have been cloned and identified. The work was completed as follows: mRNA was extracted from *Dendrobium* mutants and normal materials. The first strand of cDNA was synthesised through reverse transcription with the selected 3′ end of an anchored primer. PCR amplification was carried out with labelled substrates, using the 5′ end of a random primer and 3′ end of an anchor primer as a primer set. At present, a subtractive cDNA library has been established, and sequencing and bioinformatics analysis are in the process of being established.

6.7.2 Molecular Marker-Assisted Selection in Germplasm Resources

6.7.2.1 Concept and Significance of Molecular Marker-Assisted Breeding

Molecular marker-assisted selection is a modern breeding technology which uses the DNA marker closely linked to the target property to indirectly select the target property. This method makes the transfer of target genes possible and can not only make early accurate and stable choices but also overcome the difficulty in reusing recessive genes. Therefore, the breeding process will be sped up with improved breeding efficiency. Compared to conventional breeding, the technology can improve the breeding efficiency by about threefold. The key to the technology is the identification of DNA molecules that are closely linked to agronomic characters. As for the obvious advantages, this technology has attracted great attention in developed countries. The United States, Japan, and Western Europe have invested a lot into this work in recent years. In China, some universities and research institutes have already mastered this technology and have made a number of important achievements after recent years of research into crop breeding. Colleges of Chinese medicine and their research institutes have applied this technology to medicinal plant breeding.

The establishment of large-scale planting base for Chinese herbal medicine indicates that the production of herbal medicine in China has changed from relying on wild resources to scalable and standardised cultivation. In order to achieve the goal of high quality, high yield, and standardised production of medicinal plants, using improved varieties of medicinal herbs plays a decisive role in the production. "High content" and "high yield" are the main objectives of medicinal plant breeding. The genetic linkage map is the molecular file of organism, which can provide reference for breeders. Using restriction fragment length polymorphism analysis (RFLP), simple sequence repeat (SSR), random amplified polymorphic DNA (RAPD), amplified fragment length polymorphism (AFLP), and other molecular genetic markers to construct genetic linkage maps for important medicinal plants, essential quantitative trait loci (QTL) research on medicinal plants will be carried out. In addition, good varieties will be selected from the wild types to achieve sexual hybridisation between wild plants and domestic plants. This will become one of the important directions for medicinal plant breeding in the future.

6.7.2.2 Research on Genetic Maps of Medicinal Plant Germplasm Resources

Genetic Map

The linear arrangement of genes on a chromosome is known as the genetic map. It is based on the recombination frequency of allele in meiosis to determine its sequence and relative distance in the genome. Usually, the recombination rate is used to represent the genetic distance between genes, whose unit is the centimorgan (cM). One cM is equal to a 1% recombination rate. However, cM stands for the relative position of the gene in the chromosome instead of its actual length. There are lots of genetic mapping method, such as RFLP, RAPD, AFLP, short tandem repeats (STR), and single nucleotide polymorphism (SNP).

Parent Selection

The genetic map is constructed successively by parent selection, composition population production, genetic marker chromosomal location, and marker linkage analysis. For parent selection, the species or material with a distant genetic relationship and genetic variation should be chosen in theory; however, this should not be too large; otherwise the seed set rate and future construction accuracy of the map will be reduced. The molecular marker technique can be applied for the detection of selected material polymorphism and analysis of the result, picking out a pair or several pairs of material as the composition of the parent with certain genetic variation.

6.7.2.3 Bulked Segregant Analysis and Markers for Important Agronomic Traits

Near-Isogenic Lines (NILs)

NILs are permanently stable lines that are produced by introducing donor parent chromosome segments with certain target genes or quantitative trait loci into the genetic background of the recurrent parent. They have a high efficiency in gene mapping. However, it is nearly impossible to create NILs for some woody medicinal plants because of their long cultivation time.

Bulked Segregant Analysis (BSA)

BSA lays the foundation for the quick and efficient screening of important agronomic trait markers. This method is often used to research the marker molecule linked to important agronomic traits. It can be used in medicinal plant breeding as well. The main procedures of BSA group construction are as follows: to hybridise disease-resistant varieties with disease-susceptible varieties of a plant, the resistance gene separates in the F_2 generation. According to their resistance, the plants are divided into two groups: one is susceptible, and the other is disease resistant. Five to 10 extreme lines are picked out from each group. After DNA extraction, they are equally mixed to form resistance-susceptible DNA pools, whose polymorphism are analysed. The aim is to screen polymorphic markers and then analyse all of the separated plants, the target trait linkage, and the close linkage level.

6.7.2.4 Application of Germplasm Resource Molecular Marker-Assisted Selection

Basic Conditions for Marker-Assisted Selection

Medicinal plants for molecular marker-assisted selection should meet the following conditions:

A. Genetic distance between molecular markers and the target gene is isolated or closely linked, generally less than 5 cM to be effective for MAS.
B. A simple and efficient method of DNA automated extraction and detection should be available, which can be easily used for the analysis and operation of large groups. It is better if the molecular marker is PCR based.
C. Detection technology should be reliable, have high repeatability, and be economical and practical.

D. A multitask data-processing computer software is required. With these conditions, the breeding-assisted selection can be well carried out.

Molecular Marker-Assisted Selection Method

Introducing Favourable Genes

During cultivation, some characteristics related to the degradation of medicinal plants exist, such as decreased active components, decreased disease resistance, etc. Based on the investigation, it is necessary to select a species with strong resistance and that has been backcrossed several times to cultivated species. In addition, during each backcrossing, favourable genes and satisfying properties should be maintained so as to improve varieties. However, in practice, "linkage drag" often appears, which means the introduction of some negative genes linking with favourable genes, bring more difficulty in improving variety. Traditional backcrossing methods cannot identify recombination variety directly, due to its blindness, too many backcrosses (even more than 20), and low efficiency. On the contrary, molecular marker-assisted selection technology might identify recombination variety rapidly and directly. Using two sides' molecular markers, which are linked closely with the target favourable genes, it is possible to directly screen the recombination variety on this fragment. Generally speaking, through two or three generations, beneficial genes can be introduced without negative genes to obtain improved varieties.

Polymerisation of Favourable Genes

There are lots of favourable genes in medicinal plant germplasm resources. In the breeding process, the favourable genes of parent can be localised and transferred to a variety by hybridising or backcrossing. By this method, the favourable genes can be combined together so as to cultivate improved varieties. Molecular marker-assisted selection can overcome the shortcomings of conventional breeding and quickly combine the favourable genes together.

As for medicinal plant breeding, the use of germplasm resource molecular marker-assisted selection has been increasing recently. In the research of germplasm resources, we can learn from the experience of crop breeding, expand research on medicinal plant conventional breeding, and combine conventional breeding with molecular marker-assisted selection to cultivate good seeds of medicinal plants, serving for Chinese herbal medicine good agricultural practice (GAP) production.

6.8 Research Progress of Transgenic Medicinal Plant Resources

Because of the impact of the current environment, medicinal plant resources are in increasingly short supply, especially some rare medicinal plants. However, transgenic technology, which is used to change some genetic traits of plants, such as resistance to insect pests, etc., can improve the quality of medicinal plants and the content of active ingredients. Therefore, it plays a vital role in increasing Chinese herbal medicine resources. In principle, transgenic plants are new types of bioreactor, whose pharmaceutical products are known as transgenic plant drugs, regardless of whether the transgenic plants are medicinal plants or not.

Transgenic plants are mainly formed by *Agrobacterium*-mediated gene transfer, particle bombardment, PEG-mediated gene transfer, and germplasm system-mediated gene transfer. *Agrobacterium*-mediated gene transfer is the most popular method for creating transgenic medicinal plant resources; the others have not yet been reported in this area. Nowadays, hairy roots formed by transforming *Agrobacterium rhizogenes* root-inducing (Ri) plasmids and the crown gall formed by transforming *Agrobacterium* tumour-inducing (Ti) plasmids are used to produce effective active ingredients, which are considered a hotspot in medicinal plant biotechnology research.

6.8.1 Tissue Culture of Transgenic Medicinal Plant Resources

All over the world, hairy root tissue cultures have been established in *Ginkgo biloba*, *Taxus*, periwinkle, tobacco, *Polygonum multiflorum*, shikonin, ginseng, *Digitalis*, *Artemisia*, etc. Some important secondary metabolites obtained by this system include quinoline alkaloids, indole alkaloids, tropane alkaloids, glycosides (such as ginsenosides, beet saponins), flavonoids, quinones, polysaccharides, proteins, and several important enzymes (such as superoxide dismutase). Moreover, domestic scholars have detected some high levels of active substances in the hairy roots of *Lithospermum*, *A. annua*, *Gynostemma*, and American ginseng. In foreign countries, some hairy roots have been already applied in industry, such as *Lithospermum*, carrot, periwinkle, and so on.

Some active ingredients of medicinal plants are only synthesised in the leaves and stems; however, the crown galls are able to produce these specific compounds. For example, the root of *Psoralea* L. contains furanocoumarins, which could not be synthesised in the hairy roots, but could be in the crown galls. Crown gall tissue culture has been used to produce secondary metabolites in *Asparagus, Bidens*, periwinkle, *Cinchona*, *Digitalis*, lupine, lemon spearmint, spicy mint, *Salvia*, *T. brevifolia*, European *Taxus*, etc. *A. tumefaciens* was used to transform *Salvia* and showed that the crown gall tissue can produce the active ingredient tanshinone, exclusive to *Salvia* roots. In addition, a crown gall tissue culture of *Salvia* was induced to produce tanshinone, screening out a high-yielding strain, whose tanshinone content exceeded that in crude drugs. Thus, it may be a feasible method to produce tanshinone in this high-yielding strain with a combination of elicitors and liquid static culture. Hank et al. induced a rapid growing crown gall in a hormone-free medium, whose tumour-like tissues contained paclitaxel and its analogues, proven by mass spectrometry and ELISA. Their contents are 0.00008%–0.00004% of the dry weight.

However, the synthesis of many secondary metabolites requires the involvements of the leaves, stems, and roots; it is not possible for only the hairy roots or crown galls to produce them. Fortunately, if combining both of them, the complementation of two metabolic functions can be achieved. Subroto co-cultured genetic transformed belladonna abnormal stems and hairy roots in hormone-free medium in vitro. The scopolamine content in co-cultured stems was 3–11 times than that in intact plants (51). The ratio between scopolamine and hyoscyamine in co-culture was up to 1.9 times, which was improved significantly compared to that in belladonna hairy roots (0–0.03). Given the low side effects and high value of scopolamine, the study of co-culture conditions of hairy roots and crown galls found it possible to be used in the mass production of scopolamine. Mahagamasekera et al. found that hairy roots and crown galls from different genera of belladonna can be co-cultured, producing scopolamine as well. Nevertheless, when being cultured separately, hairy roots and crown galls of a *Duboisia* hybrid did not produce scopolamine, with only a certain amount of hyoscyamine in the hairy roots. In addition, a dual transformation system by double-transforming Ti plasmid and Ri plasmid into the hairy root culture system of *Trichosanthes* was established. The results indicated that the dual-transformed hairy root system grew more rapidly with similar protein contents. Therefore, co-culture and dual transformation methods offer a novel approach for producing some special metabolites.

6.8.2 Application of Transgenesis in the Feature Modification of Medicinal Plants

Through genetically modifying the characteristics of medicinal plants, the quality and adaptability of plants could be improved. Currently, the study on this area includes the following aspects:

6.8.2.1 Quality Improvement of Medicinal Plants

By intensively injecting genes to modify the characteristics of crude drugs, it is even possible to roll drugs and food into one. For example, due to the bitterness of *Gynostemma*, it tastes bad when taken orally. At the present, scientists hope to transfer and express the protein genes from beet into *Gynostemma*, so as to cover the bitterness. Chen introduced and expressed farnesyl pyrophosphate synthase genes from cotton into *A. annua*, obtaining a transgenic plant with an artemisinin content two to three times higher than that in the control group. The flower buds of honeysuckle have the best quality, but their blooming time in the actual production is difficult to control. When using genetic engineering technology to suppress the flowering of honeysuckle, it is possible to obtain high-quality medicine. It is also reported that the fat-soluble components (active ingredients) in *Salvia* were changed through genetic engineering technology in order to increase the active ingredients. Yahua Zhao integrated the metallothionein genes of mice into the genome of *Lycium* mediated by *Agrobacterium*, to obtain a novel zinc-rich transgenic *Lycium* variety. Although there is currently little research on this area, the method used to improve the genetic characteristics of medicinal plants is still promising.

6.8.2.2 Enhancing the Resistance of Plants

Virus disease is the most common disease in medicinal plants. The use of gene technology can improve plant disease resistance. Besides, using genetic transformation to acquire antivirus plants has succeeded in beet, *Citrus* × *aurantium* and Ningxia *Lycium*. For example, Qing Luo transferred lectin genes from Snowdrop into Ningqi No.1 strain mediated by *A. tumefaciens* LBA4404, and a transgenic aphid-resistant *Lycium* was obtained. Similarly, an enhanced aphid-resistant transgenic *Lilium* was obtained by introducing *Pinellia ternata* lectin genes into the genome of *Lilium* with *A. tumefaciens*. In addition, in terms of resistance to herbicides, Yueshi Cun introduced the acetyltransferase genes from actinomycetes into belladonna and *Scoparia dulcis* with Ri plasmids as vectors. In addition, a herbicide-resistant belladonna was cultivated successfully. Moreover, a resistant strain of woad was obtained by Tiefeng Xu, who integrated the Bar gene into tetraploid woad leaves. This transgenic plant showed significant resistance to herbicides without affecting the production of secondary metabolites.

6.8.3 Genetic Metabolic Engineering of Medicinal Plants

Relatively speaking, molecular biology research on Chinese herbal medicine began a little late, and few genes can be used at present. Thus, it limits the application of genetic engineering in improving medicinal plants. With the development of genetic engineering, determination of the key enzymes during genetic cloning and metabolism in medicinal plants is gaining increasing attention.

Heide investigated shikonin biosynthesis-related enzymes in a shikonin cell culture and primarily determined that p-hydroxybenzoic acid geranyl transferase is the key enzyme during shikonin biosynthesis. Additionally, Croteau isolated and cloned the cDNA of a diterpene cyclase synthase (taxadiene synthase) from *T. brevifolia*, whose DNA and amino acid sequences were also analysed. Croteau et al. also successfully transferred another four genes, such as taxadiene synthase, etc., into *E. coli*, resulting in the formation of taxadiene, which is only produced in *Taxus*. Furthermore, its yield for a 1 l *E. coli* culture (10^9 cells/mL) was equal to that from 7.6 kg of *Taxus* bark.

To increase the production of artemisinin, the mutant squalene synthase (SS) gene successfully replaced the wild-type gene in transgenic *A. annua* by gene targeting, adjusting the flow of metabolites. Hechun Ye cloned four key genes in the artemisinin biosynthesis pathway and transferred them by *Agrobacterium* Ti and Ri plasmids, so as to obtain a plant material with a high artemisinin content.

Nowadays, most of the patents on genes are concerned with flavonoid metabolism, including chalcone synthase 7, flavonoid skeleton hydroxylase, and glycoside transformation-related enzymes.

6.8.4 Safety of Medicinal Plant Resources

The safety of transgenic plants is currently a controversial issue. Recombinant proteins in transgenic plants are relatively stable with low expression. It is difficult to disinfect and determine their structure, and their activity depends on their structure and folding pattern; it is necessary to establish an appropriate regulation method to appraise the safety of transgenic medicinal plants.

6.9 Case Study

This case study focused on cloning and characterisation of a novel 3-hydroxy-3-methylglutaryl coenzyme A reductase gene from *Salvia miltiorrhiza* involved in diterpenoid tanshinone accumulation.

6.9.1 Material and Methods

The mature seeds of *S. miltiorrhiza* were surface sterilised by 0.1% mercuric chloride (Sigma-Aldrich, St. Louis, MO, USA) and cultured on solid, hormone-free MS basal medium. The MS medium contained 30 g/L sucrose and 8 g/L agar without ammonium nitrate for germination. Cultures were maintained at 25 °C under a 16 h light/8 h dark photoperiod with light provided by cool white fluorescent lamps at an intensity of 25 μmol m^{-2} s^{-1}.

Roots, stems, and leaves were collected from mature *S. miltiorrhiza*. The 20-day-old hairy roots were treated with MeJA for 0, 12, 24, 48, 72, and 96 h at a final concentration of 100 mM, and roots were harvested for RNA isolation. The hairy root lines were collected 30 days after inoculation. Total RNA was extracted from the tissues by Trizol method (Invitrogen, Carlsbad, CA, USA). Genomic DNA was isolated using the modified cetyltrimethyl ammonium bromide method.

5′-Rapid amplification of cDNA ends (RACE) was performed according to the manual of the SMARTTM RACE cDNA Amplification Kit (Clontech Laboratories Inc., Mountain View, CA, USA). The 5′-RACE PCR was carried out using the 5′-RACE primer and universal primer (UPM, Universal Primer A Mix). The PCR product was purified and cloned into pMD19-T vectors followed by sequencing.

After aligning and assembling the sequences, the full-length cDNA sequence of the *SmHMGR2* gene was deduced and subsequently amplified by PCR using a pair of primers. The genome sequence of the *SmHMGR2* gene was confirmed by PCR with the genome DNA as a template.

The nucleotide sequence, deduced amino acid sequence, and open reading frame (ORF) were analysed using DNAstar, and the sequence comparison was conducted through a database search using BLAST. SmHMGR2 and other HMGRs retrieved from GenBank were aligned using ClustalW. A phylogenetic tree was constructed using neighbour-joining method. Transmembrane domain was analysed by TMHMM2.0, and homology-based structural modelling was accomplished by Swiss-Model (http://www.expasy.org/).

The entire SmHMGR2 cDNA was amplified by PCR using forward and reverse primers. The PCR product was digested with EcoRI and NotI, gel purified, and ligated into the same restriction sites within the pRS406 vector. Positive clones were confirmed by PCR and subsequent sequencing analysis for the presence of the *SmHMGR2* gene and used to be transformed into *S. cerevisiae*.

The *SmHMGR2* gene was PCR amplified and the resultant PCR products were purified with a DNA purification kit (Sangon, Shanghai, China). We sequentially subcloned the PCR product into a donor vector (pDONR221) and created an entry vector. The recombination reaction between the entry and destination vectors (pH7WG2D) was performed in LR Clonase^{TM} II enzyme mix (Invitrogen, Carlsbad, CA, USA) according to the manufacturer's instructions. The *E. coli* DH5α was transformed with the product of LR reaction using the heat shock transformation method. The destination vector pH7WG2D with the ccdB (control of cell death) gene knockout was used as the control vector (pH7WG2D-Control).

The *S. cerevisiae* strain JRY2394 (*MATa, ade2, his3, met-, ura3, hmg1,* and *hmg2*) was used to examine the function of *SmHMGR2*. The *SmHMGR2* gene was cloned into the pRS406 expression vector (pRS406-*SmHMGR2*) and then transformed into JRY2394.

The vectors of pH7WG2D-*SmHMGR2* and pH7WG2D-Control were transformed into *S. miltiorrhiza* through the mediation of the *A. rhizogenes* strain ACCC10060, as described previously. Root tissues from three flasks of cultures were collected separately 10, 20, 30, 40, and 50 days after inoculation.

To detect the presence of the *Agrobacterium* rol (B, C) gene in transgenic hairy root tissues, two pairs of PCR primers were used. Primer sets for the *SmHMGR2* gene were designed by the sequence of 35S promoter with the 5′-end of the target gene and by 3′-end of the target gene with the segments of hygromycin-resistant gene. For GFP checking, the hairy root was examined using a fluorescence microscope. Transgene copy number in transformed *S. miltiorrhiza* root lines was confirmed by DNA gel blot analysis using a non-radioactive digoxigenin-11-dUTP-labelled probe as described previously.

An aliquot (200 ng) of the total RNA was used, and the first strand cDNA was synthesised with total RNA using the reverse-transcription PCR system according to the manufacturer's protocol of PrimeScript^{TM} 1st Strand cDNA Synthesis Kit (Takara, Tokyo, Japan). To estimate the relative mRNA level, a diluted series of the reference cDNA sample was used as the standard. The relative amount of the *SmHMGR2* gene was evaluated by the relative expression index of mRNA.

The contents of squalene and tanshinone in the hairy roots was determined by chromatographic methods, as described previously.

6.9.2 Results and Discussion

In recent years, there has been a remarkable progress in the understanding of the molecular regulation of diterpenoid tanshinone biosynthesis in *S. miltiorrhiza*. In the present study, the researchers attempted to further dissect the molecular biology of tanshinone biosynthesis pathways in *S. miltiorrhiza* by cloning the full-length cDNA of *SmHMGR2*. This gene was introduced into the leave explants of *S. miltiorrhiza* under the control of the strong gene promoter with the gateway vectors. The overexpression of *SmHMGR2* in transgenic hairy roots resulted in increased tanshinone and squalene production. The findings suggest that SmHMGR2 is key enzyme controlling the diterpenoid metabolic flux and could be exploited for higher tanshinone accumulation in *S. miltiorrhiza*.

The HMGR superfamily contains a number of members that responded differently to various external stimuli, including pathogen infection and exposure to xenobiotics. Distinct isoforms among HMGRs might be involved in the formation of secondary metabolites of terpenoids. Terpenoid metabolism can be elicited by jasmonic acid, or its methyl ester MeJA. The content of tanshinones in hairy roots of *S. miltiorrhiza* was significantly increased after treatment with MeJA. MeJA is not a specific inducer of the *SmHMGR2* gene; it also induces other genes contributing to terpenoid biosynthesis. It is considered that the increased content of tanshinones by MeJA is due to an integrated effect on a cluster of genes related to tanshinone biosynthesis. These results implicate that HMGR plays an important role in the production of tanshinones.

In this study, *SmHMGR2* was successfully cloned from hairy roots of *S. miltiorrhiza*. It is functionally similar to the *SmHMGR1* gene since the transcription of both *SmHMGR1* and *SmHMGR2* was responsive to MeJA treatment. However, the expression of *SmHMGR2* could be detected in all the tissues (leaves, stems, and roots) of *S. miltiorrhiza* but at different levels, with the highest expression in the leaves; this was different from *SmHMGR1*, which had the highest expression in the roots. This indicates a tissue-specific regulation of these two genes.

The multiple alignments showed that the deduced SmHMGR2 sequence had high similarity to other plant HMGRs and contained all conserved substrate-binding motifs of HMGRs. The *N*-terminal contained two transmembrane domains. The 3D model of *SmHMGR2* represented a typical spatial structure of HMGRs. However, analysis of the phylogenetic tree indicated that SmHMGR1 and SmHMGR2 were sorted into different groups, and *SmHMGR2* lacks the "RRRP" motifs. This implicates the functional difference of the two HMGR enzymes.

A detailed analysis of different transgenic lines of *S. miltiorrhiza* showed distinct phenotypic and metabolic variation. This may be due to a position effect or random integration of the transgene at non-specific sites in the plant genome, different numbers of gene copies integrated into the genome, different and specific post-translational regulation of the endogenous enzyme, and different methylation pattern and extent. These differences may exist among hairy root lines of *S. miltiorrhiza* due to infection by *A. rhizogenes* ACCC10060, leading to differences in the growth rate, aging rate, size and branches, especially the colour of hairy roots, and the content of tanshinones.

The difference in the tanshinone contents in different root lines may be attributed in part to the different growth stages of roots, since the tanshinone accumulation was high during the stationary phase (40–50 d). Actually, the growth rate of wild-type roots was slower than that of the transformed hairy roots (generally, two to three months per cycle for wild-type roots). As such, the researchers included two additional controls of hairy roots, VCK and HRCK, which could reduce the impact of the growth rate. The results support a correlation between the overexpression of the *SmHMGR2* gene and the content of tanshinones.

In transgenic tobacco, overexpressing the HbHMGR enzyme can enhance the production of sterol. The cardenolide and phytosterol levels were increased by expression of an *N*-terminal-truncated HMGR in transgenic *Digitalis minor*. Similarly, overexpression of an *N*-terminal-truncated HMGR increased the production of essential oils and sterols in transgenic *Lavandula latifolia*. The high expression of CrHMGR in these transgenic plants was associated with a higher artemisinin content. Amorphadiene production was improved by 50% through transferring the HMGR gene into engineered yeast. However, the MVA pathway can also contribute to mono- and sesquiterpene production in *L. latifolia*. This study has shown that transgenic hairy roots were morphologically distinguishable from wild-type plants, and there was a significant correlation between the increased expression level of *SmHMGR2* mRNA and the overproduction of diterpenes and triterpenes in the hairy roots of *S. miltiorrhiza*. These data suggest an important role of *SmHMGR2* in the isoprenoid biosynthesis of *S. miltiorrhiza* via the MVA pathway.

Determining the transgene copy number is an important step in transformant characterisation and can differentiate between complex and simple transformation events. The number of inserted *SmHMGR2* gene copies in the H14, H19, and H24 genomes was 2–3. This is in agreement with increased levels of *SmHMGR2* transcripts and contents of tanshinones or squalene. This provides further evidence that the overexpression of SmHMGR2 through stable gene transfer enhanced the biosynthesis of tanshinones.

Interestingly, the deducted *SmHMGR2* sequence did not contain the motif of "RRRP". The motifs rich in arginines (RRR) are considered to be important for endoplasmic reticulum retention. On the other hand, proline residues are known to strongly modify the secondary structure of proteins. Plant HMGR has been shown to insert in vitro into the membrane of microsomal vesicles, but the final in vivo subcellular localisation remains controversial. Endogenous *Arabidopsis* HMGR was found to be localised at a steady state within the endoplasmic reticulum, but also predominantly within spherical, vesicular structures located in the cytoplasm and within the central vacuole in differentiated cotyledon cells. The *N*-terminal region, including the transmembrane domain of *Arabidopsis* HMGR, was necessary and sufficient for directing HMGR to the endoplasmic reticulum and spherical structures. Further studies are needed to explore the subcellular distribution of HMGR in *S. miltimorrhiza*.

6.9.3 Examples

Example One: Production of the Antimalarial Drug Precursor Artemisinic Acid in Engineered Yeast

Malaria is a global health problem that threatens 300–500 million people and kills more than one million people annually. Disease control is hampered by the occurrence of multi-drug-resistant strains of the malaria parasite *Plasmodium falciparum*. Synthetic antimalarial drugs and malarial vaccines are currently being developed, but their efficacy against malaria awaits rigorous clinical testing. Artemisinin, a sesquiterpene lactone endoperoxide extracted from *A. annua* L., is highly effective against multi-drug-resistant *Plasmodium* spp., but is in short supply and unaffordable to most malaria sufferers. Although total synthesis of artemisinin is difficult and costly, the semi-synthesis of artemisinin or any derivative from microbially sourced artemisinic acid, its immediate precursor, could be a cost-effective, environmentally friendly, high-quality, and reliable source of artemisinin. Here, Ro et al. report the engineering of *Saccharomyces cerevisiae* to produce high titres (up to $100\,\mathrm{mg\,l^{-1}}$) of artemisinic acid using an engineered mevalonate pathway, amorphadiene synthase, and a novel cytochrome P_{450} monooxygenase (*CYP71AV1*) from *A. annua* that performs a three-step oxidation of amorpha-4,11-diene to artemisinic acid. The synthesised artemisinic acid is transported out and retained on the outside of the engineered yeast, meaning that a simple and inexpensive purification process could be used to obtain the desired product. Although the engineered yeast is already capable of producing artemisinic acid at a significantly higher specific productivity than *A. annua*, yield optimisation and industrial scale-up will be required to raise artemisinic acid production to a level high enough to reduce artemisinin combination therapies to a significantly cheaper cost.

The authors engineered artemisinic-acid-producing yeast in three steps, by (1) engineering the farnesyl pyrophosphate (FPP) biosynthetic pathway to increase FPP production and decrease its use for sterols, (2) introducing the amorphadiene synthase (ADS) gene from *A. annua* into the high FPP producer to convert FPP to amorphadiene, and (3) cloning a novel cytochrome P450 that performs a three-step oxidation of amorphadiene to artemisinic acid from *A. annua* and expressing it in the amorphadiene producer.

In summary, Ro et al. created a strain of *S. cerevisiae* capable of producing high levels of artemisinic acid by engineering the FPP biosynthetic pathway to increase FPP production and by expressing amorphadiene synthase, a novel cytochrome P_{450}, and its redox partner from *A. annua*. Given the existence of known, relatively high-yielding chemistry for the conversion of artemisinic acid to artemisinin or any other derivative that might be desired, microbially produced artemisinic acid is a viable source of this potent family of antimalarial drugs. Upon optimisation of product titres, a conservative analysis suggests that artemisinin combination therapies could be offered significantly below their current prices.

In addition to cost savings, this bioprocess should not be subject to factors like weather or political climates that may affect plant cultivation. Furthermore, artemisinic acid from a microbial source can be extracted using an environmentally friendly process without worrying about contamination by other terpenes that are produced by plants, thereby increasing the ease with which it can be produced while reducing purification costs.

Example Two: Using Tobacco to Treat Cancer

Tobacco, which has gained a reputation as a cause of cancer, may soon earn some praise rather than recrimination after being used by McCormick et al. to manufacture patient-specific vaccines against follicular B cell lymphoma.

Follicular lymphomas are a subtype of non-Hodgkin's lymphoma, the seventh leading cause of cancer-related deaths in the United States, and are a malignant disease of the lymphatic system that originates from cells of the immune system (lymphocytes). The administration of a tobacco-derived non-Hodgkin's lymphoma vaccine (a single-chain segment of an antibody protein) in a human clinical trial resulted in immune responses in more than 70% of the patients. A majority of patients showed a cellular immune response, suggesting that the vaccine specifically directs the immune system to attack cancer cells. The study not only demonstrates the safety and efficacy of the plant-made protein, but represents the first time that such responses have been observed using a subcutaneously administered antibody-based vaccine in the absence of a carrier protein (which typically boosts the immune response and has been used in all previous clinical studies). Bayer AG, a major pharmaceutical company, has acquired the supporting data from the new study, and very recently announced the opening of a production facility that will use tobacco to manufacture biopharmaceuticals, the first of which will be a candidate patient-specific antibody vaccine for non-Hodgkin's lymphoma therapy.

Although the report of McCormick et al. will undoubtedly be appreciated as an advance in immunotherapy for cancer patients, the results will likely generate even greater excitement in the plant biotechnology community. It has been almost two decades since genetically engineered plants were shown to produce monoclonal antibodies or vaccine subunits (the latter can be antigens that elicit protective antibodies). To date, however, only small biotechnology companies have used plant biotechnology to produce protein pharmaceuticals, such as glucocerebrosidase to treat Gaucher disease, lipase to treat cystic fibrosis, α-interferon, lactoferrin, and others. Meanwhile, large pharmaceutical companies have watched from the sidelines.

To understand what technical or economic forces have enticed a major player in the pharmaceutical industry into the use of plant biotechnology, one need only look at the strategy for producing a patient-specific vaccine. In the case of non-Hodgkin's lymphoma, patients are diagnosed through symptomatology, followed by excision biopsies (from either a tumour mass or lymph node where the tumorous B cells predominate). Biopsy materials are used to characterise the specific type of non-Hodgkin's lymphoma. Because each B cell bears a unique surface immunoglobulin protein, and because a malignant B cell is of colonial origin, the immunoglobulin becomes a specific marker for the tumour of a specific patient. Once the specific gene sequences encoding the individual's tumour immunoglobulin have been determined, the challenge is to find a way to obtain a portion of this very specific immunoglobulin in sufficient quantity and conformation to use it as a vaccine that will trigger the body to attack the malignant B cells bearing this immunoglobulin (and not normal B cells).

Earlier studies had shown that immunisation with a complete immunoglobulin protein (composed of four polypeptides, with regions that specify the class of antibody and specific antigen target) induced the desired immune response. However, the time between identification of the immunoglobulin gene sequences in a tumour and manufacture of the corresponding protein is slow (many months) and complex. The time delay, in particular, is a severe hindrance to the use of vaccination as a therapy for non-Hodgkin's lymphoma patients who are newly diagnosed. Rather than wait for a vaccine, these patients are more likely to start conventional (immunosuppressive) chemotherapy even with its negative side effects and uncertainty of durable remissions.

Speed of manufacture turns out to be the key to plant biotechnology's contribution to non-Hodgkin's lymphoma vaccine development. The genomic sequences of many plant viruses are known, and molecular tools are available to quickly insert new genes into a viral genome in such a way that virus replication produces a new non-viral protein as a "by-product" to virus replication. As reported by McCormick, a fragment of the tumour-specific immunoglobulin protein, called the autologous single-chain variable domain, was incorporated into a virus that infects tobacco. Because the single-chain variable domain must correspond uniquely to the immunoglobulin found on the surface of a patient's malignant B cell, a unique virus must be engineered for each patient. Once introduced into a tobacco plant, the engineered virus replicates and spreads rapidly throughout the plant, diverts normal cellular metabolism, and turns on production of the proteins encoded by the viral genome, including the single-chain variable domain gene. Even without optimising or automating production, patient-specific single-chain variable domain protein fragments could be produced under good manufacturing practice (GMP) regulations within 12–16 weeks of receiving biopsy specimens. The high-risk arena of pharmaceutical development makes it difficult to predict when (or if) a plant-made non-Hodgkin's lymphoma vaccine will ultimately be marketed, but the production facility investment commitment by a large and experienced pharmaceutical company suggests probability of success.

Academic scientists who have long promoted the use of plant biotechnology as a tool to attack global public health issues and achieve lower-cost protein drugs will learn a valuable lesson if the non-Hodgkin's lymphoma vaccine is successful: There is more to moving a pharmaceutical product into practical use than a simple "cost of goods" arguments. For example, the speed by which useful proteins can be produced in plant systems (such as in patient-specific vaccine production) is also an advantage for companies that wish to rapidly obtain GMP samples for phase I clinical trials. Moreover, the fact that each individual tobacco plant is a manufacturing unit provides an infinitely scalable manufacturing platform with low capital investment for the protein production component of biomanufacturing.

However, the value of plant biotechnology to global health should not be forgotten. Protein drugs are now widely used in the developed world, but economic barriers make most of these new biotechnology products inaccessible to all but the very wealthiest inhabitants of the developing world. It is likely that another opportunity for plant-made pharmaceuticals will come in the arena of "biosimilars" (new versions of biopharmaceutical products), or "generic" versions of existing protein drugs (the latter is often through reverse engineering, in which an existing product is produced in a redesigned manufacturing process), especially as patents on current drugs expire. Cancer therapeutics offers a major opportunity. For example, the monoclonal antibody Avastin (manufactured by Genentech) used to treat colorectal and lung cancer and approved by the US Food and Drug Administration in 2017 for treating advanced breast cancer, costs $84,700 for an average 11-month course of breast cancer treatment. It is conceivable that Avastin and a related monoclonal antibody, Herceptin (also manufactured by Genentech), prescribed for women with breast cancer with high expression of the HER2 receptor, could be manufactured using plant biotechnology with considerable cost advantages. Protein-based therapeutics is still at an early stage, and the involvement of plant biotechnology in their production is at an even earlier stage. However, the prospects are exciting.

References

1. Stanford PK. August weismann's theory of the germ-plasm and the problem of unconceived alternatives. Hist Philos Life Sci. 2005;27(2):163–99.
2. Huang HP, Li JC, Huang LQ, et al. The application of biotechnology in medicinal plants breeding research in china. Chin J Integr Med. 2015;21(7):551–60.
3. Rushforth, K. (1999). Trees of Britain and Europe. Collins ISBN 0-00-220013-9.
4. USA National Council (2015) Managing global genetic resources: agricultural crop issues and policies.
5. Schippmann U, Cunningham AB, Leaman DJ (2002) Impact of cultivation and gathering of medicinal plants on biodiversity: global trends and issues (Case Study No. 7). In Biodiversity and the ecosystem approach in agriculture, forestry and fisheries, satellite event session on the occasion of the 9th regular session of the commission on "Genetic Resources for Food and Agriculture", Rome, 12–13 October, 2002. FAO Document Repository of United Nations.
6. Guo QS, Wang CL. Retrospect and prospect of medicinal plants cultivation in china. China J Chinese Mater Med. 2015;40(17):3391–4.
7. Baranov A. Recent advances in our knowledge of the morphology, cultivation and uses of ginseng (panax ginseng c. a. meyer). Econ Bot. 1966;20(4):403–6.

8. Chen SL, Yu H, Luo HM, et al. Conservation and sustainable use of medicinal plants: problems, progress, and prospects. Chin Med. 2016;11(1):37. https://doi.org/10.1186/s13020-016-0108-7.
9. Canter PH, Thomas H, Ernst E. Bringing medicinal plants into cultivation: opportunities and challenges for biotechnology. Trends Biotechnol. 2005;23(4):180–5.
10. Ding G, Ding XY, Shen J, et al. Genetic diversity and molecular authentication of wild populations of dendrobium officinale by rapd. Acta Pharm Sin. 2005;40(11):1028–32.
11. Yuan LP, Mao CX. Hybrid rice in China—techniques and production. In: Rice. Berlin/Heidelberg: Springer; 1991. p. 128–48.
12. Liu PL, Wan Q, Guo YP, et al. Phylogeny of the genus chrysanthemum l.: evidence from single-copy nuclear gene and chloroplast dna sequences. PLoS ONE. 2012;7(11):e48970. https://doi.org/10.1371/journal.pone.0048970.
13. Meng L, Zheng G. Phylogenetic relationship analysis among Chinese wild species and cultivars of paeonia sect. moutan using rapd markers. Sci Silvae Sin. 2004;40(5):110–5.
14. Wenyuan G, Enqiang Q, Xiaohe X, et al. Analysis on genuineness of angelica sinensis by rapd. Chin Tradit Herb Drug. 2001;32:926–9.
15. Fugle GN. Laying down arms to heal the creation-evolution divide: Wipf and Stock Publishers; 2015.
16. Visarada KBRS, Meena K, Aruna C, et al. Transgenic breeding: perspectives and prospects. Crop Sci. 2009;49(5):1555–63.
17. Yi L. Transformation of antimicrobial peptide fusion gene of cecropin b and rabbit np-1 to houttuynia cordata. China J Chinese Mater Med. 2010;35(13):1660–4.
18. Choi Y, Jeong J, In J, Yang D. Production of herbicide-resistant transgenic Panax ginseng through the introduction of the phosphinothricin acetyl transferase gene and successful soil transfer. Plant Cell Rep. 2003;21(6):563–8.
19. Diemer F, Caissard JC, Moja S, et al. Altered monoterpene composition in transgenic mint following the introduction of 4S-limonene synthase. Plant Physiol Biochem. 2001;39 (7-8):603–14.
20. Suzuki KI, Yun DJ, Chen XY, et al. An Atropa belladonna hyoscyamine 6β-hydroxylase gene is differentially expressed in the root pericycle and anthers. Plant Mol Biol. 1999;40(1):141–52.
21. Nogué F, Mara K, Collonnier C, Casacuberta JM. Genome engineering and plant breeding: impact on trait discovery and development. Plant Cell Rep. 2016;35(7):1475–86.
22. Gaj T, Gersbach CA, Barbas CF. ZFN, TALEN, and CRISPR/Cas-based methods for genome engineering. Trends Biotechnol. 2013;31(7):397–405.
23. Song G, Jia M, Chen K, et al. Crispr/cas9: a powerful tool for crop genome editing. Crop J. 2016;4(2):75–82.
24. Collard BC, Mackill DJ. Marker-assisted selection: an approach for precision plant breeding in the twenty-first century. Phil Trans Royal Soc B Biolog Sci. 2007;363(1491):557–72.
25. National Research Council. Safety of genetically engineered foods: Approaches to assessing unintended health effects: National Academies Press; 2004.
26. Arora L, Narula A. Gene editing and crop improvement using CRISPR-Cas9 system. Front Plant Sci. 2017;8:1932. https://doi.org/10.3389/fpls.2017.01932.
27. Kwon YH, Smerdon MJ. Binding of zinc finger protein transcription factor iiia to its cognate dna sequence with single uv photoproducts at specific sites and its effect on dna repair. J Biol Chem. 2003;278(46):45451–9.
28. Weeks DP, Spalding MH, Yang B. Use of designer nucleases for targeted gene and genome editing in plants. Plant Biotechnol J. 2016;14(2):483–95.
29. Ainley WM, Sastry-Dent L, Welter ME, et al. Trait stacking via targeted genome editing. Plant Biotechnol J. 2013;11(9):1126–34.
30. Malzahn A, Lowder L, Qi Y. Plant genome editing with TALEN and CRISPR. Cell Biosci. 2017;7(1):21.
31. Doudna JA, Charpentier E. The new frontier of genome engineering with CRISPR-Cas9. Science. 2014;346(6213):1258096.

32. Bortesi L, Fischer R. The crispr/cas9 system for plant genome editing and beyond. Biotechnol Adv. 2015;33(1):41–52.
33. Guilinger JP, Thompson DB, Liu DR. Fusion of catalytically inactive cas9 to foki nuclease improves the specificity of genome modification. Nat Biotechnol. 2014;32(6):577–82.
34. Komor AC, Kim YB, Packer MS, et al. Programmable editing of a target base in genomic dna without double-stranded dna cleavage. Nature. 2016;533(7603):420–4.
35. Zetsche B, Gootenberg J, Abudayyeh O, et al. Cpf1 is a single rna-guided endonuclease of a class 2 crispr-cas system. Cell. 2015;163(3):759–71.
36. Jia H, Wang N. Targeted genome editing of sweet orange using cas9/sgrna. PLoS ONE. 2014;9 (4):e93806. https://doi.org/10.1371/journal.pone.0093806.
37. Hu C, Deng G, Sun X, et al. Establishment of an Efficient CRISPR/Cas9-Mediated Gene Editing System in Banana. Sci Agric Sin. 2017;50(7):1294–301.
38. Waltz E. Gene-edited crispr mushroom escapes us regulation. Nature. 2016;532(7599):293.
39. Jakočiūnas T, Bonde I, Herrgård M, et al. Multiplex metabolic pathway engineering using CRISPR/Cas9 in Saccharomyces cerevisiae. Metab Eng. 2015;28:213–22.
40. Lai MN, Wang SM, Chen PC, et al. Population-based case–control study of Chinese herbal products containing aristolochic acid and urinary tract cancer risk. J Natl Cancer Inst. 2010;102 (3):179–86.
41. Hao DC, Xiao PG. Genomics and evolution in traditional medicinal plants: road to a healthier life. Evol Bioinforma. 2015;11:197–212.
42. Zhao Y, Yin J, Guo H, et al. The complete chloroplast genome provides insight into the evolution and polymorphism of Panax ginseng. Front Plant Sci. 2015;5:696. https://doi.org/10.3389/fpls.2014.00696.
43. Chen S, Luo H, Li Y, et al. 454 EST analysis detects genes putatively involved in ginsenoside biosynthesis in Panax ginseng. Plant Cell Rep. 2011;30(9):1593–601.
44. Moses T, Pollier J, Shen Q, et al. OSC2 and CYP716A14v2 catalyze the biosynthesis of triterpenoids for the cuticle of aerial organs of Artemisia annua. Plant Cell. 2015;27 (1):286–301.
45. Chen SL, Osbourn A, Kontogianni VG, et al. Temporal transcriptome changes induced by methyl jasmonate in Salvia sclarea. Gene. 2015;558(1):41–53.
46. Vautrin S, Song C, Zhu YJ, et al. The first insight into the Salvia (Lamiaceae) genome via BAC library construction and high-throughput sequencing of target BAC clones. Pak J Bot. 2015;4 (47):1347–57.
47. Da H, Gu XJ, Xiao PG. Medicinal plants: chemistry, biology and omics: Woodhead Publishing; 2015.
48. Hao DC, Chen SL, Xiao PG, Liu M. Application of high-throughput sequencing in medicinal plant transcriptome studies. Drug Dev Res. 2012;73(8):487–98.
49. Yao X, Peng Y, Xu LJ, et al. Phytochemical and biological studies of Lycium medicinal plants. Chem Biodivers. 2011;8(6):976–1010.
50. Croteau R, Ketchum RE, Long RM, et al. Taxol biosynthesis and molecular genetics. Phytochem Rev. 2006;5(1):75–97.
51. Subroto MA, Kwok KH, Hamill JD, et al. Coculture of genetically transformed roots and shoots for synthesis, translocation, and biotransformation of secondary metabolites. Biotechnol Bioeng. 1996;49(5):481–94.

Further Reading

1. Ro D-K, Paradise EM, Ouellet M, et al. Production of the antimalarial drug precursor artemisinic acid in engineered yeast. Nature. 2006;440:940–3.

2. Arntzen CJ. Using tobacco to treat cancer. Science. 2008;321(5892):1052–3.
3. Celis C, Scurrah M, Cowgill S, et al. Environmental biosafety and transgenic potato in a centre of diversity for this crop. Nature. 2004;432:222–5.
4. Zhang JZ, Li ZM, Yao JL, et al. Identification of flowering-related genes between early flowering trifoliate orange mutant and wild-type trifoliate orange (poncirus trifoliatal. raf.) by suppression subtraction hybridization (SSH) and macroarray. Gene. 2009;430(1–2):1–104.
5. Paddon CJ, Westfall PJ, Pitera DJ, et al. High-level semi-synthetic production of the potent antimalarial artemisinin. Nature. 2013;496(7446):528–32.
6. Yin K, Gao C, Qiu JL. Progress and prospects in plant genome editing. Nat Plant. 2017;31 (3):17107.

Chapter 7
Functional Genome of Medicinal Plants

Jian Yang, Meirong Jia, and Juan Guo

As a complex organism, research on the origin, evolution, development, physiology, and genetic traits of medicinal plants has shown strong development prospects for medicinal plant functional genomics. Functional genomics is a science based on genome sequence information, which uses various genomic techniques to organically link genome sequences with gene functions (including gene networks) and phenotypes at the system level and ultimately reveal the functions of biological systems at different levels in nature (genome, transcriptome, proteome, metabolome, epigenome, etc.). Research into the genomes, transcriptomes, and proteomes of medicinal plants provides a premise and basis for comprehensively analysing various life phenomena at the molecular level. Combined with metabolomic research, it greatly promotes the application of frontier life science and technology in the field of medicinal plants, lays the foundation for clarifying the synthesis and regulation of effective components of medicinal plants, and promotes research on the interaction between genetic and environmental factors of medicinal plants.

7.1 Genomics of Medicinal Plants

The discovery of the DNA double-helix structure opened a new era of research for life science. DNA sequences, as life's most basic genetic information, determine the fundamental sources of species characteristics. The deciphering of genome

J. Yang · J. Guo (✉)
National Resource Center for Chinese Materia Medica, China Academy of Chinese Medical Sciences, Beijing, China
e-mail: yangchem2012@163.com; guojuanzy@163.com

M. Jia
Department of Plant Biology, University of California, Davis, California, USA
e-mail: meirongj@iastate.edu

L.-q. Huang (ed.), *Molecular Pharmacognosy*,
https://doi.org/10.1007/978-981-32-9034-1_7

sequences, in which all the genetic information of a species is contained, is a sign of a new era of the research and application of species. If the genome sequence of a medicinal plant with important economic and medical value is decoded, it will promote the scientific research, molecular breeding, and genetic transformation of the medicinal plant. In 1977, Frederick Sanger invented the chain-terminating sequencing method [1, 2], which represented the formation of the first generation of sequencing technology. Additionally, with the development of the second generation of high-throughput sequencing technologies and the third generation of single-molecule sequencing technology, the field has reached the post-genome era. While we are learning more about the structure of genome sequence, functional genomes have gradually become the core of these studies. With the development of sequencing technology and the decline in sequencing cost, whole genome sequencing is no longer the patent of model organisms. Studies on genome sequencing are being carried out in more and more important economic crops and medicinal plants and provide abundant data resources for medicinal plant functional genomic research. More than 360 species have had genome sequencing by now (http://www.plabipd.de/index.ep). However, different from releasing of large number of genomic data, only a small number of genes have their biological function identified. The main task of the post-genome era is to analyse the biology function of big genome data.

Genome sequences contain the origin, evolution, biological, and development information, as well as information related to physiological and genetic traits. It is the premise and foundation of analysing all kinds of life phenomena at the molecular level. Compared to model organisms in higher plants, the genetic background, basic genome data, and functional genomic research of many medicinal plants are not sufficient. As sequencing costs and time decline, the combination of different sequencing technologies has promoted genome sequencing work of increasing amounts of medicinal plants. Medicinal plant genome research greatly promoted the application of cutting-edge life science and technology in the field of medicinal plants, bridged traditional Chinese medicine (TCM) and modern life science, and discovered the "genetic code" of medicinal plants. It will establish a foundation to explain the synthesis and regulation of active ingredients in medicinal plants, accelerate variety selection in medicinal plants, and promote the scientific and commercial development of green TCM agriculture.

7.1.1 Research Contents of Medicinal Plant Genomics

Genome sequencing is just the first step in functional genomic research. The bigger challenge is to figure out the genetic information of gene sequences and its biological functions. This work depends on the correct annotation of gene sequences, the method of which depends on bioinformatics analysis. Homology is an extremely important concept in gene annotation. Homologous genes generally do not have the same nucleotide sequence, because they will be subjected to independent random mutations after their occurrence. However, homologous genes share similar sequences, and most nucleotides without mutations are at the same position. Therefore, when a new gene sequence was confirmed, we can find homologous genes of known sequences in the database according to homology and predict the function of new genes from the known homologous genes, according to the correlation of evolution. Homologous genes can be divided into two categories:

1. Orthologous genes: They are homologous genes that are distributed in the genome of two or more species that share common ancestors due to speciation events. Orthologous sequences are generally considered to have similar structure and biological functions [3]. They are highly conserved and even nearly identical, and they can even be replaced by closely related species. Moreover, they usually encode key regulatory proteins, enzymes, or coenzymes necessary for life activities [4]. In addition, many orthologous genes have similar sequence change rates, evolutionary distances, and regulatory pathways, as well as the ability to reproduce the evolutionary history of species.
2. Paralogous genes: They are homologous genes produced by gene duplication. They are often different members of a multigene family. Their common ancestral genes may exist after speciation or before species formation. The gene family is mainly produced by the replication of ancestral genes and their mutations. It is one of the ways to increase the complexity of the genome. By comparing the sequence differences among the members of the gene family, the evolutionary trajectories of genes can be traced [5].

7.1.1.1 Genome Evolution

Evolutionary analysis based on comparative genomes is an important means to understand the nature of life. Gene families are composed of a group of genes from the same ancestral gene, and identification of gene families is an important aspect of evolutionary analysis. Through the clustering of homologous genes and identification of gene family, we can obtain single- or multi-copy gene families. These gene families have a certain degree of conservation and diversity among species, which provide sequence and sequence variation information for phylogenetic studies. On the other hand, by gene family analysis, genes or gene families that are unique to a species or that undergo significant expansion or contraction in a species can be obtained. These are often associated with species-specific traits, providing a basis for medicinal plant-specific phenotypes. By using these genes and gene family analysis, phylogenetic analysis between species and time of species divergence can be further analysed to reveal the evolution of species.

Colinearity analysis refers to the homology of large fragments that take place in the same species or between two species due to replication (genomic replication, chromosomal replication, or large fragment replication) or species differentiation. Within the homologous fragment, the genes are conserved in function and arrangement order. The genes in the collinear fragment maintain a high degree of conservation during species evolution, and the massive digital mutual information is visualised by graphs through collinear analysis. The performance of the two sets of sequencing data through alignment analysis reveals the collinear relationship between them.

7.1.1.2 Genome Duplication

Gene duplication is widespread in organisms. In model organisms that have been sequenced, repetitive genes are found throughout prokaryotic and eukaryotic organisms, especially in the evolution of higher angiosperms, as they have undergone multiple multiplication processes, resulting in a larger number of repetitive genes. Gene duplication is the most important driving force for plant evolution and an important source of new functional genes. In the long-term evolution of species, the evolution of repeated genes can induce the differentiation of gene expression patterns to meet the needs of species development. Therefore, the study of repeated genes is of great significance for revealing structural changes and functional differentiation in repeated genes.

By the size of the repeat region, gene duplication can be divided into two types: (1) small-scale gene duplication, including single-gene duplication, and (2) large-scale gene duplication, including partial genomic duplication and whole genome duplication (polyploidisation) [6]. The repetition of a single-gene and part of a genome is mainly produced by unequal exchange, while the repetition of a whole genome is triggered by errors in mitosis or meiosis [7].

Gene duplication results in the presence of two or more copies of homologous gene sequences in the same genome, which may result in functional redundancy [8] and is regulated by the gene-dosage effect [9]. After natural selection, the retained repeat genes are faced with roughly the following three different fates [10]: (1) one copy retains the original function, and the other undergoes neofunctionalisation to obtain new functions; (2) both copies undergo subfunctionalisation, in which they share the function of the original gene, and their combined function covers that of the ancestral gene; and (3) one of the copies is mutated to cause non-functionalisation and thus becomes a pseudogene.

7.1.1.3 Analysis of Molecular Genetic Mechanism of Medicinal Plant Economic Traits Based on Genome-Wide Sequences

The completion of the genome-wide reference sequence provides a research basis for the molecular structure of specific forms and unique chemical formations, which are produced after long-term natural selection or manual selection. At the same time, it can also provide guidance for molecular marker-assisted breeding.

On the one hand, genetic diversity analysis is based on the whole gene resequencing. The genetic diversity determines the ability of a species to adapt to evolution. Whole genome sequencing analysis is used to construct the genetic map, and then the effective molecular markers can be screened to study the population genetics, genetic diversity, and phylogenetic evolution of the species, which can provide a basis for the mining, protection, and utilisation of medicinal genetic resources. On the other hand, molecular-assisted breeding based on a genome-wide level has application value for the genetic selection of medicinal plants.

Through genome sequencing, a large number of SNP loci are obtained, and a high-density genetic map is constructed, which provides a basis for screening the linkage of molecular markers of superior traits. By comparing genomic analysis to excavate excellent allelic variation, functional markers can be developed to accelerate the breeding process of medicinal plants and further enhance the quality traits of medicinal plants.

7.1.1.4 Assist in Sequence Assembly and the Annotation of Transcriptome Data

The development of transcriptome sequencing provides a basis for solving biological problems at different stages of growth and development, but there are different splicing forms in the transcription process; genes will generate different types of mRNA sequences according to different exon combinations during transcription, which in turn encode different protein information. This is one of the important reasons leading to the functional diversity of proteins. Genome-wide information provides the basis for comparative analysis of transcripts designed based on different biological issues. For transcriptomics studies with reference genomic sequences, the focus of bioinformatics analysis is to compare the sequence read to the genomic sequence after obtaining the transcriptome sequencing results and obtain the sequence structure, variable shear of the relevant gene, and level of transcript expression. Through horizontal comparison between different samples, the expression level of the same gene or transcript between samples can be determined, and the differential expression gene (DEG) between the samples can be found by statistical methods and further functional annotation, and classification can be performed.

7.1.2 Case Studies of Medicinal Plant Genomics

7.1.2.1 *Panax ginseng* Genome

Panax ginseng C. A. Mey is a perennial herb belonging to the family Araliaceae. It is a valuable TCM with nourishing effect and has been widely cultivated in the three north-eastern provinces. Studies have shown that ginseng is a highly heterozygous tetraploid with a large number of chromosomes. The size of the chromosomes is about 3.2 Gb, which is similar to the size of the human genome. The ginseng genome data is huge, with many genomic repeats and high complexity, which brings great challenges to research. In 2017, *GigaScience* published the results of ginseng genome sequencing. The researchers selected a genomic strain with low heterozygosity for genome sequencing. The results showed that the ginseng genome was highly repetitive, with about 62% of repeat genes, and predicted more than 42,000 proteins of ginseng. Through the combination of transcriptome data from different tissues, the biosynthesis and regulation of ginsenosides were analysed. It was found that multiple enzymes in the mevalonate (MVA) pathway of ginsenosides exist in multiple copies, such as catalysis. HMG-CoA produces MVA's rate-limiting enzyme HMGR, which has eight copies in the ginseng genome. The biosynthesis of ginsenosides requires modification with glycosyltransferases. A total of 225 glycosyltransferases are found in the ginseng genome. This is the largest family of enzyme genes in the ginseng genome. These glycosyltransferase genes have specificity for tissue expression, which provide a basis for the screening of glycosyltransferase genes in the biosynthesis of ginsenosides [11].

7.1.2.2 *Salvia miltiorrhiza* Genome

Salvia miltiorrhiza (normally known as Danshen) is one of the commonly used TCMs in China. It has the effects of promoting blood circulation and eliminating phlegm. It is widely used in the treatment of cardiovascular and cerebrovascular diseases. The main active ingredients of Danshen are tanshinone and salvianolic acid compounds. In 2016, *Molecular Plant* published the whole genome sequencing results of *S. miltiorrhiza*. The study used second-generation and third-generation sequencing techniques to systematically study the genetic information of *S. miltiorrhiza* and *S. miltiorrhiza* var. *alba*. The study showed that the Danshen genome was about 558 Mb with high heterozygosity. Moreover, 56% of the regions are repetitive sequences, which make assembly difficult. Genomic data predicts that *S. miltiorrhiza* contains more than 30,000 protein-coding genes. Its transcription factors, terpene synthase, and P450 genes are predicted and analysed. Four gene clusters of steroidal synthase and P450 were detected, suggesting that the tanshinone biosynthesis gene clusters SmCPS1 and SmCPS2 were likely to be expanded from the same CPS/CYP76AH1 gene pair. In addition, 29 genes from nine gene families related to salvianolic acid synthesis were reported. This study provided a basis for revealing the molecular mechanism of the synthesis and regulation of the main pharmacological active components of Danshen and promoting the breeding of *S. miltiorrhiza* varieties [12].

7.1.2.3 *Artemisia annua* Genome

"Qinghao" is referred to in the literature as the dried aerial parts of the annual herbaceous *A. annua*. *A. annua* is widely distributed throughout the country, and originated in hilly flats less than 400 m above sea level. Nowadays, this plant is widely distributed all over the world. The chemical composition of *A. annua* is complex, and the most widely used active component is artemisinin. Scientists at home and abroad have developed a series of derivatives based on the chemical structure of artemisinin, including artemether, artesunate, arteether, and dihydroartemisinin. Hydrogen artemisinin has become an indispensable therapeutic drug in the world market for antimalarials. In addition, though increasing our understanding of the pharmacological effects of *A. annua*, it has been confirmed to have antimalarial, antibacterial, antiparasitic, antipyretic, and immune effects. Although the pharmacological effects of *A. annua* are extensive, its mechanism of action, characteristics, and research application are still in the initial stage, and further research is needed.

In March 2018, journal *Molecular Plant* published a high-quality *A. annua* genome [13]. This study combines second- and third-generation sequencing technologies, which yielded a 1.74 Gb-sized *A. annua* genome. There are a large number of repeats in the genome of *A. annua*, and the repeat sequence is as high as 61.75%. The existence of repeats is also one of the reasons for the larger genome of Compositae. A total of 63,226 encoded protein genes were predicted. The biosynthesis and regulation of artemisinin were discussed by annotation and gene family analysis of the *A. annua* genome and transcriptome data. Studies have shown that the *A. annua* genome has a heterozygosity of 1.0–1.5% and a high-density LTR repeat. By analysing the genome of *A. annua*, it is found that there are 871 gene families unique to Compositae. The genetic code of *A. annua* provides an effective tool for the biosynthesis and regulation, metabolic engineering, genetic improvement, and variety selection of this important medicinal plant.

7.1.2.4 *Glycyrrhiza uralensis* Genome

The rhizome of *Glycyrrhiza uralensis* Fisch. is one of the sources of the Chinese medicinal material liquorice. Liquorice has many functions, such as improving liver function, anti-inflammation activity, detoxification, and relieving cough. It is often said that "10 Chinese herbal medicines have nine grasses", and liquorice is the grass, which means that liquorice participates frequently in Chinese medicine. In addition, liquorice is also an important raw material for medicine and cosmetics. In 2016, *The Plant Journal* reported the genome sequence of liquorice [14]. Researchers used second- and third-generation sequencing technology for sequencing assembly and obtained 379 Mb of the genome of liquorice; the scaffold N50 = 109 kb and 34,445 genes were predicted. Comparative analyses suggested well-conserved genomic components and synteny between *G. uralensis* and other legumes. It was observed that three genes involved in the biosynthesis of isoflavonoids formed a cluster on the genome and showed conserved microsynteny with *Medicago* and chickpea; furthermore, they annotated 257 P450s and 91 UGTs in the genome, and some of them are supposed to contribute to the structural diversity of triterpenoid saponins. The study provided an essential resource for quality improvement and engineering bioactive components through synthetic biology.

7.1.2.5 *Dendrobium officinale* Genome

Dendrobium officinale Kimura et Migo, which contains antioxidant and antitumour active compounds, has the effects of lowering blood sugar and enhancing immunity.

In recent years, it has been overexploited as a health supplement, and wild resources have been very rare; it has been listed as endangered medicinal plants. In Molecular Plant [15], the researchers used a combination of second- and third-generation sequencing to perform genome-wide sequencing. The 1.35 Gb genome of *D. candidum* was assembled, with the assembled contig N50 = 25.1 kb and scaffold N50 = 76.4 kb. The analysis showed that the repeat sequence of *D. candidum* was 63.33%. The heterozygosity rate was 0.48%. A total of 35,567 protein-coding genes were annotated, and 1462 genes were identified as specific to *D. candidum*, which were supposed to participate in plant development, light regulation, disease resistance, xylem, and cellulose synthesis. Some important biological characteristics of *D. candidum*, including cold resistance, symbiosis with fungi, and orchid genome integrity, were analysed. The biosynthesis and signalling pathways of the medicinal active ingredients of *D. candidum* were analysed. It was found that a large number of sucrose–phosphate synthetase and sucrose synthase genes related to polysaccharide production were replicated on a large scale. Ten sucrose–phosphate synthetase and 15 sucrose synthase genes were detected. At the same time, the biosynthesis of alkaloids in *D. candidum* was also discussed. This will improve understanding of the biology of *D. officinale* and ultimately facilitate modernisation of this traditional Chinese herbal medicine.

7.1.2.6 *Papaver somniferum* Genome

Papaver somniferum (known as poppy) has a very high medicinal value, and its alkaloids have calming and pain-relieving effects, including morphine, thebaine, and codeine; its outstanding efficacy has attracted much attention. In August of 2018, the poppy genome was released in journal *Science*, bringing together global attention. The study published a genome sketch of the poppy, which was sequenced using Illumina, 10× genomics, and PacBio techniques, and used nanopore and bacterial artificial chromosome (BAC) data to aid in the validation of assembly quality, resulting in 2.72 Gb of genomic information with a Contig N50 of 1.77 Mb and scaffold N50 of 204 Mb. The study compared the poppy genome with the genomes of dicotyledonous plants, such as *Arabidopsis thaliana*, coffee, and grapes, and explored the evolution of poppy. The results showed that the poppy genome had a genome-wide doubling event 7.8 million years ago, which is one of the important reasons leading to its huge genome. At the same time, the study found that there is a benzylisoquinoline alkaloid (BIA) gene cluster in a 584 kb region on chromosome 11, including a noscapine synthetic gene cluster, (S)-reticuline to (R)-reticuline catalytic enzyme gene, and four genes related to the morphine biosynthesis pathway, which are co-published in the stem to produce morphine. Analysis showed that genomic rearrangement plays an important role in the evolution of benzyl isoquinoline alkaloid anabolism [16].

7.2 Transcriptomics of Medicinal Plants

Because the cost of genomic research is enormous, and some species are completely blank in omics research, it is not possible to estimate the size of the genome, the complexity of the program, the proportion of repeats, or the number of genes involved. Firstly, with the popularity of next-generation sequencing technologies and the reduction in sequencing costs, transcriptomics has shifted from traditional model biology research to non-model organisms. The genome can be assessed at the gene level through transcriptome research. Not only can the location of genes at the evolutionary level be judged by the similarity between genes and other species, but also the genetic recognition can be made by the known species information. Therefore, transcriptomics research can provide a general understanding of the basic genetic information of a living organism at a lower cost and can estimate the number of genes in the species and the genetic background information on the species. Secondly, combined with the experimental results and scientific experimental design in the previous research work, transcriptomics research can conduct exploratory studies on important biological functions in non-model organisms. Through comparisons between different biological samples, it is possible to find information on metabolic pathways; important genes that may affect these representations, such as biological phenomena known in previous work; and information that has never been explored. Therefore, the transcriptomics study of non-model organisms can verify the experimental work at the omics level and use the discovered new phenomena to effectively guide future research.

7.2.1 *Research Contents of the Transcriptomics in Medicinal Plants*

In the post-genome era, transcriptomics, as one of the most active disciplines, is an important means of studying cell phenotype and function and an important research direction for studying gene expression, structure, and function. Unlike the genome, the transcriptome is temporal and spatial; that is, the gene expression in the same organism is not exactly the same under different growth times and environments, so the transcriptome reflects genes that are actively expressed under specific conditions, except for abnormal mRNA degradation. Through transcriptome analysis, high-throughput information about RNA levels of gene expression can be obtained, which can reveal the intrinsic relationship between gene expression and life phenomena, high-throughput characterisation of cell physiological activity, and cell metabolic characteristics, providing a scientific basis for cell modification and transformation.

A generalised transcriptome refers to the sum of all RNA transcribed from a particular cell or tissue in a specific state. RNA includes RNA that encodes proteins (such as mRNA) and RNA that does not encode proteins (such as rRNA, tRNA, microRNA [miRNA], etc.), while the narrowly defined transcript usually refers to the sum of all mRNAs encoding proteins.

The research objectives of transcriptome analysis include the following:

1. The detection of new transcripts, including unknown and rare transcripts: Using single-base resolution transcriptome sequencing technology can greatly enrich many aspects of gene annotation, including coding region boundary identification, untranslated region identification, and new transcription region identification, and has strong technical advantages in low abundance transcript discovery.
2. Gene transcription level studies, such as gene expression and differential expression among different samples: Because the change in gene expression levels in organisms is small when external stimuli or environmental changes occur, transcriptome sequencing technology can quantitatively and accurately determine the expression level of RNA, and it is possible to determine the absolute number of each molecule in the cell population and to compare the results of experiments directly.
3. Gene function annotation: Measured reads are compared with genes of an annotated function of an existing database (such as GO or KEGG) to reveal the function and biological pathway of the gene in a specific transcription state.
4. Functional studies of non-coding regions (ncRNA), such as miRNA: The study of miRNAs in transcriptomics includes three stages [17, 18] – it originated from the discovery of the first miRNA (lin-4) [19], to the determination of large-scale miRNA transcriptomes, to the current miRNA regulation of target gene expression by gene silencing. During these three stages, especially after large-scale transcriptome sequencing, a large number of miRNAs were discovered, and functional studies related to it were in full swing. Non-coding RNA, which has been considered a waste component of the genome for many years, has finally gained increasing attention. One study found that miRNA plays an important regulatory role in gene expression and growth.
5. Transcript structural variation studies, such as variable cleavage, gene fusion, RNA editing, and coding sequence polymorphism: Variations in transcript structure can reveal the diversity of the posttranscriptional expression of a gene, and alternative splicing allows a gene to produce multiple mRNA transcripts for translation into different proteins.
6. SSR and SNP tag development: Potential SNPs or SSRs were identified by comparing the sequences between the transcripts and reference genomes. The development of microsatellite markers based on transcriptome data does not require library building, cloning, or screening, which not only avoids the problem of clonal bias and the loss of microsatellites due to ineffective cloning but also improves the efficiency of microsatellite marker development. Moreover, the operation process is almost one step, which greatly saves time. In addition to extracting genomic DNA, the whole process does not require any molecular manipulation and has a low development cost, high versatility of related species, and the direct labelling of functional genes [20].

7.2.2 *Research and Application of Medicinal Plant Transcriptomics*

With the decreased cost of sequencing, medicinal parts of medicinal plants have been gradually sequenced by transcriptomes. These massive data are stored in major databases, providing a large number of components for the study of functional genes. With the completion of genome sequencing of a series of model organisms, functional genomics is in the ascendance. Reference sequences of these model organisms can be used to facilitate the study of genome-wide transcription of species, SNP differences among different individuals, and gene copy number differences. For non-model organisms, including medicinal plants, transcriptome research is of great significance in resolving many problems, such as gene evolution, secondary metabolic synthesis and regulation, genetic breeding, and ecology, due to its lack of specific characteristics in many model organisms.

7.2.2.1 Medicinal Plant Functional Gene Mining

The transcriptome represents a gene expression profile of a species or a special tissue section, which can provide more specific candidates for functional gene mining, and has become one of the important methods for the isolation and cloning of new genes and gene function research. At present, in the case that most medicinal plants are unable to perform genome-wide sequence determination, sequencing from the transcriptome is a quick way to compare, discover, and identify gene expression sequences. By constructing a cDNA library for expressed sequence tag (EST) analysis, ESTs of different tissue parts, different growth stages, and different stress treatments of medicinal plants can be used to effectively discover and identify secondary metabolite-related functional genes for the further study of gene function and activity. The compositional regulation mechanism provides the basis. Transcriptions represent the gene expression profiles of a species for a period of time or of special tissues, which can provide more definite candidates for functional gene mining.

Our laboratory used transcriptome sequencing combined with metabolome analysis to analyse the hairy roots of *Salvia miltiorrhiza* at different induction time points [21] and obtained 20,972 genes, of which 6358 showed different expressions at different induction time points; focusing on the analysis of genes related to the biosynthesis and regulation of tanshinone, a total of 70 transcription factors related to the accumulation and regulation of tanshinone were detected. This study provided the basis for the analysis of the tanshinone biosynthesis pathway. On the basis of this, by comparing transcriptome analyses, 39 genes of cytochrome P450 with upregulated gene expression and tanshinone accumulation were obtained. Through cloning and functional studies of these genes, CYP76AH1 was found. CYP76AH3 and CYP76AK1 can sequentially catalyse the metabolic intermediates in the formation of C-ring ketones of the tanshinone carbon skeleton precursor miltiradiene

[22, 23]. This research greatly promoted the analysis of the tanshinone biosynthesis pathway. Yang et al. [24] compared the root-specific accumulation mechanism of tanshinones and compared the gene expression profiles of *S. miltiorrhiza* roots and leaves by transcriptome sequencing. The two sequencing libraries produced 550 and 546 and 525 and 292 sequences, respectively. A total of 64,139 unique genes (Unigene) were obtained after splicing of the resulting sequences, including 29,883 contigs (Contig) and 34,256 single sequences (Singleton). Sequence alignments were performed using the NCBI non-redundant protein database (NR) and the Swiss-Prot database, with 32,096 unigenes (50%) obtaining functional annotations. Gene classification (GO) and KEGG analysis screened 168 unigenes on the steroid skeleton biosynthesis pathway, including 144 MEP and MVA pathways and 24 terpene synthase. Transcriptional comparison analysis showed that 2863 unigenes were highly expressed in roots, including early key enzyme genes of the tanshinone biosynthesis pathway, such as copalyl diphosphate synthase (*SmCPS*), kaurene synthase-like (*SmKSL*), and *CYP76AH1*. Other differentially expressed genes, such as cytochrome P450 oxygenase, dehydrogenase, and reductase, are presumably related to the tanshinone biosynthesis pathway, which provides a basis for further study of the tanshinone biosynthesis pathway.

7.2.2.2 Construction of Medicinal Plant Gene Regulatory Networks

In the post-genome era, along with the development of computer technology, massive experimental data provides the basis for the construction of gene regulatory networks and has revealed the interaction between a large number of genes and their products, especially the spatiotemporal mechanism of gene expression. Exploring its implicit biological laws laid the foundation for this hypothesis. In general, the expression of a gene is affected by other genes, which in turn affect the expression of other genes. These interactions affect each other to form a complex network of gene expression regulation. Therefore, almost all cellular activities are regulated by genes and their interaction networks [25].

The analysis of gene regulatory networks is an important process for understanding the function of genes and will also be an important tool for mastering the synthesis and decomposition of active ingredients in medicinal plants. The construction of metabolic pathway databases and metabolic networks through massive genetic data can help to discover unknown enzymes, conduct metabolic pathway evolution studies, and perform in vitro remodelling of metabolic pathways. Although the widely used metabolic pathway database mainly relies on model plant gene data, with the development of medicinal plant transcriptome research, the understanding of the metabolic pathway of medicinal active ingredients will form a more complete network of metabolic pathways. For example, anthraquinone alkaloids are the main active constituents of periwinkle, and vinblastine and vincristine have good antitumor activity. The biosynthesis of indole alkaloids is a complex metabolic network. Moerkercke et al. [26] constructed a detailed metabolic pathway database, CathaCyc (version 1.0), based on RNA-seq data of *Catharanthus roseus*, which contains 390 primary and secondary metabolic pathways involving 1347 synthases. These metabolic pathways include the synthesis of terpene indole alkaloids, triterpenoids, and their precursors, as well as the metabolic pathways related to their inducers, jasmonic acid hormones. This shows that RNA-seq is a useful way to construct a database of metabolic pathways. The online database CathaCyc (http://www.cathacyc.org) can be used to analyse metabolic networks and "omics" data. Using the database, combined with gene expression profile information, two complete *Catharanthus roseus* metabolite pathways were obtained, which can synthesise steroidal alkaloids and triterpenoids, and were significantly regulated by plant growth and development and environmental factors.

Artemisinin has attracted extensive attention for research on its biosynthesis, regulation, and synthetic biology due to its great antimalarial properties. Professor Xiaoya Chen's group from the Shanghai Institute of Plant Physiology and Ecology and Professor Kexuan Tang's group from Shanghai Jiaotong University have performed many studies on the regulation of artemisinin biosynthesis. Through transcriptional sequencing and functional studies, several transcription factors regulating artemisinin biosynthesis have been obtained, including AaERF1, AaERF2, AaWRKY, AaMYC2, and AaEIN3 [27–30], and have greatly promoted research into artemisinin biosynthesis regulation network and become models of active ingredient synthesis regulation in medicinal plants.

7.2.2.3 Study on the Developmental Mechanism of Medicinal Plants

With the deepening of transcriptome research into medicinal plants, individual developmental regulation and regulation mechanisms have gradually become a research hotspot. *De novo* transcriptome sequencing technology has been used to detect the changes in gene expression profiles under external stimulation to reveal the accumulation and its regulation of the active components of medicinal plants.

Glandular trichomes are places where many medicinal plants produce and store active ingredients. The development of glandular hairy trichomes is an important factor in determining the quality of medicinal materials. Professor Kexuan Tang of Shanghai Jiaotong University constructed a transcriptome library by using jasmonic acid to induce *Artemisia annua* and obtained the transcription factor protein AaHD1, of the HD-ZIP IV subfamily. Further analysis of the expression pattern of the *AaHD1* gene showed that it was similar to the initiation pattern of glandular hair development, mainly expressed at the base of glandular hairs in young leaves. Molecular biology and physiological and biochemical analyses showed that the *AaHD1* gene was induced by methyl jasmonate, and *AaJAZ8* combined with *AaHD1* could reduce the activity of the AaHD1 protein. The important function of the *AaHD1* gene in the development of glandular hairs of *A. annua* was further revealed by genetic transformation experiments. In AaHD1-overexpressing *A. annua*, the density of secretory and non-secretory glandular hairs on the leaf surface increased significantly, and the artemisinin content also increased significantly. In the plants inhibited by AaHD1 RNAi, the density of both glandular hairs decreased significantly, and the content of artemisinin also decreased significantly, which indicated that AaHD1 could participate in the regulation of both secretory and non-secretory glandular hairs of *A. annua*. Long-term treatment with methyl jasmonate showed that the percentage of increased glandular hairs in transgenic *A. annua* plants inhibited by AaHD1 RNAi was significantly lower than that in wild-type plants, which indicated that AaHD1 was an important factor involved in the regulation of glandular hair development by the jasmonic acid signalling pathway. This study is the first to reveal the developmental mechanism of jasmonic acid-regulated secretory glandular hair and proposed its regulation model [31]. Qi et al. [32] used 454-GS FLX Titanium to sequence the transcriptome of *Paris polyphylla* embryo and analysed the molecular mechanism of seed development based on gene expression profiles. The results showed that a total of 47,768 unigenes were obtained, including 16,069 contigs and 31,699 singletons. Gene function annotations found that 464 transcripts may be involved in plant hormone metabolism and biosynthesis, hormone signalling, seed dormancy, seed maturation, cell wall growth, and circadian rhythm. Further gene expression analysis indicated that there were 11 phytohormone-related genes, and five other genes during seed differentiation showed different expression profiles between the embryo and endosperm, suggesting that these genes will play an important role in the seed dormancy mechanism of *Paris polyphylla*. With the deepening of research, in addition to studies concentrated on the biosynthesis and regulation of active ingredients in medicinal plants, increasing amounts of studies will focus on the growth and development of medicinal plants and the quality characteristics of medicinal plants at the molecular level.

7.2.2.4 Development of Molecular Markers for Medicinal Plants

The development of SSR molecular marker primers is the premise of SSR molecular marker research. The EST is derived from the transcribed region of the gene and directly reflects information on the gene expression. With the rapid development of high-throughput sequencing technology and the reduction of sequencing costs, a new generation of high-throughput sequencing technology is used to sequence plants across the genome and generate rich transcriptome data, which contains a large number of EST sequences [33]. Using transcriptome sequences to develop SSR markers not only has the advantages of EST-SSR markers but also provides more comprehensive information for the development of SSR markers than EST-SSR markers, thereby improving the accuracy of genetic diversity and molecular marker-assisted breeding research.

Gai et al. [34] used the Roche 454 GS FLX platform to sequence the *de novo* transcriptome of *Paeonia suffruticosa*. A total of 625,342 ESTs with an average length of 358.1 bp were obtained, and 23,652 sequences overlap groups (Contig) and a single sequence (Singleton) were obtained after splicing, of which 15,284 were longer than 300 bp. A total of 2253 SSR loci were obtained from 454-ESTs, 149 of which were selected to design primers. A total of 121 pairs of primers could successfully extend the bands. Among them, 73 pairs of primers showed polymorphism in the PCR products. These sequences provided genetic resources for studying the physiological function of *Paeonia suffruticosa*. Jain et al. [35] obtained more than 80 million sequences of *Hippophae rhamnoides* Linn. and 88,297 unigenes by *ab initio* transcriptome sequencing, of which 7.69% had microsatellite repeats, with an average of one microsatellite per 6.704 kb. Dinucleic acid duplication accounts for the largest proportion and is mainly distributed in the 3′ and 5′ untranslated regions upstream and downstream of the coding region. AG/AG types had the highest frequency of repetition. In addition, 48.81% of unigenes with microsatellite loci were annotated by GO. Randomisation of 25 unigene-specific microsatellites for polymorphism detection of 18 sea buckthorn germplasm materials provided a basis for the further research and breeding of sea buckthorn molecular biology. Wei et al. [36] used *Houttuynia cordata* to obtain 56 million sequences from transcriptome sequencing and assembled 63,954 unigenes, of which 39,982 (62.52%) unigenes were highly homologous to non-redundant protein sequences in the NCBI database. A total of 26,122 (40.84%) unigenes have high homology to known protein sequences in the Swiss-Prot database. In total, 30,131 unigenes have obtained gene classification, and 15,363 unigenes have obtained clusters of adjacent proteins (COG). Moreover, 24,434 (38.21%) unigenes were predicted to involve 128 KEGG pathways, and 17,964 (44.93%) of the unigene sequences were highly homologous to the *Vitis vinifera* gene. In addition, 4800 cDNA SSRs were screened, and 50 SSRs were randomly selected to design specific primers for the evaluation of *Houttuynia* polymorphism. In total, 43 (86%) pairs of primers obtained PCR fragments consistent with the expected size, which will provide support for the construction of high-density genetic linkage maps and gene association analysis of *Houttuynia cordata*. Zeng et al. [37] reported a comprehensive EST database construction and EST-SSR characterisation of the traditional Chinese medicinal plant *Epimedium sagittatum* (Sieb. et Zucc.) Maxim. In this study, cDNAs of *E. sagittatum* were sequenced using 454 GS-FLX pyrosequencing technology and assembled into a total of 76,459 consensus sequences which comprised 17,231 contigs and 59,228 singlets. A total of 2810 EST-SSRs were identified from the EST database. Among them, 32 were randomly selected and tested across 52 *Epimedium* species and showed that 85.7% of them could be successfully transferred to the *Epimedium* species. This will be a powerful resource for further studies in the field, including taxonomy and molecular breeding.

7.3 Proteomics of Medicinal Plants

Proteomics refers to all proteins expressed in a genome, cell, or tissue as the research object [19]. Its purpose is to elucidate the expression and functional patterns of all proteins in organisms, including protein expression, post-translational modification, protein structure, protein–protein interactions, etc. Doing this illustrates the development of diseases and cell metabolism at the protein level. Plant proteomics is originated based on the premise of breakthroughs in genomics and high-throughput protein analysis and identification techniques. The development of plant genomics is an important prerequisite for the emergence of plant proteomics. In recent years, with the rise of whole genome and transcriptome sequencing, plant proteomics has become one of the hotspots in the post-genome era. Proteomics of different tissues and organs of plants can help us understand the mechanisms of plant development. The proteomics of plant organelles is helpful to understand the function and compartmentalisation of organelles during plant metabolism. Proteomics studies of plants responding to biological and abiotic stresses are helpful to better understand the damage mechanism of environmental stress on plants and the ecological adaptation mechanism of plants [20, 38].

Proteomics can be divided into three parts according to different research purposes: expression proteomics, structural proteomics, and functional proteomics [39–41]. Expression proteomics refers to the large-scale identification and analysis of proteins by means of bidirectional electrophoresis and other technologies. It is used for studies of protein expression abundance and modification under different conditions and then to construct expression maps in tissues or cells. It includes the whole protein expression profile of cells or tissues and differentiation analysis of protein expression profiles in different developmental stages and environments. Structural proteomics refers to the development of large-scale protein three-dimensional structures with high flux. Through analysis of the spatial structure of protein molecules, it can more accurately predict the binding sites and structural domain. By doing this, it is possible to predict protein catalytic mechanisms and pathways, protein interactions, and the interaction between protein and nucleic acid. Functional proteomics takes specific protein groups as the research object and analyses and expounds the pathways or modes of action of these proteins.

7.3.1 Research Contents of Medicinal Plant Proteomics

7.3.1.1 Preparation of Protein Samples

Protein sample preparation is the first step in proteomics research. It often employs the whole proteome of tissues or cells as the research object. The total proteome of tissues or cells is usually extracted using the trichloroacetic acid/acetone solution precipitation method. The quality of protein samples directly affects the authenticity and credibility of the experiment. Sample preparation should follow the following principles: The first step is to obtain soluble protein samples and remove nucleic acids and interfering proteins from the samples. The preparation method should be repeatable. Second, it is necessary to prevent protein aggregation, precipitation, and chemical modification during preparation. Finally, high-abundance or unrelated proteins that interfere with the results are removed as far as possible to ensure the detectability of the target protein.

7.3.1.2 Separation of Proteins

Proteins are mainly separated according to their size, isoelectric point, solubility, and specific affinity of ligands. The methods of protein separation include two-dimensional gel electrophoresis, differential gel electrophoresis, capillary electrophoresis (CE), and liquid chromatography. Two-dimensional electrophoresis, the most effective and commonly used method at present, was established by O'Farrell et al. in 1975. The first step is isoelectric focusing (IEF), which separates proteins according to the difference in their isoelectric points (pIs). The second step is sodium dodecyl sulphate-polyacrylamide gel electrophoresis (SDS-PAGE), which separates the protein by using the relative molecular weight difference of the protein (sulphate). Theoretically, each spot in the two-dimensional gel electrophoresis results corresponds to a protein in the sample. It can separate 10,000 proteins in a piece of glue simultaneously and obtain information on the isoelectric point, apparent molecular mass, and relative expression of each protein.

7.3.1.3 Identification and Quantification of Proteins

After the separation of the proteome, the isolated proteins should be identified and quantitatively analysed. The main methods used are Edman degradation, mass spectrometry (MS), isotope-coded affinity tag (ICAT) labelling, tandem affinity purification, protein chip, immunoprecipitation, and yeast hybrid/triple hybrid. Among them, mass spectrometry has become the main tool for protein analysis because of its advantages in accuracy, sensitivity, solubility, and high-throughput screening.

In 1999, Steven P. Gygi of Washington University published a paper in *Nature Biotechnology* [41], introducing a precise quantitative and simultaneous determination of individual proteins in total proteins of *Saccharomyces cerevisiae*, that is ICAT. The technology is based on isotope-coded affinity tag (ICAT) reagents and tandem mass spectrometry. ICAT technology can be divided into three steps. First, two different protein samples were labelled with heavy and light ICAT reagents. ICAT reagents and each cysteine of each protein were covalently bound and then mixed in equal amounts, and trypsin was hydrolysed into peptide segments. Second, biotin labels are used to isolate peptide segments bound to ICAT reagents, usually by affinity chromatography or capillary high-performance liquid chromatography. The identical peptide segments of the same protein labelled by light and heavy reagents have identical chemical characteristics, so they can be coeluted and appear in pairs, with a difference of 8 Da. Therefore, the elution of the final labelled peptide segment was identified by mass spectrometry and quantified according to its expression abundance.

Isobaric tags for relative and absolute quantitation (iTRAQ), a proteomic research method developed by Applied Biosystems Incorporation (ABI) based on ICAT, can simultaneously label 48 kinds of enzymatic protein samples and conduct relative and absolute quantitative proteomics research. Isotope labelling and MS (tandem time-of-flight [TOF/TOF]–MS or quadrupole time-of-flight [QTOF]–MS) are the two major material bases of this technology. The protein samples were blocked by cysteine residues, then hydrolysed by trypsinase, labelled with iTRAQ reagent, mixed with several protein samples in equal amount, purified by strong cation exchange chromatography, and analysed by LC–MS/MS. The peptide segments labelled with different reagents did not differ in their one-dimensional mass spectrometry, but existed in the form of a single peak. In two-dimensional mass spectrometry, after collision ionisation or electrospray ionisation (EI), the carboxyl group, which regulates the balance, is released by neutral particles, resulting in a relatively low-quality report. The difference in the ionic strength of the different reports indicates the relative abundance of the labelled polypeptide (the accuracy of the two-dimensional mass spectrometry is 0.1 Da), which serves as a quantitative basis for protein.

7.3.2 Application of Proteome Research in Medicinal Plants

Protein, as the executor of life function, has been researched a lot in rice, corn, soybean, wheat energy crops, castor, mulberry, and other cash crops, as well as functional bacteria, and many breakthroughs have been made. Plant proteomics research mainly focuses on the following aspects: (1) plant physiology and biochemistry, which is mainly aimed at the basic life activities of plant growth, development, and senescence; (2) tissue or subcellular proteomics, such as membrane proteomics, chloroplast, and mitochondrial proteomics; (3) cell signal transduction; (4) stress proteomics, using differential proteomics to study the effects of various stress signals on organisms as well as to study the multiple regulatory mechanisms; and (5) secondary metabolites, the active substance bases of TCM. Proteomics can be used to study the differences in protein components in different tissues of Chinese medicinal materials under different growth and development stages and soil texture, water supply, and fertiliser conditions and to provide an important theoretical basis for the physiological metabolism and regulation mechanisms.

Normally, the purpose of plant metabolism is to meet the needs of growth and development, such as the synthesis of new cells and cell division. Environmental stress inhibits active growth and affects photosynthesis. Instead of synthesising new cell components, cells synthesise a large number of stress defence components. At the same time, the regulation of transcription, protein, and metabolism levels reached a new growth balance. There are usually several ways for plants to cope with stress. (1) Because the process of stress relief is accompanied by the biosynthesis of stress defence components, stress defence requires high energy, so stress usually upregulates carbon metabolism pathways, especially glycolysis pathways, to ensure an adequate energy supply. (2) One of the characteristics of various stresses is the production of a large number of reactive oxygen species, which induces the upregulation of antioxidant enzymes, especially those involved in the ascorbic acid–glutathione cycle or glutathione S-transferases (GSTs). (3) Low temperature, high heat, drought, salt, and mechanical damage will produce dehydration stress. Water shortage will lead to the accumulation of osmosis-related substances (such as betaine, proline, and sugar) and hydrophilic COR/LEA proteins, which have the function of protecting and binding water. (4) Stress factors, such as extreme temperature, can lead to protein misfolding, resulting in the loss of protein function. Therefore, plants express chaperone proteins, such as heat-shock proteins and binding proteins, in the process of stress defence. (5) Heavy metal stress can induce the upregulation of complex protein expression, especially phytochelatin and glutathione derivatives, in order to enhance the detoxification ability of plants. However, there are significant differences in stress tolerance among different species or even between different genotypes of the same species. Proteomic analysis shows that tolerant plants can more effectively adjust several defence systems, ensure an adequate supply of metabolites (such as photosynthates and ATP), and maintain high productivity to eliminate stress.

Professor Jingkui Tian of Zhejiang University has done a lot of work in the study of medicinal plant proteomics. By using proteomics to assist transcriptome and functional research, the synthesis mechanism of active compounds of medicinal plants was expounded. *Lonicera japonica* is a well-known medicinal plant that is widely used in TCM. In order to explore the molecular mechanism of the production of its active ingredient, a label-free proteomics technique was used in combination with polyethylene glycol fractionation and combinatorial peptide ligand library analysis to remove high-abundance proteins and enrich low-abundance proteins. Proteins related to the oxidative pentose phosphate pathway, signalling, hormone metabolism, and transport were highly enriched. Specifically, 28 proteins related to secondary metabolism were identified, which provide references for biopathway analysis of the active ingredient [42] (R: variations in metabolites and proteome in *Lonicera japonica* Thunb. buds and flowers under UV radiation). *Catharanthus roseus* produces a variety of indole alkaloids of significant pharmaceutical relevance. Zhu et al. used proteomics technique to investigate the potential stress-induced increase of indole alkaloid biosynthesis. It was found that 21 proteins were down regulated and 66 proteins were upregulated upon binary stress. The proteins related to tricarboxylic acid cycle and the cell wall were increased. Specifically, 10-hydroxygeraniol oxidoreductase involved in biosynthesis of indole alkaloids was upregulated. This study illustrated the probable mechanism of increased alkaloid content in *C. roseus* under stress [43].

Biotic and abiotic stresses can inhibit plant growth, resulting in decreased crop productivity. However, moderate adverse stress can promote the accumulation of valuable natural products in medicinal plants. Elucidating the underlying molecular mechanisms thus might help optimise the variety of available plant medicinal materials and improve their quality. In a study by Wang et al., *Salvia miltiorrhiza* hairy root cultures were employed as an in vitro model of the Chinese herb Danshen. A comparative proteomics analysis was performed. By comparing the gel images of groups exposed to the stress of yeast extract (YE) combined with Ag^+ and controls,

64 proteins were identified that showed significant changes in abundance for at least one time point after treatment. It was found that YE and Ag^+ stress induced a burst of reactive oxygen species and activated the Ca^{2+}/calmodulin signalling pathway. The expression of immune-suppressive proteins increased. Epidermal cells underwent programmed cell death. Energy metabolism was enhanced, and carbon metabolism shifted to favour the production of secondary metabolites, such as lignin, tanshinone, and salvianolic acids. The tanshinone and salvianolic acids were deposited on the collapsed epidermal cells, forming a physicochemical barrier. Together, the defence proteins and these natural products enhanced the stress resistance of the plants. This study sheds new light on quality formation mechanisms of medicinal plants and provides reference for studying how the planting environment affects the efficacy of herbal medicines [44].

7.4 Epigenetics of Medicinal Plants

Epigenetics studies the changes in heritable gene expression or cell phenotype by certain mechanisms without altering the DNA sequence [45]. The concept of "epigenetics" was proposed by developmental biologist Conrad Waddington in the 1940s and was originally used to describe the interactions between genes and the interaction of genes with the environment, which was further defined as changes in biological phenotype, morphology, or molecular hierarchy without changing the coding sequence, upstream of the gene, or the context of a promoter region [46]. Epigenetics changes phenotypic traits through DNA methylation, histone modifications, non-coding RNA changes, and loss of imprinted genes. Epigenetic research is a hotspot in model plant research, but in view of the long culture period and immature genetic transformation system, it has been rarely researched and applied in medicinal plants.

7.4.1 Research Contents of Medicinal Plant Epigenetics

7.4.1.1 DNA Methylation

The most well-studied epigenetic mechanism is DNA methylation, in which the 5′ carbon position of cytosine on the CpG dinucleotide in genomic DNA is covalently bonded to a methyl group under the action of DNA methyltransferase [47]. DNA methylation is also the most prevalent epigenetic modification found in eubacteria, archaea, and all eukaryotes. The establishment and maintenance of DNA methylation are achieved by DNA methyltransferases; the sequences are highly similar in bacteria, plants, and animals, indicating their common origin [48].

In plants, DNA methylation has two main roles: To protect the genome against the intrusion of the transposon and to participate in the regulation of gene expression [49]. In some advanced plants, such as corn, transposons are abundant in the genome. Since the transposition effect is harmful to most organisms, methylation of the transposon can inhibit its transposition activity, thereby protecting the organism from the threat of transposition [50]. In addition, DNA methylation results in transcriptional silencing in the promoter region, thereby regulating the expression of related genes, which is also closely related to histone modification and chromatin condensation in related regions [51].

7.4.1.2 Histone Modification

Chromatin is composed of histones and DNA, and the covalent modification of histones is another important mechanism regulating gene expression. Histones are a class of highly conserved proteins whose *N*-terminus is located on the surface of nucleosome octamers. In eukaryotes, histones provide the major packaging structure for chromosomal DNA, and each histone protein coats a DNA fragment of approximately 146 bp in length to form nucleosomes. Each nucleosome contains two copies of four different histones: H2A, H2B, H3, and H4. These histone amino acid residues are subject to extensive covalent modification. Covalent histone modifications include methylation, acetylation, phosphorylation, and ubiquitination. The coding of specific histone modifications can determine the corresponding chromatin structure, thereby affecting gene transcription, DNA repair and replication, and chromosome rearrangement physiological processes [52, 53].

The most mature studies in histone modification are based on the acetylation of the *N*-terminus of histones, that is, the acetylation of the ε-amino group in the evolutionarily conserved lysine residue at the *N*-terminus. It is generally believed that the acetylation of histones can reduce the surface charge of positive histones, facilitate the dissociation of DNA and histone octamers, and thus relax the nucleosome structure, allowing various transcription factors and synergistic transcription factors to bind to DNA. Site-specific binding activates the transcription of genes; however, studies have reported the opposite. Experiments have shown that overexpression of deacetylase (reduced acetylation status) in the model plant *Arabidopsis* can lead to the overexpression of certain pathway genes, while the gene expression of other pathways is inhibited. Moreover, in *Arabidopsis*, the acetylation state can act as a regulatory element at different developmental stages [54]. Other epigenetic modifications with epigenetic effects include methylation, phosphorylation, and ubiquitination, but the mechanisms are still unclear [53, 55].

7.4.1.3 Non-coding RNA

Analysis of human genome sequencing results showed that two-thirds of the more than three billion base pair genes were reverse transcribed, but in the end, less than 2% of the sequences were used to encode proteins. In addition to rRNA and tRNA involved in mRNA translation, these non-coding RNAs include narrow non-coding RNAs: miRNA, small interfering RNA (siRNA), and long non-coding RNA (lncRNA). These large amounts of non-coding RNA are thought to play an important regulatory role in the process of biological growth and development and have great research prospects.

miRNA molecules were first discovered in *Caenorhabditis elegans* larvae in 1993 [56]. They are about 21–25 nucleotides in length and are derived from endogenous hairpin RNA, often targeting genes that are similar but not identical to their sequences, acting as transcriptional repressors at the transcriptional and post-transcriptional levels. The horizontal negative regulation of gene expression is widely involved in plant growth and development, stress response, and signal transduction [41, 57]. In plants, some miRNA molecules control the polarisation of the front/back of the leaf and the individual development of the flower, regulating heterochronic shifts and other developmental pathways [58].

Small interfering (siRNA) is a double-stranded RNA of 20–25 nucleotides in length, obtained by cleavage of longer double-stranded RNA [59]. The siRNA can form a perfectly paired double strand with the homologous target gene RNA, which mediates the Argonaute protein to cleave the target gene RNA to degrade, thereby silencing the target gene expression. siRNA is mainly involved in post-transcriptional regulation, transposon silencing, antiviral activity, chromatin remodelling, and other biological processes [60].

lncRNAs are RNA molecules 200–100,000 bases in length. The promoters can also bind to transcription factors. Local chromatin histones also have characteristic modification and structural features. The conservatism is less than 10% in different species, and the expression abundance is low, showing strong specificity in tissues and cells. Our understanding of lncRNA is still in its infancy. Recent studies have shown that lncRNA is involved in the regulation of plant growth, spring flowering, and other growth and development processes [38].

7.4.2 Research Methods of Epigenetics in Medicinal Plants

With the development of modern molecular biology technology, recently developed tools can be combined with traditional methods to study the complex interactions between genotypes, epigenetic modifications, and the environment and to test their respective contributions to phenotypic differences and species evolution [61]. A set of mature molecular biological research methods have been developed for genotype and epigenetic modification. However, in order to clarify the role of epigenetic modification in plant phenotypic differentiation, these molecular biological methods need to be combined with field investigations.

7.4.2.1 Study on Phenotypic Differences

As far as traditional plant taxonomy is concerned, phenotypic differences arise from studies on the external morphology and internal anatomical characteristics of plants.

The first problem to be solved is to determine whether the phenotypic differences of the selected research objectives are continuous or discrete (discontinuous). Discontinuous phenotypic differences are often caused by relatively independent reasons. Phenotypic differences with discontinuous changes can be selected for subsequent research. By comparing and analysing various morphological, anatomical, and physiological traits, we can determine whether a well-defined and distinct group has been formed or whether extreme phenotypes are closely linked to the individual's traits in the group.

As far as medicinal plants are concerned, the phenotypic differences are mainly focused on the analysis of their effective chemical constituents, which have clear quantitative characteristics. For a single chemical component, it can be easily distinguished by setting a threshold. For many chemical constituents, phenotypic differences can be determined by fingerprint similarity analysis, principal component analysis, and cluster analysis.

In addition, we need to measure and observe environmental characteristics (such as the microclimate, geology, soil, and biological interactions) to determine the limiting environmental factors in the field operations and study the phenotypic differences and combine these environmental factors with the characteristics of plant anatomy, morphology, physiology, and chemical composition.

7.4.2.2 Research Methods of Epigenetics

Epigenetic differences are mainly focused on DNA methylation, histone covalent modification, and the expression of miRNAs.

1. Detection of genomic DNA methylation

The core goal of methylation detection technology is to distinguish methylated and unmethylated cytosine in genomic DNA. The methods can be divided into four categories:

(a) Methylation-sensitive amplification polymorphism (MSAP): Based on restriction endonuclease digestion, an amplified fragment length polymorphism (AFLP) technique was developed. One or more enzymes were used to restrict the cleavage of unmethylated DNA, followed by two rounds of PCR amplification. The methylation status of genomic CpG loci could be obtained by counting and analysing the amplified bands. This method, combined with chip and capillary sequencing [62], has been used to detect genome-wide methylation in a variety of organisms, but it is limited to the CpG loci recognised by endonucleases.
(b) Bisulphite sequencing PCR: This method relies on the bisulphite transformation of genomic DNA. After bisulphite treatment, the unmethylated cytosine (C) of genomic DNA is converted to uracil (U), and the methylated C remains unchanged [63]. Subsequently, methylated and unmethylated cytosine can be distinguished by sequencing. It is worth mentioning that the emergence of a new

generation of sequencing technology makes whole genome sequencing single-base analysis possible. The methylation profiles of *Arabidopsis* genomic DNA (about 120 Mb) were analysed by bisulphite combined with a new generation of sequencing technology [64].

(c) Immunologically based methylated DNA immunoprecipitation (MeDIP or mDIP) method: Genome methylation or unmethylated fragments were enriched by immunoprecipitation using 5-methylcytosine-specific antibodies or proteins containing methyl-binding domains to identify [65].

(d) High-resolution melting (HRM): Based on the 'melting' characteristics of DNA in solution, DNA templates with different methylated cytosine contents were identified by melting point analysis at different melting temperatures [66].

2. Detection of histone covalent modification

Histone modification includes acetylation, methylation, phosphorylation, and ubiquitination. There are few methods for histone modification. At present, the most commonly used technique is chromatin immunoprecipitation (ChIP). ChIP was proposed by O'Neill and Turner [67] as a technique for studying the interaction between proteins and DNA *in vivo*. Its basic principle is to fix the DNA–protein complex in the nucleus of living cells, cut the chromatin into small chromatin fragments within a certain length by ultrasound or enzyme treatment, and finally precipitate the complex by immunological methods to dissociate the DNA fragment from the protein. By purifying and detecting the target DNA fragments, the information on protein–DNA interactions can be obtained. In practical applications, ChIP can be combined with biochip [68] or sequencing technology [69], and DNA binding sites or histone modifications can be analysed with high-throughput screening in the whole genome or large genome regions.

3. Detection of miRNAs

Because mature miRNAs are only 21–25 bases in length, they are not expressed in cells. These characteristics make the detection of miRNAs completely different from the traditional detection of RNA. At present, there are three main methods for the detection of miRNAs: northern blot, real-time fluorescent quantitative PCR, and DNA chip.

(a) Northern blot is a classical semi-quantitative miRNA detection method. It does not require much material for experimental instruments. Its basic principle is to separate the total RNA samples by gel electrophoresis, transfer the miRNA part to the special membrane by electrospinning, and then use specific probes to hybridise and detect them.

(b) Quantitative real-time fluorescent PCR (RT-PCR) is a method of quantitative and qualitative analysis of the unknown initial content template by adding fluorescent groups into the PCR reaction system and detecting the fluorescent signals generated in each cycle of the PCR amplification reaction in real time. Quantitative RT-PCR is very sensitive to RNA detection and can achieve the detection ability of single copy. However, miRNAs are very small, and their

21–25 nucleotide lengths are comparable to those of primers, an issue which needs to be solved using a special primer-plus-tail [70] or stem-ring method [71].

(c) The DNA chip method is used to fix antisense DNA probes on the chip and use hybridisation to detect fluorescent-labelled miRNAs [72]. DNA chip can integrate thousands of densely arranged molecular arrays, which can analyse samples quickly with high-throughput screening. The efficiency is hundreds of times higher than that of the traditional northern blot.

4. Verification of epigenetic modification

If analysis at the molecular level suggests that plant phenotypic differences are not derived from differences in DNA sequences, but are related to epigenetic modification, then it is necessary to verify whether these phenotypic differences can be steadily inherited. Only when these phenotypic differences can be steadily inherited can it be confirmed that epigenetic modification plays a decisive role in plant evolution and germplasm generation, rather than simply phenotypic plasticity (fluctuation of gene expression under environmental effects). To verify whether it is possible to stably inherit phenotypic differences can be completed through mutual transplantation experiments; plants with phenotypic differences should be reproduced in the same growth environment for at least three to five generations to observe whether the phenotypic differences can be retained [61]. If it cannot be retained, the original phenotypic differences are likely to be caused by phenotypic plasticity.

7.4.3 Application of Epigenetics in the Study of Medicinal Plants

Epigenetics is involved in regulating plant disease resistance, stress resistance, growth, and morphogenesis and plays an important role in the biosynthesis of active ingredients in medicinal plants.

7.4.3.1 Epigenetics Regulates the Biosynthesis of Active Components in Medicinal Plants

The active components of medicinal plants are mainly secondary metabolites. At present, most of the studies focus on the analysis of secondary metabolic pathways. With the analysis of pathways, the excavation of functional genes, DNA methylation, and non-coding RNA regulation of secondary metabolic pathways have also received widespread attention. Researchers used 5-azacytidine to demethylate *Dendrobium* tissue-cultured seedlings and analysed the changes in growth and expression of active components and related genes. The results showed that the content of polysaccharides and alkaloids in *Dendrobium* seedlings treated with methylation significantly increased. It is speculated that DNA demethylation can regulate the secondary metabolic pathway of *Dendrobium*, but there is no further experimental test [73]. Through the transcriptome sequencing and functional annotation of ginseng, 14 probable miRNAs were found, and their targets were predicted to be transcription factors and transporters related to ginseng growth, development, and defence [74]. Vashisht et al. sequenced and analysed the transcriptome of *Coptis chinensis*, identified 18 miRNAs, and predicted that the target gene of micro4995 was the enzyme involved in cinnamic acid synthesis in the phenylpropanoid pathway. They predicted that cinnamic acid was one of the necessary substrates for the synthesis of active components of *Coptis chinensis*. It was speculated that the miRNAs were involved in the regulation of biosynthesis [75].

7.4.3.2 Epigenetic Involvement in Regulating the Growth and Development of Medicinal Plants

Studies have shown that epigenetic modification is involved in regulating plant morphogenesis, growth, and development by regulating plant hormone metabolism, transport, and signal transduction. It has been reported that epigenetics regulate flower development, fruit development and maturation, and heterosis. Researchers studied the genetic and epigenetic modification of *Salvia miltiorrhiza* populations in different regions. It was found that there were high similarities in genetic background and epigenetic diversity among individuals in *Salvia miltiorrhiza* populations. The DNA methylation patterns between cultivated and wild *S. miltiorrhiza* were not completely consistent [76]. This study provided a basis for the evaluation and protection of *Salvia miltiorrhiza* germplasm resources, introduction and domestication, and breeding of excellent varieties. Ginseng has been widely planted in northeast China. The differences between cultivated and wild ginseng and their molecular mechanisms have always been a research hotspot. By comparing the genetic information on wild and cultivated ginseng, researchers have found that compared to wild ginseng, cultivated ginseng has a lower degree of cytosine methylation. Many studies have shown that the degree of cytosine methylation is related to plant development and differentiation, such as morphological modification, variation, flowering time, stress response, etc. [77]. The lower cytosine methylation and different methylation forms to those of wild ginseng were associated with phenotypic changes in cultivated ginseng [78].

7.4.3.3 Epigenetic Regulation of Stress Response in Medicinal Plants

Plants can adapt to the environment through gene recombination or mutation in their long-term evolution, but these changes are slow and irreversible. Epigenetic modifications, such as DNA methylation, play an important role in the face of sudden environmental changes. Most epigenetic modifications will be erased after the stress disappears, but some of them are stable and can be inherited to offspring, so that offspring can acquire stress resistance. In model plants, such as *Arabidopsis thaliana* and rice, epigenetic regulation of stress-response-associated gene expression to adapt to environmental changes has been reported. The *Arabidopsis thaliana* *AtHKT1* gene encodes a sodium ion transporter, which plays an important role in salt tolerance. The DNA methylation level in the promoter region of the *AtHKT1* gene regulates its expression, and changes in the expression of the *AtHKT1* gene become sensitive to salt stress [79]. miRNAs play an important role in regulating plant responses to abiotic stresses. miRNAs often affect methylation modification through the RNA-mediated DNA methylation pathway, which in turn affects the stress response. Huang et al. found that SIAGO4A, as an important factor in the RNA-mediated DNA methylation pathway, participated in the response of tomato to drought and high salt stress [80]. Although epigenetics has been extensively studied in model plants and important crops, few studies have been reported on medical plants because of their complex genetic backgrounds, long growth cycles, and immature genetic transformation systems. It is believed that with the development of technology, more epigenetic studies will be applied to regulating the synthesis of medicinal active ingredients and the genuine research of TCM in the future.

7.5 Metabolomics of Medicinal Plants

Over the past two decades, extensive sequencing work has been carried out on a large number of organisms. At the same time, analytical techniques used to analyse different cell products have also made great strides. These techniques, known as "omics", have become an important tool for studying biological systems and their response to environmental stimuli or disturbances. Among them, the rapid development of metabolomics usually uses complex analytical techniques in parallel with statistically and multivariate analysis methods to quantitatively analyse all metabolites in the organism, to find the relative relationship between metabolites and physiological and pathological changes, and then to extract data for interpretation. The research objects are mainly small molecular substances with relative molecular weights of less than 1000. Metabolomics plays an important role in disease diagnosis, drug development, food application, and biology through the comprehensive analysis of metabolites.

Medicinal plant metabolism is a relatively narrow concept. It mainly studies changes in external factors, such as climate change, nutrient stress, and environmental stress, especially the effects of metabolite changes. As one of the most powerful tools in modern Chinese medicine research, metabolomics has theoretical significance and application value for the analysis of biosynthetic pathways of medicinal active ingredients, drug efficacy and safety, Chinese medicine resources, and quality control.

7.5.1 Research Contents of Medicinal Plant Metabolomics

Metabolites are generally thought to provide the most direct reading of the physiological state of a biological system. Mass spectrometry uses different chromatographic separation techniques, such as liquid–gas phase or nuclear magnetic phase, to play an important role as the primary tool for the simultaneous determination of large amounts of metabolites. Despite the increasing sophistication and sensitivity of analytical techniques, there is currently no single technology that can solve all the problems in metabolomics research due to the diversity of compounds and the vast differences in their content. In response to the above problems, a large number of complementary combinations have been established to extract, detect, quantify, and identify metabolites as much as possible [81].

Metabolomic studies are usually divided into four levels (see Table 7.1): (1) target analysis, (2) metabolic profiling, (3) metabolomics, and (4) metabolic fingerprinting [82]. Among them, target analysis has been widely used, for example, to measure and quantify a small group of known metabolites and to analyse large or small groups of known or unknown compounds. Metabolomic analysis quantifies compounds as much as possible using a combination of techniques, such as liquid chromatography–tandem mass spectrometry (LC–MS/MS), gas chromatography–mass spectrometry (GC–MS), and nuclear magnetic resonance (NMR). Metabolic fingerprinting is distinguished by giving a metabolic "signature" or mass spectrometric feature of the target sample and comparing it to a large number of samples. When a significant difference between the samples is captured, the biological correlation of the compound can be explained, and the time spent on the analysis can be greatly reduced.

The basic steps of metabonomics research include sample collection, preparation, detection, and analysis and data conversion and data analysis. Good metabonomics

Table 7.1 Metabolomics classification [83]

Classification	Definition
Target compound analysis	Quantitative study of proprietary metabolites
Metabolic profiling	Qualitative or quantitative studies of a group of related compounds or proprietary metabolic pathways
Metabolomics	Qualitative or quantitative study of all metabolites
Metabolic fingerprint	Classify samples by rapid, holistic analysis

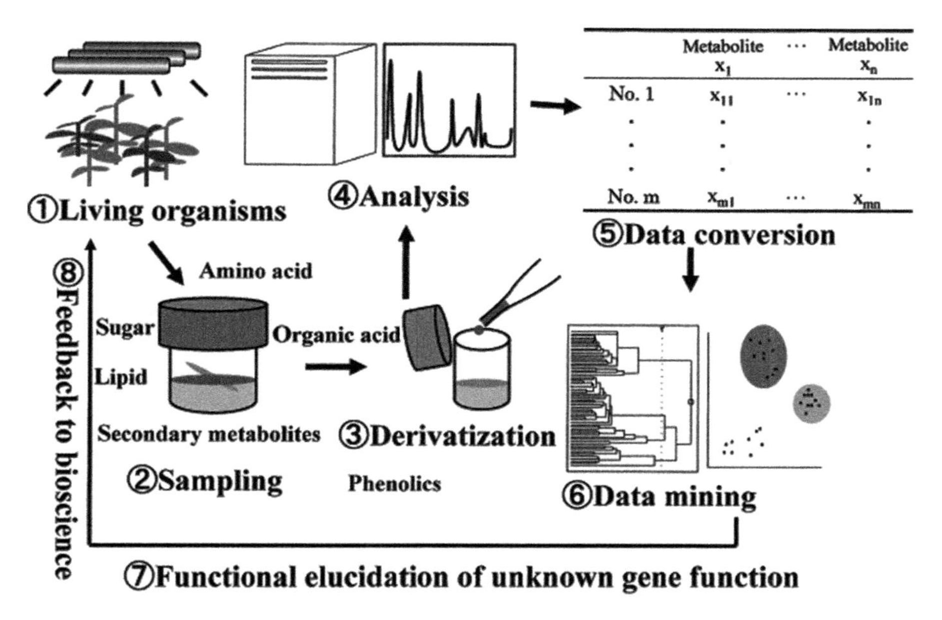

Fig. 7.1 Metabolomics research system [83]

analysis should keep all experimental conditions consistent and minimise experimental errors as much as possible (see Fig. 7.1). The following technical issues in each step are elaborated in detail.

7.5.1.1 Sample Collection

Sample collection needs to take into account the time, region, location, and growth stage of medicinal plants to reduce the impact of sample differences on metabolomics analysis results. After sample collection, it is necessary to inactivate medicinal plants, such as freezing in liquid nitrogen, in order to reduce redox reaction and the degradation of various hydrolytic enzymes. Strict experimental design and sample collection are the first steps to the success of metabolomics experiments. Sample acquisition requirements are as follows: (1) control the basically identical plant growth environmental conditions (if each experiment cannot be completed under completely identical conditions), and ensure that the growth environmental conditions of different treatments or materials in the same experiment are identical; and (2) set up experimental repetitions, generally four to six times, which will further eliminate the errors of environment and experimental operation and obtain statistical data.

7.5.1.2 Sample Preparation

When the external environment changes, the small molecule metabolites in plants also change rapidly. Therefore, in the extraction process, it is necessary to ensure that metabolites are not affected by physical or chemical substances as much as possible. In addition, samples should be preprocessed according to the characteristics of target metabolites and selected research methods, such as the number and content of metabolites to be examined, and appropriate analytical methods and equipment should be selected. For target analysis, reasonable internal standard compounds and purification methods are particularly important; for metabolic spectrum analysis, it is necessary to take into account the characteristics of each metabolite, such as water solubility, plasmin–nucleus ratio, and so on, to cover the scope of metabolites as much as possible. It is also very important to repeatedly explore the extraction and separation efficiency through preliminary experiments. In order to maintain a high extraction efficiency, plant samples need to be uniformly pulverised. Therefore, the advantages of a ball mill are more obvious than that of an agitator when a sample has more rigid tissues.

At present, most of the analytical objects of high-throughput metabolomics are hydrophilic small molecules. In addition, since most primary secondary metabolites are not volatile, it is necessary to derivatise the samples after extraction, using methods such as esterification, acylation, ionisation, and alkylation, in order to obtain more samples for analysis. The stability of reagents, reaction conditions, and derivatives needed to be assessed correspondingly, and the specificity and efficiency of derivation should also be verified.

7.5.1.3 Detection and Analysis

The quality of metabolomics data is usually determined by resolution and quantity. Quantitative analysis of metabolites by chromatography or electrophoresis coupled with mass spectrometry is a common way to obtain metabolomics data. In the process of data acquisition, separation may have a greater impact on both resolution and quantification. Ideally, obtaining high-resolution and accurate quantitative data is very important for metabolomics research, but in the actual operation process, the cost of optimising the system is very high. Therefore, for different research purposes, specific considerations are needed for resolution and quantification.

Separation techniques of metabolites include GC, LC, and CE. Detection and identification techniques include IR spectroscopy, UV spectroscopy, NMR, and MS. The combination of separation and identification techniques constitutes a common separation and identification method in high-throughput metabolomics, such as GC–MS, HPLC–MS, and CE–MS. These techniques have their own characteristics; for example, GC–MS has more comprehensive atlas data and is easier for qualitative analysis; LC–MS is efficient, is fast, and has unique advantages in the analysis of compounds with high polarity and poor thermal stability; CE–MS is more convenient for the separation and identification of charged metabolites. In the separation and identification of compounds, it is often necessary to determine suitable separation and quantitative techniques according to the properties of the sample and selectivity and sensitivity of the instruments used.

7.5.1.4 Data Conversion

In metabolomics, multivariate analysis is needed to elucidate complex linear and nonlinear relationships. The original data obtained from chromatographic or electrochemical analysis need to be converted into matrix tables for multivariate analysis through peak recognition and integration. Data from different sources are usually recorded in their proprietary format, so all matrix tables need to be adjusted to the same data format for analysis. Spectral data and data obtained from GC–MS or LC–MS generally contain a certain degree of perturbation and thus need to be compared and preprocessed. Data preprocessing is also an extremely necessary step for data mining in the later stage. Preprocessing generally includes the following aspects: (1) data denoising, (2) baseline correction, (3) resolution optimisation, (4) data normalisation, and other processing methods, such as peak separation, are also commonly used [84].

7.5.1.5 Data Analysis

Because metabolomics information obtained by MS and NMR has a large sample size, complex data information, and high correlation among variables in a multidimensional data matrix, it is often impossible to extract data information using traditional single-variable analysis methods. Therefore, pattern recognition based on multivariate data analysis is the most commonly used data analysis method in metabolomics. Multivariate analysis methods generally include principal component analysis, hierarchical cluster analysis, factor analysis, and canonical analysis. The above analysis is mainly used to characterise the data structure and the significant trends contained in the data. Principal component analysis (PCA), hierarchical clustering, and self-organising map (SOMs) are the most commonly used multivariate analysis methods.

1. Principal component analysis

PCA is the analysis of data transformed from the original data axis to the main axis. Principal component analysis is very effective in reducing the dimensions of large data sets, and it is also helpful to find valid signals from noise data. In most cases, data obtained from GC–MS, LC–MS, or CE–MS need to be processed for PCA analysis. The target metabolite is used as an independent variable, and the corresponding content of the metabolite is used as a dependent variable. In addition to using principal component identification to find useful information, PCA analysis of metabolic data sets can provide two sets of data, scores and loads. Among them, the load is used to evaluate the contribution of each metabolite to the overall information of the metabolome, which is useful for understanding the level difference of each metabolite between samples [83].

2. Hierarchical cluster analysis

Hierarchical cluster analysis (HCA) is a type of cluster analysis method based on multivariate distances for each pair of data points. With HCA analysis, data is usually not split in one step. Instead, a single cluster containing all targets is split through a series of splits until a better grouping of the data is completed. In general, the HCA method is accurate for the analysis of data sets with a large number of data points, but it is very time-consuming. Now, researchers are more inclined to use K-means clustering (KMC) or batch-learning self-organising analysis (batch-learning self-organising map, BL-SOM) [83].

3. Self-organising analysis

Self-organising maps (SOM) are a non-cluster exploration data analysis method. SOMs can be used in metabolomics in addition to other omics studies, such as genomes and transcriptomes [85]. The original SOM algorithm was very time-consuming, and the improved BL-SOM significantly improved the efficiency, especially the need for computing power; BL-SOM can be performed on a normal computer. Due to the shortcomings of PCA for nonlinear/discontinuous data structure analysis and the fact that HCA may give error information, SOM or KMC should be analysed for the change in metabolite content, but SOM cannot provide information to distinguish clusters. Therefore, special attention should be paid to the actual use of SOM for analysis [83].

7.5.2 Application of Metabolomics in the Study of Medicinal Plants

The identification of TCM ingredients plays a crucial role in ensuring the safety and efficacy of TCM in clinical use. The quality and content of active ingredients in medicinal plants are also significantly different depending on the species, medicinal site, production area, and cultivation period. In this section, we describe the application of metabolomics in the aspects of drug species, medicinal sites, cultivation and harvesting, and metabolome and evolution.

7.5.2.1 Identification of Medicinal Plant Species

Different medicinal plant species may contain the same components, but the content may vary widely, which in turn affects the therapeutic effect. LX Duan et al. [86] studied two *Astragalus membranaceus* varieties, *Astragalus membranaceus* (Fisch.)

Bge. var. *mongholicus* (Bge.) Hsiao and *Astragalus membranaceus* (Fisch.) Bge., which are highly similar in their morphologies, chemical compositions, and DNA sequences by metabolomics. The above two plants were successfully distinguished by AFLP and GC–TOF/MS fingerprinting. Differences in water-soluble sugar, fatty acid, proline, and polyamine contents reflect the extent to which plants adapt to different growing environments. After multivariate and univariate analysis, three AFLPs and eight metabolites were found to be used as AFLP gene fragments and metabolic markers, respectively, to distinguish the above two species. Xie G et al. [87] used 25 ultra-performance liquid chromatography (UPLC)-QTOFMS and multivariate analysis techniques for the metabolic profiling of five different drugs, *Panax ginseng* (Chinese ginseng), *Panax notoginseng* (Sanchi), *Panax japonicus* (Rhizoma Panacis Majoris), *Panax quinquefolius* L. (American ginseng), and *P. ginseng* (Korean ginseng). PCA analysis indicated that these five different *Panax* species can be divided into five different phytochemical groups. Chemical markers, such as ginsenosides Rf, Rb1, and Rb2, explained the differences between different species, which were confirmed by TOF–MS and reference compounds. For the two often confused Chinese herbal medicines, *Epimedium wushanense* and *Epimedium koreanum*, Wang LB et al. [88] used pattern recognition-assisted fingerprints to analyse their secondary metabolites. PCA and HCA analysis of the HPLC chromatographic data revealed that each of the two major clusters formed contained one species. They grouped the entire data into training and test sets. Supervised pattern recognition, soft independent modelling by class analogy (SIMCA), and a backpropagation artificial neural network (BP-ANN) were used for analysis. Although SIMCA failed to identify one of the samples, the BP-ANN accurately predicted the entire test set. Therefore, fingerprint recognition analysis assisted by pattern recognition technology can play a great role in distinguishing and identifying drug species. Sun H et al. [89] designed UPLC–QTOF–HDMS combined with ultra-definition mass spectrometry to study the composition of the roots of two *Aconitum* plants. From the data obtained, it was found that *Aconitum carmichaelii* Debx (CHW) and *Aconitum carmichaelii* Debx (SFZ) had 32 metabolites, while *Aconitum kusnezoffii* Reichb (CW) and CHW had 13 metabolites. Among them, the Junggar aconitine, carbaryl, and iso-talazine found in SFZ and CHW did not exist in CW. *Artemisia annua* has long been used as an antimalarial drug in China, and *Artemisia afra* has played a similar role in southern Africa. While measuring the antimalarial activity of *Artemisia afra*, Liu NQ et al. [90] also studied the metabolic differences between the two species by multivariate analysis. The hydrophilic metabolites in the crude extract were identified by one-dimensional and two-dimensional NMR spectroscopy. Twenty-four semi-polar components were found in *Artemisia afra*, including three new phenylpropanoids, caffeic acid, and quinic acid. The PCA analysis score sheet clearly shows the difference in chemical composition between the two. Among them, the principal component 1 (PC1) of the *X*-axis and the principal component 2 (PC2) of the *Y*-axis explain the reason for the differences of 41.8% and 18.0%, respectively. In the quality control of TCM, NIR spectroscopy successfully identified *Eleutherococcus senticosus*. The spectral data was processed by different pattern recognition techniques, and the classification success rates obtained by SIMCA and PLS-DA techniques were as high as 84% and 92%, respectively. The determination of the mixture prepared in the laboratory revealed that depending on the degree of proximity to Araliaceae, if about 5% of the foreign substance was found, it was identified as a defective product [91]. The use of HPLC fingerprints to distinguish between Chinese herbal medicines has also been successfully applied [92]. Analysis of 13 plant species (Japanese Angelicae, JA; Szechwan Lovage Rhizome, SL; and Cnidium Rhizome, CR) similar to Angelica (CA) found that the content of ligustrazine A content in CA was much lower than that in SL and CR, which can be used as a chemical marker to distinguish these species. In addition, JA can be distinguished from CA, SL, and CR based on chromatographic behaviour and/or the coniferate content. In addition to the difference in the content of certain substances, the chromatographic behaviour of SL and CR is very close, and in chemical taxonomy, the two are highly correlated. A similar study was conducted by Tistaert C et al. [93]. Radix Bupleui is a commonly used TCM collected from *Bupleurum* (*Bupleurum chinense* and *B. scorzonerifolium*). Tian RT et al. [94] analysed the quality of multiple batches of *Bupleurum chinense* samples and commercially available samples by HPLC–evaporative light-scattering detector (ELSD) and high-performance thin layer chromatography (HPTLC) analysis of the main active ingredient, saikosaponin. Preprocessed data is processed by chemometrics (e.g. K-nearest neighbour [KNN] and artificial neural network [ANN]) for similarity and pattern recognition analysis. KNN improves performance for processing HPTLC fingerprints, especially for those other algorithms that are difficult to handle. These two chromatographic fingerprinting methods show a good complementarity to the quality control of medicinal materials. Experiments have shown that roots obtained from different species in the genus *Bupleurum* can be classified. At the same time, the study also found that *Bupleurum chinense* obtained from the main medicinal materials distribution centres in the country usually belong to *Bupleurum chinense*. The content of its main active ingredient, saikosaponin, varies greatly. On this basis, Qin XM et al. [95] used ^{1}H NMR combined with multivariate analysis of 67 *Bupleurum* samples to distinguish the plant sources used and based on the metabolic spectrum on the quality of the production and cultivation methods. It was found that the contents of arginine, citric acid, sucrose, saikosaponin b1 and b2, volatile oil, and fatty acid in *B. scoreonerifolium* were higher than those in *B. chinense*, while the contents of saikosaponins a, c, and d were lower. By comparing the content of saikosaponin, it was found that the high content of samples collected from Shaanxi Province showed the effect of production area on the quality of *Bupleurum*. In addition, no significant changes were found between wild-type and artificially cultivated-type plants. The ^{1}H NMR method can be used to detect the saikosaponin and distinguish the skeletons of two kinds of saikosaponins, which is suitable for species differentiation and quality control in *Bupleurum*.

In addition, other metabolomics techniques are also of great significance for the quality control of Chinese herbal medicines. Direct analysis in real time–mass spectrometry (DART–MS) can generate a new mass spectrometer ion source. In the study of liquorice, the m/z 339 peak was derived from a species-specific compound, liquorice chalcone A $[M + H]^+$. Liquorice can be distinguished from other species by the detection of liquorice chalcone A [96]. GC–MS combined with similarity analysis (SA) and PCA can distinguish *Scutellaria barbata* D. Don and counterfeit [97]. By applying SA and HCA, *Aconitum kusnezoffii* can be separated from the dopants, and HCA and chemical taxonomic analysis are associated with good reliability [98]. *Curcuma* spp., such as *Curcuma wenyujin* Y. H. Chen et C. Ling and Curcuma longa L., can also be distinguished by HPLC–diode array detection (DAD)–MS [99].

7.5.2.2 Identification of Different Medicinal Parts of Medicinal Plants

The content of active ingredients in different parts of medicinal plants usually differs significantly, so the use of unused medicinal parts will eventually affect the therapeutic effect of the drug. For a long time, different medicinal parts have been used as different drugs in the process of using medicinal plants. For different medicinal parts of *Panax notoginseng*, Dan M et al. [100] used UPLC-ESI-MS for metabolic profiling. PCA analysis of UPLC-ESI-MS data revealed that the components between the flower buds, roots, and rhizomes of *Panax notoginseng* can be clearly distinguished. The saponins causing this change were found by the corresponding loading weight; the retention time on the UPLC-QTOF–MS was confirmed by the tandem recipe and saponin standard and further explained the factors that cause different metabolic phenotypes of *Panax notoginseng*. They believe that UPLC–ESI–MS/MS combined with multivariate analysis can not only assess the chemical composition of different parts of plants but also have a certain value for explaining the mechanism of intrinsic phytochemical diversity. In addition, Jurišić Grubešić R et al. [101] determined the total polyphenols and tannic acid content in different parts of *Papaver* plants grown in Croatia, such as the leaves, stems, and flowers, and found that the total polyphenol content was different. The difference in the parts is more obvious. Among them, the total polyphenol content in the leaves was as high as 10.5%, while those of the stems and flowers were only 4.34% and 5.56%, respectively. The content of tannic acid in the stem is between 0.28% and 1.00%, and in leaves and flowers, it is 2.26% and 2.21%, respectively. The metabolomics study of mulberry leaves using ^{1}H NMR technique also received good results [102].

7.5.2.3 Classification of Different Medicinal Plants

The environment in which medicinal plants are cultivated, such as temperature, humidity, soil, and climate, plays a decisive role in plant growth. Kong WJ et al. [103] established a simple, sensitive, and reliable quantitative and chemical fingerprinting method for the quality control and analysis of Rhizoma coptidis. Five active alkaloids were quantitatively analysed by ultraperformance liquid chromatography with photodiode array detector (UPLC-PAD). There was a good regression ($R > 0.9992$) and recovery (98.4%–100.8%) in the assay range, and the detection and quantitation limits were 0.07 and 0.22 μg/mL, respectively. According to different production areas, the UPLC fingerprints of *Coptis chinensis* were compared using SA, PCA, and HCA. The results showed that the above chemometric studies were successfully used to classify the samples of *Coptis chinensis* by source. In addition, the study also found that the five marker components can be further used for quality control and precise differentiation during the quantitative analysis of *Coptis*. As a fast and direct method of characteristic spectral acquisition, NIR spectroscopy is applied in the analysis of the origin of *Corydalis yanhusuo* [104]. The NIR spectra of two different sources of *Corydalis* powder were microwave pretreated. The training set of the *Corydalis* spectral data was simulated using four different chemometric methods, least-squares support-vector machine (LS-SVM), radial basis function artificial neural network (RBF–ANN), partial least-squares discriminant analysis (PLS-DA), and KNN. Based on spectral recognition and prediction criteria, although LS-SVM has a success rate of up to 95%, there are no significant statistical differences between the four different analytical methods.

There are many studies on the classification of medicinal plants by metabolomics research. For example, based on HPLC fingerprints, HCA analysis was used to classify the origins of *Isatis indigotica* and *Ganoderma lucidum* [105, 106]. The origin was identified by GC–MS analysis of the essential oils of Cinnamon Cortex and *Artemisia capillaris* herba [107, 108]. The successful application of NIR spectroscopy combined with DA and PLS-DA analysis to distinguish the source of Radix *Scutellaria baicalensis* is also noteworthy [109].

7.5.2.4 Different Harvesting Periods of Medicinal Plants

For the quality control of medicinal plants, the planting period is another important factor that determines the duration of time for plants to obtain nutrients from the soil. Ginseng is an important medicinal plant on a global scale, but the current stage of falsification of the breeding age has become a serious problem in the market. Kim N et al. [110] established a UPLC–QTOF–MS method for the metabolic analysis of 60 ginseng samples that were 1–6 years old (*Panax ginseng*). Yang SO et al. [111] performed a similar metabolomics study on ginseng of 2–6 years old under the guidance of good agricultural practices (GAP) standards by NMR technology. The HPLC–heuristic evolving latent projection (HELP)–PCA analysis strategy is also very helpful for harvesting season and timing optimisation [111]. Studies have shown that July is the best season for harvesting green Mandarin tangerine peel (Pericarpium Citri Reticulatae Viride), while winter is more suitable for harvesting orange peel. (Pericarpium Citri Reticulatae).

In summary, medicinal plants derived from different species or different parts of the same plant usually show different effects. Pharmacological effects and clinical applications vary depending on the amount and type of active ingredient. Species diversity greatly increases the difficulty of quality control of medicinal plants. In addition, unused habitats, climatic conditions, and incubation cycles have also resulted in changes in the chemical composition. The improper and unreasonable use of medicinal plants can lead to treatment failure and even toxic reactions. The effective identification and differentiation of medicinal plants play a crucial role in the quality control of TCM. Many plants are very similar in morphology, with little difference in composition, and it is often difficult to distinguish effectively. Metabolomics, through the use of various analytical tools combined with multivariate statistical analysis, provides powerful tools and approaches for differentiating these complex systems. At the same time, information obtained by the Metabolomics Institute not only helps to understand the metabolic network of plant and their response to changes in the external environment but also provides new ideas for the natural characteristics of plant phenotypes.

7.6 Systematic Biology of Medicinal Plants

Organisms are complex systems. The genetic central dogma believes that the flow of information is from DNA to mRNA to protein. The enzyme protein catalyses the reaction of metabolites and finally aggregates and interacts to produce a variety of biological phenotypes. DNA plays a vital role as a carrier of life information. Genomics is the earliest developed bio-omics. Genomics research has led to the rapid development of life sciences, which has greatly promoted the rapid development of transcriptomics, proteomics, metabolomics, and phenotypes. Complex biological phenomena can be elucidated comprehensively and systematically through molecular mechanism research and pathway and network integration. How to integrate the massive biological data that is accelerating and accumulating at present and what effective analysis strategies are used to reveal the essential characteristics of biological activities and reveal the laws of life as a whole are the key issues facing the post-genome era. The emergence and development of systematic biology provides new ideas and methods for a deeper understanding of the universal laws of life.

7.6.1 *Methods and Technology Platforms for Systems Biology*

Systems biology is the study of the composition of all components (genes, mRNAs, proteins, metabolites, small biomolecules, etc.) in a biological system, the interrelationship between these components under specific conditions, and the analysis of this biological system at a certain dynamic process in time. At present, a wealth of biological data has been accumulated for the study of various components of biological systems. Systems biology can combine isolated interactions at the genetic and protein levels, as well as various metabolic pathways and regulatory pathways, to predict the overall function from known components to illustrate the biology as a whole.

The research ideas and research methods of systems biology are essentially the construction of multiple information fusion and system models. Classical molecular biology research is a vertical study that uses one or more methods to study one or several genes and proteins. Genomics, proteomics, and various other omics are horizontal studies that study thousands of genes or proteins simultaneously in a single way. Systems biology research integrates horizontal and vertical research into a "three-dimensional" study. Systems biology is not meant to replace other fields to clarify specific problems at the genetic and protein levels but to find connections based on the results of molecular biologists or geneticists. Systems biology is the search for connections or special interactions between life-level activities at different levels (genetic, transcription, protein, metabolic, etc.) or at the same level to create complex systems that exhibit emergent behaviours and patterns. This complex system presents new behaviours that cannot be reflected by a single system. Systems biology is not a simple accumulation of "omics" data, but rather some mathematical models to choose from them. These mathematical models can not only simulate the behaviour of biological systems but can also be used to predict the future behaviour of the system once it is stimulated and externally interfered.

Systems biology mainly uses bioinformatics and computational biology to manage, count, and analyse related data and to perform computer simulations of biomolecules and processes. The research consists of at least three steps: step 1, integrate all theoretical and experimental data at different levels of biological systems; step 2, based on the integration of the above data, propose a model that can describe the system working together at different levels; and step 3, the proposed model is used to predict what might happen in the future of the system, to guide the experiment and understand the biological process, and ultimately to realise and understand the process mechanism of the system from genotype to phenotype. Data mining is a process of knowledge discovery. Its emergence is the inevitable development of life science research. Systems biology uses high-throughput testing technology and advanced computer hardware, software algorithms, and mathematical methods to represent the ideas and directions of biological integration research. Systems biology's interpretation of biological systems is based first on the integration of data on the different properties of biological systems. Large amounts of component data are collected from the laboratory (wet data) and public data resources (dry data), the perfect integration of wet and dry data. It is true systems biology.

7.6.2 The Opportunities and Challenges of Systems Biology in Molecular Pharmacognosy Research

Molecular pharmacognosy is an interdisciplinary subject between pharmacognosy and molecular biology and is closely related to the development of molecular biology. The dominant idea of the biological research as a whole has always been

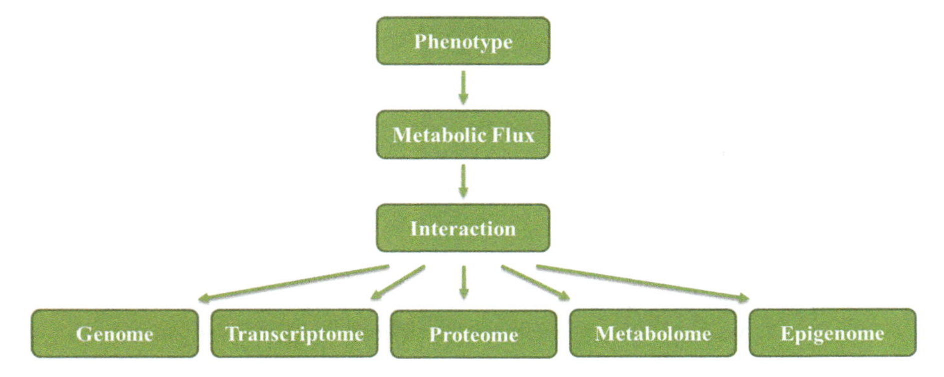

Fig. 7.2 Hierarchy and relevance of various omics

the idea of reductionism. The biology of the genome era has evolved from simple observation and description of a biological phenotype to molecular level research. It is a huge leap. However, when genome sequencing was completed, it was found that for a system, the whole is not a part simple addition but a combination of components of different natures.

At present, the genomes, transcriptomes, metabolomes, and other data of medicinal plants have been gradually accumulated. To fully understand medicinal plants and their systems, it is obviously not successful to rely on these scattered data alone. Systems biology can realise the integration from gene to cell, to organisation, to individual levels, and the whole is greater than the sum of parts. Systems biology can realise the identification of various molecules (nucleic acids, proteins, small molecule compounds, etc.) in medicinal organisms and their interaction research to the construction of pathways, networks, modules, and road maps of life activities, presenting new properties between different parts and levels (Fig. 7.2).

Because systems biology is characterised by holistic research, and bioinformatics technology is an important method for linking "reduction" and "system", it has a lot in common with TCM theory, so systems biology can provide new ideas and methods for TCM research. The organic combination is both an opportunity and a challenge for Chinese medicine to re-examine and develop its own characteristics.

References

1. Sanger F, Nicklen S, Coulson AR. DNA sequencing with chain-terminating inhibitors. Proc Natl Acad Sci. 1977;74(12):5463–7.
2. Maxam AM, Gilbert W. A new method for sequencing DNA. Proc Natl Acad Sci. 1977;74 (2):560–4.
3. Feng C, et al. Assessing performance of orthology detection strategies applied to eukaryotic genomes. PLoS One. 2007;2(4):e383.
4. Pan Z, et al. Reviews in comparative genomic research based on orthologs. Hereditas. 2009;31 (5):457–63.

5. Ohta T. Evolution by gene duplication revisited: differentiation of regulatory elements versus proteins. Genetica. 2003;118(2–3):209–16.
6. Hakes L, et al. All duplicates are not equal: the difference between small-scale and genome duplication. Genome Biol. 2007;8(10):1–13.
7. Li X, et al. Origin and evolution of new genes. Chin Sci Bull. 2004;49(13):1219–25.
8. Zhang J. Evolution by gene duplication: an update. Trends Ecol Evol. 2003;18(6):292–8.
9. Edger PP, Pires JC. Gene and genome duplications: the impact of dosage-sensitivity on the fate of nuclear genes. Chromosom Res Int J Mol Supramol Evol Asp Chromosom Biol. 2009;17 (5):699.
10. Sun H, Ge S. Review of the evolution of duplicated genes. Chin Bull Bot. 2010;45(1):13–22.
11. Xu J, et al. Panax ginseng genome examination for ginsenoside biosynthesis. GigaScience. 2017;6(11): gix093–gix093.
12. Xu H, et al. Analysis of the genome sequence of the medicinal plant *Salvia miltiorrhiza*. Mol Plant. 2016;9(6):949–52.
13. Shen Q, et al. The genome of Artemisia annua provides insight into the evolution of Asteraceae family and artemisinin biosynthesis. Mol Plant. 2018;11(6):776–88.
14. Mochida K, et al. Draft genome assembly and annotation of Glycyrrhiza uralensis, a medicinal legume. Plant J. 2017;89(2):181–94.
15. Yan L, et al. The genome of Dendrobium officinale illuminates the biology of the important traditional Chinese orchid herb. Mol Plant. 2015;8(6):922–34.
16. Guo L, et al. The opium poppy genome and morphinan production. Science. 2018;362 (6412):343–7.
17. He S, et al. MicroRNA-encoding long non-coding RNAs. BMC Genomics. 2008;9(1):1–11.
18. Mattick JS. The functional genomics of noncoding RNA. Science. 2005;309(5740):1527–8.
19. Blackstock WP, Weir MP. Proteomics: quantitative and physical mapping of cellular proteins. Trends Biotechnol. 1999;17(3):121–7.
20. González-Díaz H, et al. Proteomics, networks and connectivity indices. Proteomics. 2008;8 (4):750–78.
21. Gao W, et al. Combining metabolomics and transcriptomics to characterize tanshinone biosynthesis in Salvia miltiorrhiza. BMC Genomics. 2014;15(1):73.
22. Guo J, et al. CYP76AH1 catalyzes turnover of miltiradiene in tanshinones biosynthesis and enables heterologous production of ferruginol in yeasts. Proc Natl Acad Sci. 2013;110 (29):12108–13.
23. Guo J, et al. Cytochrome P450 promiscuity leads to a bifurcating biosynthetic pathway for tanshinones. New Phytol. 2016;210(2):525–34.
24. Lei Y, et al. Transcriptome analysis of medicinal plant Salvia miltiorrhiza and identification of genes related to Tanshinone biosynthesis. PLoS One. 2013;8(11):e80464.
25. Van Someren EP, et al. Genetic network modeling. Pharmacogenomics. 2002;3(4):507–25.
26. Alex VM, et al. CathaCyc, a metabolic pathway database built from Catharanthus roseus RNA-Seq data. Plant Cell Physiol. 2013;54(5):673–85.
27. Ma Y-N, et al. Jasmonate promotes artemisinin biosynthesis by activating the TCP14-ORA complex in *Artemisia annua*. Sci Adv. 2018;4(11):eaas9357.
28. Yu Z-X, et al. The jasmonate-responsive AP2/ERF transcription factors AaERF1 and AaERF2 positively regulate artemisinin biosynthesis in Artemisia annua L. Mol Plant. 2012;5 (2):353–65.
29. Shen Q, et al. The jasmonate-responsive AaMYC2 transcription factor positively regulates artemisinin biosynthesis in Artemisia annua. New Phytol. 2016;210(4):1269–81.
30. Tang Y, et al. AaEIN3 mediates the downregulation of artemisinin biosynthesis by ethylene signaling through promoting leaf senescence in Artemisia annua. Front Plant Sci. 2018;9:413.
31. Yan T, et al. HOMEODOMAIN PROTEIN 1 is required for jasmonate-mediated glandular trichome initiation in Artemisia annua. New Phytol. 2017;213(3):1145–55.
32. Qi J, et al. Mining genes involved in the stratification of Paris Polyphylla seeds using high-throughput embryo transcriptome sequencing. BMC Genomics. 2013;14(1): 358–358.

33. Simon SA, et al. Short-read sequencing technologies for transcriptional analyses. Annu Rev Plant Biol. 2009;60(1):305.
34. Gai S, et al. Transcriptome analysis of tree peony during chilling requirement fulfillment: assembling, annotation and markers discovering. Gene. 2012;497(2):256–62.
35. Jain A, Chaudhary S, Sharma PC. Mining of microsatellites using next generation sequencing of seabuckthorn (Hippophae rhamnoides L.) transcriptome. Physiol Mol Biol Plant. 2014;20 (1):115–23.
36. Lin W, et al. Transcriptome analysis of Houttuynia cordata Thunb. by Illumina paired-end RNA sequencing and SSR marker discovery. PLoS One. 2014;9(1):e84105.
37. Zeng S, et al. Development of a EST dataset and characterization of EST-SSRs in a traditional Chinese medicinal plant, Epimedium sagittatum (Sieb. Et Zucc.) Maxim. BMC Genomics. 2010;11:94.
38. Wang Y-D, Wang X, Wong Y-s. Proteomics analysis reveals multiple regulatory mechanisms in response to selenium in rice. J Proteome. 2012;75(6):1849–66.
39. Schmid MB. Structural proteomics: the potential of high-throughput structure determination. Trends Microbiol. 2002;10(10):s27–31.
40. Aggarwal K, Lee HK. Functional genomics and proteomics as a foundation for systems biology. Brief Funct Genomics. 2003;2(3):175–84.
41. Lesley SA, et al. Structural genomics of the *Thermotoga maritima* proteome implemented in a high-throughput structure determination pipeline. Proc Natl Acad Sci. 2002;99(18):11664–9.
42. Zhu W, et al. Variations of metabolites and proteome in Lonicera japonica Thunb. Buds and flowers under UV radiation. Biochim Biophys Acta (BBA) – Proteins Proteomics. 2017;1865 (4):404–13.
43. Zhu W, et al. Binary stress induces an increase in indole alkaloid biosynthesis in Catharanthus roseus. Front Plant Sci. 2015;6:582.
44. Wang Y, et al. Comparative proteomic analysis of the response to silver ions and yeast extract in Salvia miltiorrhiza hairy root cultures. Plant Physiol Biochem. 2016;107:364–73.
45. Adrian B. Perceptions of epigenetics. Nature. 2007;447(7143):396–8.
46. Jablonka E, Lamb MJ. The changing concept of epigenetics. Ann N Y Acad Sci. 2010;981 (1):82–96.
47. Finnegan EJ, et al. DNA methylation in plants. Annu Rev Plant Physiol Plant Mol Biol. 1998;49(1):223–47.
48. Loïc P, Wen-Hsiung L. Evolutionary diversification of DNA methyltransferases in eukaryotic genomes. Mol Biol Evol. 2005;22(4):1119–28.
49. Chan S, Henderson I, Jacobsen S. Gardening the genome: DNA methylation in Arabidopsis thaliana. Nat Rev Genet. 2005;6(5):351–60.
50. Bernard A, Emilie C, Rachel M. Environmentally induced phenotypes and DNA methylation: how to deal with unpredictable conditions until the next generation and after. Mol Ecol. 2010;19(7):1283–95.
51. Chiang PK, et al. S-Adenosylmethionine and methylation. FASEB J Off Publ Fed Am Soc Exp Biol. 1996;10(4):471.
52. Turner BM. Histone acetylation and an epigenetic code. BioEssays. 2000;22(9):836–45.
53. Jenuwein T, Allis CD. Translating the histone code. Science. 2001;293(5532):1074–80.
54. Tian L, et al. Reversible histone acetylation and deacetylation mediate genome-wide, promoter-dependent, and locus-specific changes in gene expression during plant development. Genetics. 2004;169(1):337–45.
55. Tariq M, Paszkowski J. DNA and histone methylation in plants. Trends Genet. 2004;20 (6):244–51.
56. Lee RC, Feinbaum RL, Ambros VR. The C. elegans heterochronic gene lin-4 encodes small RNAs with antisense complementarity to lin-14. Cell. 1993;75(5):843–54.
57. Bartel DP. MicroRNAs: genomics, biogenesis, mechanism, and function. Cell. 2004;116 (2):281–97.

58. Kidner CA, Martienssen RA. The developmental role of microRNA in plants. Curr Opin Plant Biol. 2005;8(1):38–44.
59. Hammond SM, et al. An RNA-directed nuclease mediates post-transcriptional gene silencing in Drosophila cells. Nature. 2000;404(6775):293–6.
60. Zamore PD, et al. RNAi: double-stranded RNA directs the ATP-dependent cleavage of mRNA at 21 to 23 nucleotide intervals. Cell. 2000;101(1):25–33.
61. Flatscher, R, et al. Environmental heterogeneity and phenotypic divergence: can heritable epigenetic variation aid speciation? Genet Res Int. 2012;2012:698421–698421.
62. Rollins RA, et al. Large-scale structure of genomic methylation patterns. Genome Res. 2005;16(2):157–63.
63. Frommer M, et al. A genomic sequencing protocol that yields a positive display of 5-methylcytosine residues in individual DNA strands. Proc Natl Acad Sci U S A. 1992;89(5):1827–31.
64. Cokus SJ, et al. Shotgun bisulphite sequencing of the Arabidopsis genome reveals DNA methylation patterning. Nature. 2008;452(7184):215–9.
65. Nair SS, et al. Comparison of methyl-DNA immunoprecipitation (MeDIP) and methyl-CpG binding domain (MBD) protein capture for genome-wide DNA methylation analysis reveal CpG sequence coverage bias. Epigenetics. 2011;6(1):34–44.
66. Wojdacz TK, Alexander D. Methylation-sensitive high resolution melting (MS-HRM): a new approach for sensitive and high-throughput assessment of methylation. Nucleic Acids Res. 2007;35(6):e41.
67. Oneill LP, Turner BM. Immunoprecipitation of chromatin. Methods Enzymol. 1996;274:189–97.
68. Miura H, Tomaru Y. ChIP on chip for transcriptional regulatory network analysis. Tanpakushitsu Kakusan Koso Protein Nucleic Acid Enzyme. 2004;49(17 Suppl):2710.
69. Nix DA, Courdy SJ, Boucher KM. Empirical methods for controlling false positives and estimating confidence in ChIP-Seq peaks. BMC Bioinf. 2008;9(1):1–9.
70. Shi R, Chiang VL. Facile means for quantifying microRNA expression by real-time PCR. BioTechniques. 2005;39(4):519–25.
71. Mestdagh P, et al. High-throughput stem-loop RT-qPCR miRNA expression profiling using minute amounts of input RNA. Nucleic Acids Res. 2008;36(21):e143.
72. Liu CG, et al. An oligonucleotide microchip for genome-wide microRNA profiling in human and mouse tissues. Proc Natl Acad Sci U S A. 2004;101(26):9740–4.
73. Ni Z, et al. Effects of 5-azacytidine on bioactive components of Dendrobium. J Zhejiang Agric Sci. 2014;7:1018–20.
74. Li C, et al. Transcriptome analysis reveals ginsenosides biosynthetic genes, microRNAs and simple sequence repeats in Panax ginsengC. A. Meyer. BMC Genomics. 2013;14(1):245.
75. Vashisht I, et al. Mining NGS transcriptomes for miRNAs and dissecting their role in regulating growth, development, and secondary metabolites production in different organs of a medicinal herb, Picrorhiza kurroa. Planta. 2015;241(5):1255–68.
76. Zhang Y, Chu H, Zhang J. Comparison of population genetic and epigenetic diversity of Salvia miltiorrhiza in Qinba Mountains. Acta Agric Boreali-Occiden Sin. 2012;10:142–8.
77. Boyko A, et al. Transgenerational adaptation of Arabidopsis to stress requires DNA methylation and the function of dicer-like proteins. PLoS One. 2010;5(3):e9514.
78. Li M-R, et al. Genetic and epigenetic diversities shed light on domestication of cultivated ginseng (Panax ginseng). Mol Plant. 2015;8(11):1612–22.
79. Baek D, et al. Regulated AtHKT1 gene expression by a distal enhancer element and DNA methylation in the promoter plays an important role in salt tolerance. Plant Cell Physiol. 2011;52(1):149–61.
80. Huang W, et al. SlAGO4A, a core factor of RNA-directed DNA methylation (RdDM) pathway, plays an important role under salt and drought stress in tomato. Mol Breed. 2016;36(3):28.

81. Villas-Boas SG, et al. Metabolome analysis: an introduction, vol. 24. Hoboken: John Wiley & Sons; 2007.
82. Fiehn O. Metabolomics – the link between genotypes and phenotypes. Plant Mol Biol. 2002;48(1–2):155–71.
83. Fukusaki E, Kobayashi A. Plant metabolomics: potential for practical operation. J Biosci Bioeng. 2005;100(4):347–54.
84. Trethewey RN. Metabolite profiling as an aid to metabolic engineering in plants. Curr Opin Plant Biol. 2004;7(2):196–201.
85. Hirai MY, et al. Integration of transcriptomics and metabolomics for understanding of global responses to nutritional stresses in Arabidopsis thaliana. Proc Natl Acad Sci U S A. 2004;101 (27):10205–10.
86. Duan L, et al. Use of the metabolomics approach to characterize Chinese medicinal material Huangqi. Mol Plant. 2012;5(2):376–86.
87. Xie G, et al. Ultra-performance LC/TOF MS analysis of medicinal Panax herbs for metabolomic research. J Sep Sci. 2008;31:1015–26.
88. Wang L, Wang X, Kong L. Automatic authentication and distinction of Epimedium koreanum and Epimedium wushanense with HPLC fingerprint analysis assisted by pattern recognition techniques. Biochem Syst Ecol. 2012;40:138–45.
89. Sun H, et al. UPLC–Q-TOF–HDMS analysis of constituents in the root of two kinds of aconitum using a metabolomics approach. Phytochem Anal. 2013;24(3):263–76.
90. Liu NQ, et al. Metabolomic investigation of the ethnopharmacological use of Artemisia afra with NMR spectroscopy and multivariate data analysis. J Ethnopharmacol. 2010;128 (1):230–5.
91. Lu G, et al. Development of high-performance liquid chromatographic fingerprints for distinguishing Chinese Angelica from related umbelliferae herbs. J Chromatogr A. 2005;1073(1):383–92.
92. Qin X, et al. Metabolic fingerprinting by 1HNMR for discrimination of the two species used as Radix Bupleuri. Planta Med. 2012;78(09):926–33.
93. Tistaert C, et al. Dissimilar chromatographic systems to indicate and identify antioxidants from Mallotus species. Talanta. 2011;83(4):1198–208.
94. Tian RT, Xie PS, Liu HP. Evaluation of traditional Chinese herbal medicine: Chaihu (Bupleuri Radix) by both high-performance liquid chromatographic and high-performance thin-layer chromatographic fingerprint and chemometric analysis. J Chromatogr A. 2009;1216 (11):2150–5.
95. Qin X, Dai Y, Liu N, et al. Metabolic fingerprinting by 1HNMR for discrimination of the two species used as Radix Bupleuri. Planta Med. 2012;78(09):926–33.
96. Fukuda E, et al. Identification of Glycyrrhiza species by direct analysis in real time mass spectrometry. Nat Prod Commun. 2010;5(11):1755–8.
97. Pan R, et al. Development of the chromatographic fingerprint of Scutellaria barbata D. Don by GC–MS combined with Chemometrics methods. J Pharm Biomed Anal. 2011;55(3):391–6.
98. Zhao Y, et al. An expeditious HPLC method to distinguish Aconitum kusnezoffii from related species. Fitoterapia. 2009;80(6):333–8.
99. Wu H. Studies on warm and cold nature of JiangHuang and YuJin based on metabonomics. Chinese Academy of Chinese Medical Sciences: Beijing; 2011.
100. Dan M, et al. Metabolite profiling of Panax notoginseng using UPLC–ESI-MS. Phytochemistry. 2008;69(11):2237–44.
101. Grubesic RJ, et al. Spectrophotometric method for polyphenols analysis: Prevalidation and application on Plantago L. species. J Pharm Biomed Anal. 2005;39(3):837–42.
102. Fukuda E, et al. Application to classification of mulberry leaves using multivariate analysis of proton NMR metabolomic data. Nat Prod Commun. 2011;6(11):1621.
103. Kong W, et al. Quantitative and chemical fingerprint analysis for quality control of Rhizoma Coptidischinensis based on UPLC-PAD combined with chemometrics methods. Phytomedicine. 2009;16(10):950–9.

104. Lai Y, Ni Y, Kokot S. Discrimination of Rhizoma Corydalis from two sources by near-infrared spectroscopy supported by the wavelet transform and least-squares support vector machine methods. Vib Spectrosc. 2011;56(2):154–60.
105. Zou P, Hong Y, Koh H. Chemical fingerprinting of Isatis indigotica root by RP-HPLC and hierarchical clustering analysis. J Pharm Biomed Anal. 2005;38(3):514–20.
106. Chen Y, et al. Discrimination of Ganoderma lucidum according to geographical origin with near infrared diffuse reflectance spectroscopy and pattern recognition techniques. Anal Chim Acta. 2008;618(2):121–30.
107. Gong F, et al. Gas chromatography-mass spectrometry and chemometric resolution applied to the determination of essential oils in Cortex cinnamomi. J Chromatogr A. 2001;905 (1):193–205.
108. Guo F, et al. Analyzing of the volatile chemical constituents in Artemisia capillaris herba by GC–MS and correlative chemometric resolution methods. J Pharm Biomed Anal. 2004;35 (3):469–78.
109. Li W, et al. Classification and quantification analysis of Radix scutellariae from different origins with near infrared diffuse reflection spectroscopy. Vib Spectrosc. 2011;55(1):58–64.
110. Kim N, et al. Metabolomic approach for age discrimination of Panax ginseng using UPLC-Q-Tof MS. J Agric Food Chem. 2011;59(19):10435–41.
111. Yang SO, et al. NMR-based metabolic profiling and differentiation of ginseng roots according to cultivation ages. J Pharm Biomed Anal. 2012;58(1):19–26.

Chapter 8
Molecular Mechanisms and Gene Regulation for Biosynthesis of Medicinal Plant Active Ingredients

Lei Zhang, Hexin Tan, and Philipp Zerbe

Abstract This chapter introduces the basic principles and methods for the studies on molecular mechanisms and gene regulation for the biosynthesis of medicinal plant active ingredients, including the biosynthetic pathway, functional genes related to biosynthesis, and manual gene regulation for the biosynthesis of active ingredients in medicinal plants. In addition, the authors explain the problems in gene cloning, stability, and transportation and propose development prospects and research directions.

8.1 Overview on the Studies of Molecular Mechanisms and Regulation of the Biosynthesis of Medicinal Plant Active Ingredients

The use of medicinal plants in the prevention and treatment of diseases has been ongoing for thousands of years. Today, about 25% of all prescription drugs come from medicinal plants. With the thought of returning to the natural becoming more and more popular, medicinal plants are attracting increasing attention. Scientists have conducted many successful studies and achieved encouraging outcomes in studies on the classification, cultivation, ecology, active ingredient analysis, action mechanism analysis, and processing methods of medicinal plants. Many of the active ingredients in medicinal plants can be directly used as drugs; and those that cannot frequently become the raw materials of chemical semisynthesis and the targets for structural transformation, so the active ingredients of medicinal plants are the focus of many studies.

L. Zhang (✉) · H. Tan
School of Pharmacy, Second Military Medical University, Shanghai, China
e-mail: leizhang100@163.com; hexintan@163.com

P. Zerbe
Plant Biology Department, College of Biological Sciences, University of California, Davis, Davis, CA, USA
e-mail: pzerbe@ucdavis.edu

L.-q. Huang (ed.), *Molecular Pharmacognosy*,
https://doi.org/10.1007/978-981-32-9034-1_8

The chemical structures of the active ingredients of medicinal plants usually contain multiple chiral centres with a relatively complex structure; it frequently requires a very long time from the discovery to elucidate the chemical structure of many active ingredients. Take morphine as an example, which contains five asymmetric centres: its chemical structure was determined in 1952, 146 years after the isolation of the compound. The structural complexity of the active ingredients in medicinal plants reflects the magical mechanism of biosynthesis and indicates the extent of the difficulty of relevant studies. Recently, the genome of opium poppy and key genes in the morphine synthesis pathway have been reported, which will greatly promote morphine biosynthesis and related research [1]. In the late 1950s, researchers fed medicinal plants with a radio-labelled precursor, which started the experimental biology research stage for the biosynthetic pathway of the active ingredients in medicinal plants [2, 3]. Later, with the emergence and use of mass spectrometry, spectroscopy, NMR spectrometry, and other advanced techniques, in combination with the radio-labelled precursor feeding experiments, studying the biosynthetic pathway of the active ingredients in medicinal plants has become much easier.

Since the 1970s, with the use of cultured cells in studies on the biosynthetic pathway of pharmaceutical active ingredients, especially the enzyme involved in the biosynthetic pathway of the active ingredient, studies on active ingredients in medicinal plants have reached the protein level. It is estimated that, from then till now, more than 100 enzymes involved in alkaloid biosynthesis have been successfully isolated and identified. In the late 1980s, scholars shifted the focus to the research field of the cloning, isolation, characterisation, and gene regulation of the genes related to the biosynthesis of medicinal plant active ingredients. Gene regulation studies on the active ingredients have been conducted in tens of plants, such as periwinkle, poppy, *Lithospermum*, sweet wormwood herb (*Artemisia annua* L.), yew, *Berberis*, henbane, *Datura*, belladonna, black poison, tobacco, *Camptotheca acuminata* Decne, etc.

Since the 1990s, encouraged by the advances in molecular biology research techniques and research results, studies on the gene regulations of active ingredients in medicinal plants have achieved further development. Take the anti-malaria drug artemisinin for example: in 2016 alone, there were still an estimated 216 million new cases of malaria, 445,000 deaths, and nearly one billion people living in areas with a high risk of the disease [4]. Artemisinin and its semisynthetic derivatives are a group of drugs used against malaria. It was discovered in 1972 by Tu Youyou, a Chinese scientist who was awarded half of the 2015 Nobel Prize in Medicine for her discovery [5]. Besides its antimalarial activity, many other therapeutic effects of artemisinin on diseases such as cancer [6, 7], tuberculosis [8], and diabetes [9] have been reported. Currently, the supply of artemisinin-based combination therapies (ACTs) is reliant on the agricultural production of artemisinin; in the future, the market source of artemisinin may depend on a balance between plant extraction and biosynthesis [10]. The genome of *Artemisia annua* has been reported [11], and in the last two decades, metabolic engineering has been demonstrated to be a useful approach to increasing the artemisinin content in *A. annua*. Previous studies employed several metabolic engineering strategies to enhance artemisinin production, including the overexpression of artemisinin biosynthetic pathway genes [12, 13], overexpression of transcription factors (TFs) that can enhance the expression of artemisinin biosynthetic genes [14–19], and overexpression of the ADP-FPS fusion gene to stimulate substrate channelling [20]. In addition, a synthetic biology strategy for artemisinin biosynthesis, in which the complete biosynthetic pathway of artemisinic acid, the precursor of artemisinin, is introduced into tobacco plants, has also been reported [21, 22].

In recent years, with the development of sequencing technology, the genomes and transcriptome of many plants have reported. Up to now (January 2019), the genome sequences of 361 plant species have been published (http://www.plabipd.de/index.ep). These have greatly promoted the research of medicinal plants at the molecular level. The biosynthetic pathways and gene regulation details of some active ingredients with extremely complex structures in medicinal plants have been gradually revealed, and primary success in the manual control of some active ingredients has been acquired, indicating a promising development prospect in this research field.

8.2 Significance of Studies on Molecular Mechanisms and Regulation of the Biosynthesis of Medicinal Plant Active Ingredients

The contents of alkaloids, saponins, flavonoids, glycosides, terpenes, and other active ingredients in medicinal plants are usually minimal. It is very difficult to significantly elevate the contents of the targeted active ingredients through planting and cultivation; however, if we regulate the key genes for biosynthesis at the molecular level to promote their expression, the contents of the target products could be elevated. Furthermore, through regulation at the molecular level, the proportions of various ingredients could also be altered, and some toxic ingredients could also be reduced or even be removed completely, which is beyond the reach of ordinary cultivation practices. Metabolic engineering based on such regulation could produce new leading compounds for the development of new drugs and be used to enhance the ability of plants to resist diseases, increase their contents of beneficial chemical components, and change their taste and colour.

Since the 1980s, plant cell engineering, which produces active ingredients in medicinal plants through plant cell culture technology, has greatly developed. In the middle of the 1990s, more than 400 kinds of plant have been studied in this field, and more than 600 kinds of secondary metabolites have been isolated from the cultured cells; however, among them, only a few cell cultures, such as those from ginseng and *Lithospermum*, have reached the industrial production scale. The reason for this is mainly because very low contents of active ingredients are contained in the cultured plant cells; even if occasionally high contents are found, unstable yields might also become a problem. Therefore, clarification of the mechanisms of the biosynthesis of active ingredients at the molecular level and regulating biosynthesis at the gene level will increase the gene expression of key enzymes in the process of active ingredient biosynthesis and improve and stabilise the production of active ingredients in suspension to harvest cell lines of high and stable production. This could solve the problems of low output and instability in active ingredients of suspension culture plant cells and make the industrial production of active ingredients using medicinal plant cell culture technology possible. Studies on gene regulation of the active ingredients in medicinal plants may provide unprecedented possibilities in purposely regulating the active ingredient contents in medicinal plants; this line of study could help to realise the goal of mass-producing medicinal plant active ingredients by cell cultivation, so it has a very high significance. In addition to technically ensuring the purposeful regulation of the contents of active ingredients in medicinal plants, genetic engineering studies will provide unprecedented possibilities in improving the quality of herbs and cultivation of new varieties. Regulation at the molecular level can also be used to alter the proportions of the relevant ingredients in medicinal plant, to reduce or remove some undesirable components (toxic ingredients) to resolve the toxic effects of traditional Chinese medicine (TCM), and to technically ensure the quality of TCM. Therefore, studies on the molecular mechanism and gene regulation of biosynthesis of the active ingredients in medicinal plants have great significance, both in theory and in production practices.

Because the synthesis of the active ingredients in medicinal plants is realised through a number of enzymatic reaction steps and completed in the specific differentiated cells, even with molecular biological measures, there are still many difficulties in fully clarifying the biosynthetic mechanism. An important prerequisite for target regulation of the active ingredients in medicinal plants is an understanding of the biosynthetic pathway of the active ingredient, particularly the key steps. In consideration of the not yet clear understanding of the molecular mechanisms of biosynthesis for most of the active ingredients in medicinal plants, it would be much more difficult to purposely regulate the active ingredients in medicinal plants. Even so, in the past few years, considerable advances have been achieved in studies on the molecular mechanisms and the regulation of active ingredient biosynthesis in medicinal plants. In the subsequent sections, we will first introduce the basic principles and methods for studying the gene regulation of active ingredients in medicinal plants and then, with terpenes as the example, describe the progress of gene regulation for this class of compounds.

8.3 Basic Principles and Methods for Studies on Molecular Mechanisms and Gene Regulation for the Biosynthesis of Medicinal Plant Active Ingredients

The active ingredients contained in traditional drugs are secondary metabolites; their biosynthetic pathways are very complicated, often involving dozens of enzymes during the reaction. Thus, identifying key enzymes in the formation of particular product has become a key step in the utilisation of genetic engineering technology to produce active ingredients of medicinal plants. To find out the genes for key enzymes, the selection of an appropriate carrier and promotion of the expression of exogenous or endogenous genes have become the most important problems.

The study of molecular mechanisms and the regulation of biosynthesis for medicinal plants generally start with research on the available biosynthetic pathways of the active ingredients in medicinal plants. After this, research is performed on the extraction, analysis, purification, and characterisation of each enzyme involved in the enzymatic reactions of biosynthesis; this is followed by cloning and separation of the corresponding genes and researching their expression characteristics. Finally, manual regulation of these relevant genes, especially the key enzymes, is performed with transgenetic technology. In the following sections, we will individually discuss the basic theories and methods involved in the above-mentioned four aspects.

8.3.1 Biosynthetic Pathways of Active Ingredients in Medicinal Plants

Understanding the biosynthetic pathways of active ingredients in medicinal plants is a prerequisite for carrying out genetic regulation on the active ingredients. In the 1970s, the magical effect of isotopes was discovered. People begin to use isotopes to label possible or identified precursors and then use the labelled precursors to feed the whole plant and study the biosynthetic pathway of the active ingredients in medicinal plants. Later, it was discovered that utilisation of isotope-labelled possible or identified precursors to the cultured plant cells might be a more suitable method. Although the biosynthetic capacity and overall plant biosynthetic capacity of cultured cells may not be identical, the cells can be cultured in sterile conditions under manual control and proliferation; therefore, it has many advantages over experiments on the whole plant. On one hand, research findings from cultured cells might be used as the reference for studies on the biosynthetic pathway using the whole plant. On the other hand, research findings on the biosynthetic pathway of active ingredients in cultured plant cells could be directly used in the industrial production of targeted active ingredients using cell culture.

In the following paragraph, we will introduce the basic steps of the biosynthetic pathway of active ingredients in medicinal plants:

In carrying out studies on the biosynthetic pathway of active ingredients in medicinal plants, the first step is to infer and label the intermediates and precursors. Firstly, the possible intermediates and initial precursors in the biosynthetic pathway of the product should be inferred according to the structure of the end product first, and then the intermediates and initial precursors should be chemically synthesised and labelled with ^{3}H, ^{13}C, ^{14}C, etc. Isotopically labelled compounds (such as the possible precursors) might have the following sources: the commercial product (if market available), enough relevant compounds that are isolated from plants or cultured cells and isotopically labelled, or product that is manually synthesised (chemical synthesis) and isotopically labelled. The database isoMETLIN can be used for isotope-based metabolomics [23]. Isotope labelling is a very practical method for researching biosynthetic pathways [24].

The second step is to directly feed plants or cultured cells with the isotope-labelled compounds. Then, radioactive high-performance liquid chromatography (radio-HPLC), spectroscopy, mass spectrometry, and nuclear magnetic resonance (NMR) technology can be used to analyse the extracts from plants or plant cells. Through radio-HPLC analysis of the extracts from plants of cultured cells fed with isotopes for some time, the retention time and quantities of all the peaks with radioisotope incorporation will be recorded; according to the known standards or through mass spectrometry and NMR identification of the collected peaks, the structure and properties of the compounds represented by radioisotope incorporation peaks will be determined. Note that the peak must be a pure peak containing only one compound. In the case of a pure peak incorporated with radioactive isotopes corresponding to a compound with the same structure as the final product, the fed compound is unquestionably the precursor of the final product, because the radioactivity of the fed compounds can be incorporated into the end product.

After being fed with isotopically labelled compounds, with the exception of the peaks of the final product, other peaks with radioisotope incorporation may be intermediate products produced in the biosynthesis pathway of the final product, or compounds outside the biosynthesis pathway of the final product. How can we determine the peaks with radioisotope incorporation corresponding to the intermediate products of the final product? In other words, how can we determine the peaks with radioisotope incorporation representing compounds involved in the biosynthesis of the target final product? The methods for these determination processes are the same as previously described methods, i.e. isolation of an adequate amount of compounds represented by peaks with radioisotopes, labelling the compound with radioisotopes or labelling the manually synthesised compound with isotopes, and then feeding relevant plants or cultured cells with the labelled compound. If the radioactivity of the fed compound can be effectively incorporated into the final product, then the fed compound is an intermediate product of the target final product.

Similarly, the isotope-labelled exact precursor of the target product can also be used to feed cultured plant cells, which can then be used in combination with modern analytic techniques, such as spectroscopy, mass spectrometry, NMR, and especially radio-HPLC, to carry out research in this area. After analysis with radio-HPLC, we can observe the peak retention time and number of radioisotopes incorporated and also determine the structure and properties of the compound at the peak with radioisotope incorporation. This can be performed in accordance with the known standards or through identification with mass spectrometry and NMR on the collection of the peaks. One of the peaks with radioisotope incorporation will have corresponding compounds, with the same structure as the end product. Similarly, after feeding with the isotope-labelled specific precursor, with the exception of the final product peaks, other peaks with radioisotope incorporation may be the intermediate products produced in the biosynthesis pathway of the final product, or compounds outside the biosynthesis pathway of the final product. Next, similarly, it should be determined whether the compounds corresponding to the peaks with radioisotope incorporation between the precursor and final products are the intermediate product, that is whether it participates in the biosynthesis of the end products. If the radioactivity of the fed compound can be effectively incorporated into the final product, then the compound used for feeding is certainly the intermediate product of the target final product.

After determining the chemical structure of the final product and some intermediates, there are many difficulties in acquiring the complete profile of the biosynthetic pathway of an active ingredient of medicinal plants (of course, the final product acquired after feeding with a specific precursor must be the active ingredient). These difficulties are mainly reflected in the sites of various intermediates in the biosynthetic pathway. Generally, it is believed that the compound with stronger radioactivity after the plant or cultured cells are fed with the labelled intermediate product is the next intermediate product in the biosynthetic pathway. Though the principle for studying the biosynthetic pathways of active ingredients in medicinal plants is simple, specific training is required on the isotope labelling technique and the identification of the compound structure. Further detailed descriptions on relevant techniques can be found in the relevant books and articles.

Many of the active ingredients in medicinal plants are secondary metabolites. The biosynthetic pathways of active ingredients in higher medicinal plants are still unclear, but the primary profiles of the biosynthetic pathways of various types of secondary metabolites have been described in many published articles. According to the biogenesis pathway, the active ingredients of medicinal plants are generally divided into three classes: terpenes, aromatic compounds, and alkaloids. The active ingredients of medicinal plants can be subdivided into seven categories: phenylpropanoids, quinones, flavonoids, tannin, terpenes, sterols, and glycosides and alkaloids. The main biosynthetic pathways are as follows: (1) acetate–malonate pathway – produces fatty acids, phenolic compounds, and anthraquinone; (2) mevalonate pathway – produces terpenoids; (3) cinnamic acid pathway – produces phenylpropanoids, coumarins, lignin, and flavonoids; (4) amino acid pathway – produces alkaloids; and (5) the combined pathway of the above pathways – produces natural products with more complex structures. Regarding the biosynthesis of the above-mentioned compounds, there are two biosynthesis pathways for isopentenyl pyrophosphate (IPP), an important intermediate product in the biosynthetic pathway of terpenoid: the first is the mevalonate (MVA) pathway in the cytoplasm, which exists in almost all organisms in nature; the second is a newly discovered pathway – the 1-deoxy-D-xylulose 5-phosphate (DXP) pathway, which exists in plastids. Currently, it is considered that the enzymes in the MVA pathway, which are associated with the synthesis of triterpenes, carotenoids, and steroidal in higher plants, exist in the cytoplasm; however, the synthesis of monoterpenes and diterpenes occurs in plasmids through the DXP pathway; for the synthesis of sesquiterpene, there is no clear conclusion yet. Phosphoenolpyruvate, produced in the 4-phosphate erythritol glycolytic pathway, will form 7-phosphate ketoheptose by condensation. Then, through a series of transformation, it will enter into the shikimic acid pathway; subsequently, aromatic amino acids will be formed through many branches, and finally, various aromatic active ingredients will be produced. Alkaloids are produced by the transformation of amino acids produced in the Krebs cycle, and indole alkaloids are synthesised by the transformation of tryptophan produced in the shikimic acid pathway.

8.3.2 Cloning of Genes Related to the Biosynthesis of Active Ingredients in Medicinal Plants

Studies on the characteristics of clones of the genes related to the biosynthesis of active ingredients in medicinal plants are the prerequisite of the artificial regulation of the relevant target ingredient and thus have very high significance. According to the results of the previous studies, the most commonly used separation methods for the genes relevant to the biosynthesis of active ingredients in medicinal plants are summarised as follows:

1. Initial separation of protein sequence

This is the "original" method of gene isolation. The method uses separated and purified proteins. First, the amino acid sequence of a polypeptide fragment in a protein is obtained, and then the corresponding nucleotide sequence is designed and synthesised. In the past, the above nucleotide sequences were used as molecular probes to hybridise the chromosomal gene or cDNA library, and the corresponding gene would be isolated. Now, however, the corresponding genes can be easily found on the National Center for Biotechnology Information (NCBI https://www.ncbi.nlm.nih.gov/) database by BLAST searching for the relevant nucleic acid sequences. Amino acid sequences can also be used to design degenerate nucleotide primers, after which the PCR technique is used to amplify a nucleotide fragment, using it as screening gene or the probe of the cDNA library. So far, there have been some gene sequences isolated through the initial separation of the protein sequence. However, due to the low protein yield of the genes relevant to biosynthesis, the separation and purification of these proteins become very difficult; thus, this separation method is very poor in its ability to separate genes in medicinal plants.

2. Homologous gene sequence method

According to the homologous conservation of gene sequences, the molecular probe prepared by the isolated gene forms a kind of organism that can be used to separate the homologous genes in medicinal plants. For example, P450 commonly exists in a variety of eukaryotic organisms; some P450 genes in plants have been isolated successfully by using the cDNAs or their conserved region to prepare a molecular probe and then hybridising the cDNA library or gene library. The homologous sequence hybridisation method can be used to separate the homologous genes from the known genes in other medicinal plants.

Additionally, the conservative region of the gene can be used to design a PCR primer of dozens of nucleotides, after which the PCR technique can be used to amplify the corresponding fragment in the chromosomal DNA or cDNA of the plant. Then, with the obtained PCR product as the molecular probe to hybridise and screen the gene bank, the clones of the gene to be isolated will be harvested. Compared to homologous hybridisation, this technology not only reduces the length required for homologous sequences but also accelerates the process of gene isolation. This homologous sequence gene PCR separation technology is widely used. Many P450 genes involved in the hydroxylation of active ingredients in medicinal plants are isolated by this method.

For genes that are expressed under induction by fungi or jasmonic acids and involved in the biosynthesis of active ingredients in medicinal plants, we can design a PCR primer through the conserved region of gene sequences and then amplify the cDNA fragment related to the medicinal plant with reverse transcriptase polymerase chain reaction (RT-PCR) [25]. Then, with the obtained PCR product as the molecular probe to hybridise and screen the gene bank, the clones of the gene to be isolated can be harvested.

3. Protein functional complementation cloning method

Some genes involving the most basic biological functions are very conservative in the process of biological evolution. Such genes are commonly called "housekeeping genes". Some of the proteins produced by housekeeping genes have the same functions in different species, which can be exchanged among them. According to the principle of proteins with the same function, the gene isolation method of protein functional complementation was developed in *E. coli* or yeast. For example, Corey EJ et al. [26] expressed the cDNA of *Arabidopsis thaliana* in a yeast mutant lacking lanosterol synthetase, which confirmed a full-length cDNA clone encoding *Arabidopsis thaliana* cycloartenol synthetase (with similar function to lanosterol synthetase).

4. Expression separation by differential hybridisation

Many genes have unique expression patterns. Different gene expression levels can be found under different growth conditions, tissues, or developmental stages and have been utilised to develop the separation method by differential hybridisation. With this separation method, assuming two different sources, A and B, firstly, construct a cDNA gene bank using mRNA from source A, and then prepare molecular probes by using cDNA from source A and mRNA from source B. In the next step, the above two probes are used to hybridise the cDNA gene bank of source A, respectively, to screen the differential clones. Finally, with the cloned cDNA as a probe, compare the expression pattern of the hybridised mRNA with the expression pattern of the targeted gene; if conformity in the expression patterns can be confirmed, the cDNA is that of the gene to be cloned [4]. The P450 gene CYP72A1, involved in the synthesis of vinca alkaloids, is separated in this way [27]. This practice shows that many limitations exist in the separation of the target gene with expression by differential hybridisation: first, the sensitivity of differential hybridisation is relatively low, especially for a low abundance of mRNA. Because the hybridisation probes used in differential hybridisation are the cDNA group reverse-transcribed from mRNA, and the similarity of the probes to the nucleotide sequence of the target gene is very low, it is very difficult to detect the cDNA clones from a low abundance of mRNA. Second, in differential hybridisation, lots of hybridisation filters have to be screened, and lots of plaques or cloned fragments have to be identified; this is time-consuming and laborious. Moreover, regarding the membrane between the two sets of parallel transfers, the retained quantity of DNA is frequently different, so the acquired hybridisation signal intensities will not be consistent; it is necessary to repeat dot-blot hybridisation for further positive clone identifications. Thus, poor reproducibility is another defect of differential hybridisation screening [28].

5. mRNA differential display separation

mRNA differential display was established in 1992 by Liang Pardee and was intended to be used to search for genes of specific expression by comparing mRNA in different cells. The basic strategy of mRNA differential display is as follows: a specific small fragment of mRNA is to be amplified through reverse transcription and PCR amplification, and then the DNA sequence analysis gel (6–8% polyacrylamide gel) is used to conduct synchronisation separation, showing the amplified product for comparison. Firstly, using 5′-T-(*n*)*MN*-3′ (3′-end fixed primer, in which $n = 10$–20, $M =$ dA/dC/dG, and $N =$ dA/dC/dT/dG) as the 3′-end primer, the mRNA to be compared is reverse-transcribed to obtain the single-stranded cDNA. After reverse transcription, the single-stranded cDNA will be used as a template, and PCR will be immediately performed with the above-mentioned fixed primer as the 3′-end primer and a random primer of 10–13 bases as the 5′-end primer. Using different combinations of the twelve 3′-end primers and twenty-five 5′-end primers, 15,000 different genes can be analysed at a 95% confidence interval, which basically includes all of the genes expressed in a single cell [29].

The basic procedures of mRNA differential display include the following:

(1) Separate the total mRNA from the material to be compared, and synthesise the first strand of cDNA with the selected 3′-end fixed primer by reverse transcription.
(2) Perform PCR amplification with the pairs of random primer consisting of a 5′-end random primer and a 3′-end fixed primer under the conditions of adding radio-labelled dNTP.
(3) Perform electrophoresis separation through adding the amplified sample into the denatured DNA sequencing gel. After X-ray film exposure, the DNA

amplification bands from the differentially expressed mRNA in the material to be compared will be displayed.
(4) Cut the DNA band related to differential expression out of the sequencing gel, and recover the amplified DNA bands.
(5) Because the amount of DNA in the gel band is very small, it cannot be directly used for cloning; thus, it is necessary to perform the second PCR amplification with the same primers. Only after achieving a certain amount of DNA, recombinant cloning is to be performed.
(6) Perform southern and northern hybridisation blotting of the specific DNA clones with the genomic DNA and total mRNA, respectively, to verify the specificity of its expression; then, with this fragment as a probe, screen full-length cDNA or genomic clones from the cDNA library or gene library [2].

mRNA differential display technology has the features of high sensitivity, simplicity, and intuition and the ability to compare multiple samples simultaneously; it has become very popular among researchers of molecular biology. However, after several years of practice, the mRNA differential display technology also exposes some obvious shortcomings, such as a high false-positive ratio and small fragments. Recently, through the efforts of scholars, these two shortcomings have basically been overcome.

6. Separation by expression library

Clone cDNA in the expression vector, and then introduce into *E. coli* host cells to build a cDNA expression library. Prepare antibodies from the purified protein, and then screen the gene clone from gene expression cDNA library with the prepared antibodies.

As for nucleic acid hybridisation screening, from the expression library screening, the first step is to perform in situ reproduction of the plaques or bacterial colonies onto a nitrocellulose membrane. When cultured phages develop into plaques, each plaque contains a lot of the fusion protein produced by the phage, and in the reproduction process, they will be transferred to the membrane. After properly treating the filter membrane, the transferred proteins from each plaque will be exposed. Perform incubation with the antibodies containing the target protein (the first antibody). After an appropriate time, rinse the membrane to remove the unbound antibodies, and then perform incubation again after adding a second antibody. The second antibody can be radio-labelled, coupled with biotin, or combined with a certain enzyme (such as alkaline phosphatase). Through a combination of the first and second antibody, based on observable markers (such as radiolabels or colour), we can determine the site of the positive clone, which can be specifically identified by the first antibody. Finally, compare the film (or the X-ray negatives) with the original plate control, find the positive clone, and perform sequencing and identification of the isolated recombinant phage DNA [2]. Lois et al. used this method to successfully separate the casbene synthase gene from the expression library [30].

7. Database cloning

Database cloning is a very useful method of gene isolation. It involves the use of expression sequence tags (ESTs) to isolate the gene; it is also known as the expression sequence tag method. ESTs are actually a part of the exon.

The basic steps of database cloning include the sequencing of cDNA clones by using a DNA sequencer or manual sequenator to acquiring full-length cDNA. Then, the sequence of related cDNA cloning can be compared to that of related gene sequences in Genebank for homology, providing an initial determination on the nature of the gene. The relevant full-length cDNA can be constructed into a suitable expression vector, and protein expression can be achieved in a heterologous expression system, such as *E. coli*, yeast, or insect. Finally, in combination with the relevant substrate, we can identify the functions of the protein expressed according to the relevant gene. In recent years, with the development of sequencing technology, gene transcription databases can be easily obtained by transcriptome sequencing, or directly downloaded from the website, such as NCBI (https://www.ncbi.nlm.nih.gov/) or PLEXdb (http://www.plexdb.org/).

8. Gene position cloning

Gene position cloning is also called map-based cloning, which is an efficient way to separate the unknown genes of the coding products.

Gene position cloning requires yeast artificial chromosome (YACs) as the carriers to build a YAC bank containing DNA of large fragments and DNA probes closely linked to the same target genes, which should be in the genetic map distance of hundreds of kilobases. In addition, gene position cloning needs to build a high-density DNA restriction fragment length polymorphisms (RFLP) or random amplified polymorphic DNA (RAPD) molecular markers map. The so-called RFLP molecular marker determines the length variety in DNA fragments produced by cutting DNA molecules using specific restriction endonucleases enzymes. DNA restriction fragments are separated through agarose gel electrophoresis (AGE) and stained by ethidium bromide (EB). After these steps, the visual specific electrophoresis bands translated from RFLP can be seen under ultraviolet light.

The elementary steps of gene position cloning are as follows: (1) position the target genes in the chromosomes, and determine a pair of closely linked RFLP or RAPD molecular markers on both sides of the target genes; (2) using the most closely linked pair of two-sided molecular markers containing the target gene as probes, clone and separate the specific genomic fragments located between the markers with the genome walking technique; (3) using sequencing, genetically modified complementary experiments, site-directed mutagenesis, and gene silencing (such as RNA interference, RNAi), determine the functions of the gene.

The starting point of the chromosome walking technique is the identified molecular marker (RFLP or the known gene clone), which is adjacent to the separated target gene as closely as possible. Take the RFLP marker study as an example: draw up the restriction map marked and cloned by the starting RFLP, and subclone the restriction fragment closest to the target gene. This restriction fragment is marked as the molecular hybridisation probe using a radioisotope, and the new clone (one-step cloning) is chosen from the genome banks for overlapping the starting point clone. The above-illustrated steps are repeated to get a second new clone (two-step cloning). Continue on with this (three-step, four-step, etc.), and new clones more closely linked to the target gene will be obtained; chromosome walking is continued until the clone of the target gene is obtained. Because this method takes small steps to get closer to the target gene through overlapping sequences, it is called chromosome walking. Although the mechanism of this method is easy, the practice is complicated [2].

9. Gene chip

Gene chip is an advanced technology developed in the 1980s, which is also called DNA chip or DNA microarray. Through the in situ synthesis, point contact, or ink-jet method, a large amount (the reticular density per square centimetre usually reaching thousands or even tens of thousands of molecules) of DNA molecules (target) are fixed onto a buttress, such as a membrane or a glass/silicon slice. Because the matrix is similar to a computer's chip, it is called gene chip or DNA chip.

After the hybridisation of the gene chip and marked molecule sample, all the tested genes could be analysed and examined qualitatively and quantitatively by testing the hybridisation signal intensity of every probe molecule. Therefore, the transcriptome can be studied at a larger scale, even at the genome level (the models and laws of genetic expressions). The chip with cDNA cloning fixed in the support holders (membrane, glass, or silicon slice) is called a cDNA chip, and the related technique is called the cDNA chip technique. The elementary steps of cDNA chip analysis are illustrated in the Methods section.

10. Proteomics technology

Although tens of thousands of gene expressions can be tested by functional genome technologies (DNA microarray analysis and serial analysis of gene expression [SAGE]), the real levels of gene expressions and proteins are often not correlated due to the transportation, degeneration, translation regulation, and procession after the translation of mRNA. Thus, we cannot obtain the complete information. One of the ways around this is to study proteins that are the products of the related gene expressions. Some species, individuals, organs, tissues, and even all the proteins of the cells are studied together. Corresponding to the genome, this is called the proteome.

The proteome was firstly proposed by Wilkins and Williams from Australia in the meeting of dimensional electrophoresis held in Siena, Italy, in 1994. The subject aimed at the study of proteomes is proteomics, which is a new study field. The main method for proteomics studies is as follows: using two-dimensional polyacrylamide gel electrophoresis (2D-PAGE) to separate the complicated components of proteins; scanning tens of thousands of protein spots tested through silver staining, fluorescent staining, and autoradiography; and then obtaining and analysing the data from the images using special computer software. In addition, the protein spots recollected from the gel need to be identified through the analysis of amino acid composition, mild protein sequencing, and mass-spectrometry techniques, to elucidate the nature of the proteins, the changes in expressions, and the process after the translation.

The proteomics technique includes the elementary steps listed below:

(1) Extraction of the proteins: the preparation of protein samples is carried out through acetone precipitation, trichloroacetic acid (TDA)-acetone precipitation, etc. Sometimes, organelles (such as microsomes) need to be separated and purified to enrich the organelle proteins.
(2) Dimensional electrophoresis of the proteins: using the IPGphor system for isoelectric focusing (IEF) electrophoresis, after balancing, conducting vertical SDS-PAGE electrophoresis, and staining the gel by silver staining, fluorescent staining, and Coomassie Brilliant Blue (CBB) staining.
(3) Scanning of the protein spots and image analysis: the stained gel is scanned by a light density spectrometer to digitalise the images, and the data is analysed by PDQUEST, LIPS, HEMES, and GEMINI software. The differential proteins are screened out initially.
(4) Identifying the candidate proteins by their mass spectra: using the peptide mass fingerprint (PMF) to identify the proteins. The PMF technique uses enzymes (commonly used pancreatin) to process the proteins separated from dimensional electrophoresis in gel or membrane for enzymolysis in the C-ends of arginine or lysine. MALDI-TOF mass spectrometry is used for testing, and the proteins database is searched to identify the proteins by comparing with the masses of peptides in the database.
(5) Mild sequencing of the proteins: the proteins separated through the gel are blotted on PVDF membrane, and after being stained and cut, the proteins are sequenced by the protein sequencer. The N-end sequencing method is used to identify the proteins.

Besides the methods listed above, other methods for separating the genes are developed on the basis of the interactions of large biological molecules, such as yeast one-hybrid, yeast two-hybrid, mammalian two-hybrid, bacterial two-hybrid, yeast three-hybrid, reverse two-hybrid, and reverse one-hybrid systems.

8.3.3 Manual Regulation of the Biosynthesis of Medicinal Plant Active Ingredients

8.3.3.1 The Fundamental Principles of Manually Regulating the Biosynthesis of Medicinal Plant Active Ingredients

The main methods for improving the metabolic engineering of the plant's secondary metabolites are as follows: (1) improving the flow of synthesis of the target compounds; (2) increasing the amount of production cells; and (3) reducing the catabolism of the target products. The fundamental principles of these three methods are illustrated as follows:

1. Improving the flow of synthesis of the target compounds

In the synthesis of the target compound, many factors can control the flow, such as rate-limiting enzymes, feedback inhibition, and competition. As for the control of rate-limiting enzymes, the protein-coding genes controlled by the strong promoter are introduced to improve the levels and increase the flow. In addition, the heterologous genes coding enzymes of similar functions are introduced from other plants or microbes to increase the flow. To overcome the feedback inhibition, a gene coding a similar enzymes is used, and this enzyme is not sensitive to the feedback inhibition. This can also be achieved by protein engineering. In these efforts, the expression of a functional protein in the cells is needed.

Through analyses of the nature of the biosynthesis of medicinal plant active ingredients and the enzymatic activities and reaction dynamics of the related genes' heterologous expressions, we can determine which enzymes and genes are essential for biosynthesis of the active ingredients and which step is the regulation step. With this knowledge, we can control the related enzymes and genes to improve the biosynthesis levels of the target active ingredients. The expression amount or activities of the essential enzymes are increased through some elicitors. Meanwhile, some inhibitors are used to inhibit the key enzymes in the branched approach of biosynthesis, or the expression amount or activities of the enzymes participating in the degeneration of the target active ingredients, to further improve the levels of biosynthesis of the active ingredients.

The choice of a regulation step depends on the total flows of primary and secondary metabolism. If the total flow is high but the output of the target compounds is low, the regulation target is the secondary metabolism, taking the production of vinblastine in the culture of *Catharanthus roseus* as an example. If the amount of secondary metabolites is low, the prime regulation target will refocus the enzymes from the primary to the secondary metabolism.

The competition approach can be inhibited through antisense genes. Genes can overcome the influence of the competition approach through improving the enzymatic levels of catalysing the target products or introducing an enzyme with similar functions which has a high affinity to the target products. A totally different concept is to introduce a gene to code an antibody, which can inhibit the activities of enzymes that compete with target products for the same substrate in the metabolism of the target products.

Obviously, these efforts need a comprehensive understanding of the related secondary metabolism for all the midbodies, enzymes, and protein-coding genes and the metabolic regulation and principles at different levels. However, a few approaches to the secondary metabolism are understood. The flavonoid-anthocyanidin metabolic approach is best known and has been used to change the colours of flowers. For most metabolic approaches, the study level is still left at the midbodies. In recent years, further studies have been performed at the enzyme level, such as the study on the metabolic approaches of terpenoid indole alkaloid, isoquinoline, tropane, and some terpenoids. The first step is cloning the coding genes. The following sections will illustrate this.

2. Increasing the amount of production cells

Different tissues of the plants produce different secondary metabolites. Only a small portion of cells participate in the synthesis of the target compounds. Some of the plant cell cultures reveal that not all the cells synthesise the target compounds. The study on *C. roseus* cell cultures shows that the amount of anthocyanidin depends on the cells producing the compounds and that the levels of anthocyanidin in the cells that do synthesise it are similar. Through the method of increasing the production cells by genetic regulation, the production rate of anthocyanidin can increase by five to six times in the process of *C .roseus* cell culture. However, we know little about the process of a cell producing a secondary metabolite. This subject needs more basic studies on the process of cell differentiation. This is a meaningful future research topic for studying the difference in production of secondary metabolites in different cells.

3. Inhibiting catabolism

In some reports, the final products degenerate in the cell cultures. In a study on *C. roseus* cell cultures, in the last loop of growth, the rate of alkaloid degeneration was similar to the rate of synthesis [31]. If the degeneration metabolism is to be prevented, it is essential to find the related enzymes. However, the reversion of chemical degeneration is inevitable.

8.3.3.2 Basic Methods of Manually Regulating the Target Active Ingredients of Medicinal Plants on the Basis of Gene Levels

1. Transformation technologies of medicinal plants

Using transgenic technologies for the manual control of biosynthesis on the basis of the gene level, another basis is to build ideal transformation technologies. Plant transgenic technologies include the agrobacterium-mediated method, polyethylene glycol (PEG) induction method, electrisation method, viral vector method, liposome method, microinjection method, gene gun method (or micro-bombardment method), DNA solution direct-culture method, pollen-mediated transformation method, pollen tube track method, electrophoresis method, laser-mediated transformation method, and sonoporation method [32]. For medicinal plants, the most commonly used method is the agrobacterium-mediated approach.

There are two kinds of agrobacterium in the medicinal plant transgene, *Agrobacterium tumefaciens* and *Agrobacterium rhizogenes*. The former induces tumours in plants, and the latter induces hairy roots. The transgenic mechanisms of both are similar and are based on a kind of plasmid (Ti or Ri plasmid). This plasmid has one fragment that can be integrated into the plant genome and be expressed as T-DNA. The area of this T-DNA is 20 kb, and it contains the genes for controlling the differentiation of stems and roots and opine synthetic enzyme genes. In addition, at both ends of the T-DNA area, there is a terminal repeat sequence of around 25 bp, which is called either RB or LB. RB is essential for the integration of T-DNA into plant chromosomes. A series of genes in the Vir area of the Ti plasmid are helpful for the integration of T-DNA into the plant chromosomes.

In the upstream and downstream regions of the exogenous genes, a suitable promoter (such as CaMV-35S) and terminator are connected, and the connected sequences are inserted in the T-DNA area of the Ti plasmid, in which the oncogene is knocked out. Because the Ti plasmid is about 200 kb, it is difficult to insert the exogenous genes directly into the T-DNA area. Therefore, in practice, the exogenous genes are cloned to the intermediate carriers in *E. coli*, and the intermediate carriers with exogenous genes are transferred into *A. tumefaciens*. There are two kinds of carrier systems that use this method: cointegrate vector systems and binary vector systems. The oncogene sequence in the T-DNA area of the cointegrate vector Ti plasmid is replaced by the DNA of the pBR322 plasmid. When the pBR322-derived intermediate carriers with the exogenous genes enter *A. tumefaciens* from *E. coli*, the exogenous genes can be integrated into the T-DNA of noncarcinogenic Ti plasmids because of the homologous recombination of intermediate carriers and the pBR322 sequence in noncarcinogenic Ti plasmids. The common cointegrate vector systems include the pGV3850, pGV2260, and pMON220 systems. Binary vector systems comprise two compatible Ti plasmids. One is a shuttle plasmid, which belongs to the multifunctional cloning vector plasmid and is replicated in both *E. coli* and *A. tumefaciens*. This plasmid has the following features: (1) in its T-DNA, there are cloning points and plant selectable marker genes; (2) the areas outside the T-DNA area contain bacteria selectable marker genes; (3) it can be transported from *E. coli* to *A. tumefaciens*. The other is a Ti-derived plasmid, such as pGV2260, which can provide functional areas of trans-toxicity to help T-DNA integrate into the plant chromosomes. Binary vector systems can transport more than 40 kb DNA into the plants. Either system needs to transport the intermediate vectors from *E. coli* to *A. tumefaciens*. To finish this process, there are two methods. One is triparental mating, taking the pMON220 system as an example. The triparental mating method needs an *E. coli* with the intermediate plasmid pMON220, an *E. coli* with the transportation plasmid pRK2013, and an *A. tumefaciens* with noncarcinogenic Ti plasmid pTiB6S3SE. The steps are as follows: first, transport the pRK2013 plasmid into the *E. coli* with the pMON220 plasmid; second, code transportation proteins RK2 and movement proteins combining the pMON220 plasmid to help it transfer into *A. tumefaciens*; third, using the LIH sequence, homologous recombination is performed in pTiB6S3SE plasmid. The other method is to use $CaCl_2$ to prepare competent cells of *A. tumefaciens* to change the permeability of the cell membranes and to transfer the intermediate vectors into *A. tumefaciens* using the freeze-thawing method [29].

Besides the usage of *A. tumefaciens*, *A. rhizogenes* can be used for the transformation of exogenous gene. The Ri plasmid of *A. rhizogenes* can induce the growth of hairy roots in plants, and the hairy roots can differentiate into the complete plants by the effect of plant hormones. In *A. rhizogenes*, the cointegrate vector system containing the Ri plasmid $pRiA_4$ is used. The pCGN529 plasmid (containing exogenous genes) can form a cointegrant with the triazole-linked analogue of DNA (^{TL}DNA) of the $pRiA_4$ plasmid due to its ^{TL}DNA segment. In the transformational plant tissues, the integrant $pRiA_4$:pCGN529 induces the formation of hairy roots. These uncertain roots can be cut and placed in a medium containing kanamycin to select the cells with recombinant T-DNA, and a suitable plant hormone can be added into the medium to induce the development of fertile plants [2]. In practice, it is less common to use *A. rhizogenes* than *A. tumefaciens*. People use wild *A. rhizogenes* to induce the plants to develop hairy roots for useful secondary metabolites [33, 34]. In recent years, this topic is a hotspot in the field, and many valuable results have been obtained.

There are many ways for *Agrobacterium* to transform the plants. The most-used technology is the leaf disc method, which takes the leaves as the recipient cells with hereditary consistence and develops the complete plants using tissue cultures. The steps are as follows: first, the bacteria-free leaf slices and the selected soil *Agrobacterium* are cultured together for 24–36 h and are then transferred to the selected medium containing antibiotics for culturing. In this medium, the untransformed cells are killed by the antibiotics, and the transformed cells will develop into complete plants [29].

2. Several approaches to manually regulating the biosynthesis of medicinal plant active Ingredients

To raise the biosynthesis levels of the active ingredients in medicinal plants, based on current knowledge, we can use several approaches, listed in Fig. 8.1 [35].

First, some key genes in the biosynthesis of active ingredients are imported into the plant to increase the expression amount; this increases the biosynthesis of the target active ingredients.

Second, antisense technology can be used. Based on the principle of base complementarity, DNA or RNA fragments (or their chemical modification products) of certain components of manual synthesis or biosynthesis inhibit or close some gene expressions. Through this technology, antisense RNA fragments can be introduced into the plant cells to inhibit the expressions of key enzymes, catalysing a branch of metabolism. Thus, the amount of target compound can be increased, and the synthesis of other compounds is inhibited. At present, this method has produced successful results. For example, Mol JNM et al. [36] successfully regulated the activities of cinnamyl alcohol dehydrogenase in the hairy roots of *Linum flavum* and inhibited the synthesis of the lignin molecule in the branched metabolism to increase the synthesised amount of anticancer compound 5-methoxy podophyllotoxin.

Third, RNAi technology can be used to inhibit the related gene expressions to increase the biosynthesis of the active ingredient. There are three kinds of RNA: tRNA (transfer RNA), rRNA (ribosome RNA), and mRNA (messenger RNA). RNA is commonly a single-stranded linear molecule, and there are also double-stranded and single-stranded loops. In 1983, RNA molecules with branched chains were found. According to the early genetic points, RNA appeared to be affiliated with DNA in the hereditary process of biological characters, which took simple functions that RNA transfers DNA's genetic information from the cellular nucleus to ribosomes synthesising proteins as the messenger and is a template of protein biosynthesis. With deeper studies of RNA, it was found that RNA is not only a messenger

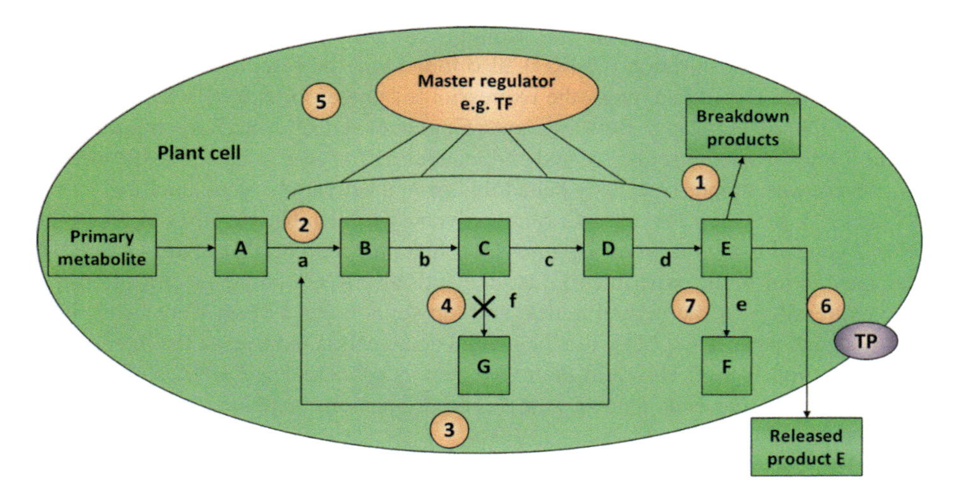

Fig. 8.1 Hypothetical biosynthetic pathway of the secondary metabolite E. Genes a–d are involved in the biosynthesis of the desired metabolite E. Gene e is involved in further metabolism of the compound E to F. The function of gene f can be blocked using RNAi or antisense techniques. TF is a transcription factor with a role as a master regulator. TP is a transporter protein that is involved in transporting compound E from inside the plant cell to the vacuoles and extracellular spaces. (*1*) Decrease the catabolism of the desired compound. (*2*) Enhance the expression or activity of a rate-limiting enzyme. (*3*) Prevent feedback inhibition of a key enzyme. (*4*) Decrease the flux through competitive pathways. (*5*) Enhance expression or activity of all genes involved in the pathway. (*6*) Compartmentalisation of the desired compound. (*7*) Conversion of an existing product into a new product

of hereditary information but also a controller of gene expression, especially for small molecules with nucleotide lengths of 21–28 kb. RNA can close the genes or change the expression levels and even have catalytic effects on enzymes, which were previously unknown functions.

Similar studies have been reported. Dated back to the 1920s, it was known that once plants were infected by a wild virus, they could produce resistance to the virus by changing the expression of related genes. This is the phenomenon of virus-induced gene silencing (VIGS). At the beginning of the 1980s, Napoli C et al. [37] imported the key gene of synthesising anthocyanin-chalcone synthase to petunia to make the flower a deeper purple. In contrast to their expectations, some flowers became mottled, even white. The test results showed the expressions of both the imported exogenous chalcone synthase and the endogenous chalcone synthase of the petunia had been suppressed. This gene silencing phenomenon of the transgene suppressing the related endogenous gene and its expression was called the co-suppression of plant genes.

Several years later, Satan et al. found a similar phenomenon [38]. They intended to import the replicase gene of the replication of tomato virus X to tobacco, with the purpose of blocking the life circle of the virus and controlling its growth. The results showed that some tobacco presented the antivirus characters and the others presented the opposite. Waterhouse et al. [38], from the Commonwealth Scientific and Industrial Research Organisation (CSIRO), also found that the combination of sense and antisense RNA could suppress the expression of potato virus Y (PVY) and create anti-PVY in potato plants. Besides plants, Cogoni et al. [39] from Italy imported *al-1* or *al-3* into *Neurospora crassa*, and endogenous *al-1* or *al-3* in cells transformed by *N. crassa* was suppressed. They called this gene inactivation "gene quelling".

In 1995, Su Guo et al. from Cornell University used antisense RNA to block the *Par-1* gene of *Caenorhabditis elegans*. Outside of their expectations, the sense RNA (as the control in the experiment), just like antisense RNA, could specifically block the expressions of the related genes, and the suppression effects were better than those of antisense RNA. How did the plant's endogenous virus resist the infection of the exogenous virus? How did the exogenous genes and sense RNA suppress the expressions of the endogenous genes? These questions remained mysteries for a long time.

In February 1998, Andrew Fire from Washington Carnegie Institute, Craig Mello from the University of Massachusetts, and other researchers first revealed the truth. They found, in various organisms, that when endogenous or exogenous double-stranded RNA were imported into the cells, mRNA with the same origins as the double-stranded RNA would be degenerated, and the corresponding gene expression would be suppressed. What's more, the suppression effects could survive until the next generation [40]. Because this is the suppression of gene expressions at the RNA level, they called this double-stranded RNA-mediated posttranscriptional gene silencing (PTGS) "RNA interference" (RNAi). Following this, in various organisms, such as fruit fly, *Arabidopsis*, and mouse, double-stranded RNA-mediated RNAi was found. RNAi has since been found in fungi, plants, invertebrates, and mammals.

At this point, plant virus-induced gene silencing (VIGS), co-suppression of plant cells, gene quelling in fungi, and RNA interference widely existing in the biosphere have been gradually harmonised because the same basic molecular mechanism, namely, 21–23 nt RNA-mediated posttranscriptional gene silencing (PTGS), was discovered. Actually, RNA interference is a naturally occurring mechanism in eukaryotic organisms, monitoring and defending exogenous and movable genetic materials. Therefore, RNAi technology was formed to use this mechanism purposefully and create specific gene silencing. A series of major breakthroughs have occurred in studies on small RNA molecules represented by RNAi technology in the past 3 years, and in the annual selection of "the world's top 10 scientific and technological breakthroughs" of "SCIENCE" magazine, RNA studies have occupied an important position and won an award in 2002.

Unlike antisense RNA technology and co-suppression, RNA interference (RNAi) has the following four important advantages:

(1) RNAi target sequences are selective. After double-stranded RNA corresponding to exon sequences of a gene is imported into biological cells, mRNA of the corresponding target gene can be specifically degraded, thus inhibiting the expression of the gene. However, double-stranded RNA importing intron or promoter sequences of the gene show no interference effect.

(2) RNA interference is highly specific. Small RNA molecules can specifically degrade mRNA of a single endogenous gene corresponding to its sequence, thereby specifically inhibiting the expression of target genes.
(3) RNA interference shows a high efficiency in inhibiting the expression of genes; only a very small amount of double-stranded RNA molecules (the number is far less than that of endogenous mRNA) can completely inhibit the expression of the corresponding genes, and even the substoichiometric level of a few dsRNAs in each cell is still sufficient to inhibit target gene function. Phenotypes generated by RNAi can achieve the extent of the lack phenotype of mutant.
(4) RNAi has a magnifying effect. RNAi effects can be enlarged; the RNAi effect can still be maintained after 50–100 times cell proliferation. The effect of RNA interference inhibiting gene expression can also go through cell boundaries, delivering and maintaining efficiency through long distances among different cells. Interference effects in *C. elegans* can even be passed to offspring. In plants, the gene expression silencing effect of RNAi can be passed from local areas to the whole plant.

RNAi technology has been widely used in the identification of biological gene function and the screening and breeding of new genes. In plants, Professor Waterhouse from the CSIRO performed a 100% inhibition of expression of the *Arabidopsis δ12-desaturase* gene (FAD2) using the technology; the activity of FAD2 proteins was significantly decreased, and one of the transgenic strains was passed through five generations, after which the inhibitory effect of the phenotype was still very stable. This is the first example in the world to use RNAi technology to successfully make plant seed traits into stabilised genetic changes. In addition to the work of Waterhouse, RNAi technology has been widely used in global research into plant gene function and genetic breeding: Chuang CF et al. successfully, effectively, and specifically inhibited functions of multiple genes of *A. thaliana* using the intron hairpin RNAi technology invented by Waterhouse, and the effect of RNA inhibiting gene expression can be passed down to future offspring [41]; Schweizer P et al. produced effective interference in the function of the *dihydroxyflavonol-4-reductase* gene encoding maize and barley at the single-cell level using the RNAi method, and the accumulation of red anthocyanin related to cells was decreased [42]; Wang E et al. successfully identified the function of two genes that play an important role in the metabolism of terpenes using RNAi (one gene is responsible for cyclising geranyl pyrophosphate (GPP) into cembratriene-ol (CBT-ol), and the other gene is responsible for transforming CBT-ols into CBT-diols [43]); Ogita et al. successfully inhibited the *CaMXMT1* gene involved in caffeine biosynthesis using RNAi technology and reduced 70% of the caffeine content in transgenic plants, which makes the future production of decaffeinated coffee beans possible [44]. Recently, Davuluri et al. inhibited the regulatory gene *DET1* in tomato photomorphogenesis using RNAi, so that carotenoid and flavonoid contents in tomato fruits were greatly improved [45]. It is no wonder that the Nobel Laureate Director and Professor Phillip A Sharp from the McGovern Institute for Brain Research, Massachusetts Institute of Technology (MIT), praised RNAi technology as the most important and exciting scientific breakthrough of the past 15 years and even of the past few decades. The discovery of the key role of double-stranded RNA in a variety of gene silencing phenomena indicates the arrival of a new era of RNA.

The content of the target product can be increased by controlling the expression in regulatory genes of biosynthetic genes of the effective ingredients of medicinal plants (such as transcription activator). Because of the plasticity of secondary metabolites in plants, it is often very difficult to improve target products through the artificial regulation of key biosynthetic genes of the effective ingredients. In this case, we can regulate regulatory genes of the related gene to achieve our objective.

8.4 Problems

8.4.1 Problems in Gene Cloning

First of all, cloning a gene from the secondary metabolic pathway is very complicated. Though a variety of molecular biology means are available for gene cloning, such as transposon tagging and differential screening, these methods are often unsuccessful in the cloning of secondary metabolic pathway genes. This is mainly due to the lack of understanding of secondary metabolic pathways and their intermediates.

Presently, many studies have been conducted along the pathway: the identification of metabolic intermediates and separation, purification, and identification of metabolic enzymes followed by the cloning of genes. The low level of most secondary metabolic synthetases is a very complex factor in the study. G10H, a cytochrome P450 enzyme, is a good example. The enzyme shows only a very low yield after purification. Obtained antibodies are not sufficient to be selected for the cloning of expressed proteins. Using a probe designed to combine PCR with the conserved region of the cytochrome P450 enzyme, 16 very closely related genes were obtained from *C. roseus*. It is unrealistic to express all of these cDNAs to identify the genes encoding G10H. Schroeder and his colleagues used the differential screening of the *C. roseus* cDNA library to isolate a putative G10H cDNA clone, but the expression of cDNA in different systems has shown no G10H activity presently [46]. The cloning of genes encoding NADPH has been successful.

8.4.2 Problems in Stability

According to reports, with the reproduction of plants from generation to generation, transferred genes gradually become silent. In plant cell culture processes, the fastest-growing cells should be screened continuously. Under such continuous selection pressure, the degree of stability of the transgene remains unclear. The continuous growth of cells in a selective medium is very expensive for industrial production. Initial results have shown that the transgenic cell line selected in the culture of transgenic *C. roseus* cells is relatively stable [47].

8.4.3 Problems in Transportation

The synthesis of secondary metabolites involves different organelles or even different tissues and organs, so the transportation of intermediates plays an important role. For physical and mechanical transportation, the factor affecting the biosynthesis rate is mainly the diffusion rate, while metabolic engineering can only be beneficial to increasing the concentration gradient. If the transportation is selectively active, transport proteins are involved in the transportation process, or the transportation is driven by the pH gradient, genes encoding transport proteins or genes controlling pH will require overexpression [48]. The storage of alkaloids in the vacuoles of *C. roseus* cells is a good example. The low pH value in the vacuoles makes them an ion trap for alkaline alkaloids. In this case, it is necessary to identify pH-regulatory genes, which ultimately affect the transfer and promote the accumulation of alkaloids [49].

8.5 Prospects for Development and Future Direction of Efforts

Metabolic engineering has wide application prospects in the modification of secondary metabolic pathways, and there will be many successful examples. In particular, a study in which only one step of response changing can lead to a high yield of target compounds is likely to become the first commercial application. For example, scopolamine is produced by *Atropa belladonna* L. cell culture, but it is to produce hyoscyamine under normal circumstances. It can also be envisaged to allow plants to produce entirely new compounds, for example, by the method of introducing oxidases, which have broad substrate specificities.

In short, we have entered a new area of plant culture and can assume the eventual factory production of plants which have a diversity of comprehensively useful chemical compositions. Moreover, we can also change the secondary metabolic pathways in plants so that they have special qualities in terms of colour, taste, toxicity, health, and resistance to diseases and insects. Studies on the genetic engineering of secondary metabolites in medicinal plants to obtain secondary products using transgenic technology have shown very attractive development prospects and have become the focus of attention for many scholars. Nevertheless, the biosynthesis pathways of effective ingredients in many important medicinal plants have not been thoroughly studied yet. There have been few enzymes and genes, especially key enzymes and genes involved in the biosynthesis of active ingredient for separation and identification, and it can be said that there is little understanding of the regulatory mechanism of biosynthesising genes of active ingredients in most medicinal plants. In order to make better use of molecular biology techniques to achieve the purpose of the artificial regulation of active ingredients, we should focus on the following aspects in the future while further clarifying the biosynthetic pathways of effective ingredients:

The secondary metabolic pathways of most medicinal plants are not yet very clear, so the study of metabolic pathways should be strengthened. More key enzymes and their genes for the biosynthesis of specific active ingredients should be isolated and identified. The total flow of increasing a metabolic pathway may require a genetic engineering series where each gene encodes another rate-limiting step. Therefore, it is necessary to study the overall regulation of alternative methods and strive to find key steps to control the overall metabolic pathway. For example, studies on the regulation of expressions of pyruvate decarboxylase (*PDC*) genes and strictosidine synthase (*STR*) genes in *C. roseus* cells show that both were governed by the downregulation of auxin and upregulation of inducers. It can be seen that a gene is likely to control the expression of the two genes at least and may also control a greater part of metabolic pathways and even the entire metabolic pathways of alkaloids.

Tissues and the development of the regulatory mechanism of biosynthetic gene expression of active ingredients should be further studied. We should further make clear the various cis- and trans-acting factors that play spatial and temporal regulation roles in biosynthetic genes of active ingredients. Conclusions can be drawn from all of the above experiments; the gene cloning of secondary metabolites is feasible and can show overexpression, but limitation needs to be overcome. Therefore, the identification and cloning of regulatory genes may eventually lead to greater success to improve the content of secondary metabolites. This is because these genes can control a large part of secondary metabolism if not the whole process. Of course, we should start the study from genes encoding special steps.

Promoters with the function of tissue-specific expression need to be separated to achieve the directional expression of active ingredients, in particular tissues or cells. The development of a new type of bioreactor and the establishment and perfection of efficient cell culture techniques are also needed.

Studies on functional genomics and systems biology of plant secondary metabolites are being carried out. As we all know, most previous studies on the plant quality focused on the clarification of the biosynthetic pathways of secondary metabolites in related plants and the cloning, transformation, and regulation of related genes, such as paclitaxel in *Taxus* and arteannuin in *Artemisia apiacea*. However, these studies remain at the stage of studying individual genes one by one. Studies conducted simultaneously on multiple genes, and even all genes and proteins, as well as their interaction network of biosynthetic pathways of secondary metabolism in plants, such as paclitaxel in *Taxus* and arteannuin in *Artemisia apiacea*, are expected to fundamentally reveal the mechanism and regulation rhythm for biosynthesis of secondary metabolites in plants. This could also clarify plant quality and reasons for the formation of resistance to achieve the purpose of artificially controlling plant quality and breeding new varieties. This will be the new trend in the development of research areas for the biosynthetic pathways of secondary metabolites. It can be expected that functional genomics and systems biology techniques, including gene chip technology and proteomics and metabolomics technologies, will be widely used in the study of the biosynthesis and regulation of plant secondary metabolites.

We have every reason to believe that in the near future, we will obtain more new varieties of transgenic medicinal plants in which the secondary metabolite content is increased or the ratio between secondary metabolite compositions is changed and use them in productive practice. In addition, we may also obtain better cell lines through genetic engineering technology and conduct the large-scale production of effective ingredients so that more large-scale cultivations of plant cells tend toward industrialisation, thus providing further protection of human health.

8.6 Case Study

CYP76AH1 Catalyses Turnover of Miltiradiene in Tanshinone Biosynthesis and Enables Heterologous Production of Ferruginol in Yeasts

Tanshinones are abietane-type norditerpenoid quinones found in the commonly used Chinese medicinal herb *Salvia miltiorrhiza* Bunge. This group of diterpenoids mainly includes dihydrotanshinone I, tanshinone I, tanshinone IIA, and cryptotanshinone [50]. These compounds have exhibited diverse pharmacological activities, including antibacterial, antioxidant, anti-inflammatory, cytotoxic, neuroprotective, cardioprotective, antiplatelet, and antitumour effects [50, 51]. However, the content of tanshinones in *S. miltiorrhiza* is low. Biotechnology strategies are a good way to efficiently increase the yield of tanshinones from the cultured hairy roots. Here, the authors introduce one representative study on the cloning of CYP76AH1, which catalyses the turnover of miltiradiene in tanshinone biosynthesis and enables the heterologous production of ferruginol in yeasts.

8.6.1 Material and Methods

1. Isolation of Total RNA and qRT-PCR Analysis

For CYP76AH1 expression analysis, hairy roots were induced with Ag^+ and sampled after 6, 12, 24, 48, 72, and 120 h. Total RNA was extracted from hairy roots using TRIzol reagent (Invitrogen) following the manufacturer's directions. First-strand cDNA was synthesised with RevertAid First Strand cDNA Synthesis Kit

(Fermentas) using oligo (dT)18 primer. Quantitative real-time PCR was performed using the SYBR Premix Ex Taq (Takara Bio) and an Applied Biosynthesis 7500 Real-Time machine. The primers used were 5′-TCGTGGATGAGTCGGCAAT-3′ and 5′-TGAGTATCTGAGTTCCCT-3′. Actin was used as the endogenous control to normalise the expression value. At least three independent experiments were performed for each analysis (tissues and time points) [52].

2. Isolation and Quantification of Ferruginol

Dried and powered hairy roots were extracted with ethyl acetate, with a fivefold concentration of these extracts used for ferruginol quantification by GC analysis. GC analyses were carried out using an Agilent GC7890 system with HP-5 column (320 μm × 0.25 μm × 30 m) and flame ionisation detection, with quantification by comparison with an authentic standard for ferruginol, which was obtained from WUXI APPTEC Co. Samples (1 μL) were injected in the splitless mode at 280 °C. The GC oven temperature was programmed to increase at 6 °C/min from 150 to 220 °C, at 3 °C/min from 220 to 230 °C, and at 20 °C/min from 230 to 280 °C [52].

3. cDNA Cloning and Heterologous Expression of CYP76AH1 in Yeast

The 5′- and 3′-ends of the targeted CYP and SmCPR2 were cloned by RACE (Invitrogen) according to the manufacturer's directions. Full-length cDNA was cloned from cDNA isolated from induced hairy roots 12 h after induction with Ag^+, using PrimeStar DNA polymerase (Takara Bio). The ORF region of CYP76AH1 was subcloned into yeast epitope-tagging vector pESC-His via EcoRI and SpeI digestion with PCR amplification. The pESC-His-CYP76AH1 construct was verified by complete gene sequencing, which was transformed into the yeast strain WAT11U. Transformants were selected on a synthetic drop-in medium−His (SD−his) containing 20 g/L of glucose and grown at 28 °C for 48 h. The resulting recombinant strain was initially grown in a SD−His liquid medium with 20 g/L of glucose at 28 °C for about 48 h for an OD600 of 2–3. Cells were centrifuged and washed three times with sterile water to remove any residual glucose. The cells were then resuspended in the yeast peptone galactose (YPL) induction medium (10 g/L yeast extract, 20 g/L bacto peptone, and 20 g/L galactose) and grown overnight at 28 °C to induce recombinant protein expression. ORFs of SmCYP-2, SmCYP-10, SmCYP-11, SmCYP-18, and SmCYP-20 were also subcloned into pESC-His as performed for CYP76AH1, and transformants were generated and confirmed by sequencing [52].

8.6.2 Results

1. Intermediacy of miltiradiene in tanshinone biosynthesis

Whereas the planar conformation of the distal ring in miltiradiene is suggestive, a role for miltiradiene in tanshinone biosynthesis remained hypothetical. This was investigated by stable isotope labelling. To generate labelled miltiradiene, the authors used a previously developed *Escherichia coli*-based metabolic engineering system that enables the coexpression of SmCPS and SmKSL with a GGPP synthase [53], along with the upregulation of several key enzymes in the endogenous methylerythritol-5-phosphate (MEP)-dependent isoprenoid precursor pathway. Accordingly, growth of the resulting recombinant *E. coli* cells in optimised minimal media enabled the production of labelled miltiradiene from $^{13}C_6$-glucose. Using this methodology, fully ^{13}C-labelled miltiradiene was generated [52]. Whereas it was possible to observe the incorporation of ^{13}C-labelled miltiradiene into the oxidised intermediate (ferruginol), endogenous production levels were still sufficient to preclude the detection of any other fully ^{13}C-labelled tanshinones (Fig. 8.2). This confounding endogenous metabolism was further suppressed by the addition of the MEP pathway inhibitor, fosmidomycin, as well as AMO1618, along with feeding with increased amounts of labelled miltiradiene. Under these conditions, fully ^{13}C-labelled cryptotanshinone was observed, demonstrating that the labelled miltiradiene could undergo substantial elaboration and serve as a precursor to compounds found in the late stage of the proposed tanshinone biosynthetic pathway (Fig. 8.2).

2. CYP candidate gene discovery

To find candidate CYPs for tanshinone biosynthesis, the authors moved beyond the previously reported EST database [54] and used the next-generation sequencing of mRNA from induced hairy roots to generate an extensive transcriptome (accession no. SRX224100), resulting in 25,793 isotigs ranging from 100 to 1,100 nt in length. Analysis of this data set revealed the presence of ~300 CYP isotigs. Moreover, using an RNA-seq approach to examine the change in transcriptome upon induction with Ag^+, 14 CYP genes were selected as initial candidates for investigation on the basis of their significant increase in transcript levels.

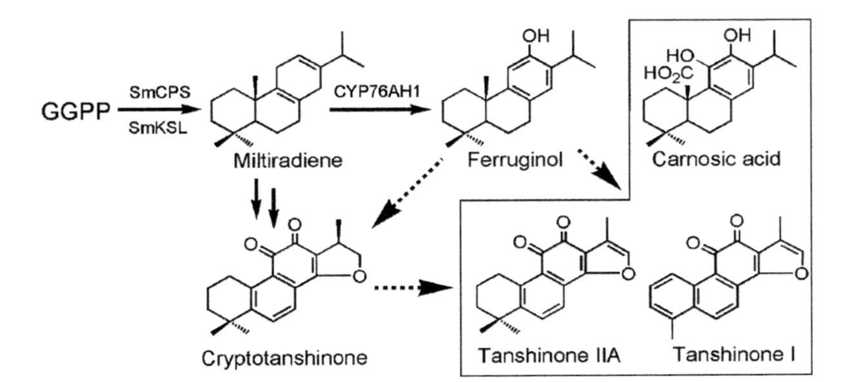

Fig. 8.2 Partial pathways for tanshinone biosynthesis and structures for some representative tanshinones. Solid arrows indicate the established relationships, and dashed arrows indicate hypothetical relationships [52]

Given the accumulation of tanshinones in the rhizome [55], and previous tissue-specific transcription demonstrated for other diterpenoid metabolism [56–59], the author hypothesised that tanshinone biosynthetic enzymes would be specifically expressed in the rhizome. Indeed, consistent with previous investigations [60], quantitative real-time (qRT)-PCR analysis demonstrated that SmCPS and SmKSL mRNA levels were higher in the rhizome than in the aboveground tissues of *S. miltiorrhiza*. Hence, the relative expression of each of the 14 inducible CYPs was similarly analysed, and transcripts for 6 of the 14 candidate CYPs were found to be more abundant in the rhizome. These six CYPs (GenBank accession nos. CYP76AH1 [JX422213], SmCYP-2 [JX422214], SmCYP-10 [JX422215], SmCYP-11 [JX422216], SmCYP-18 [JX422217], and SmCYP-20 [JX422218]) were then cloned to enable biochemical assays using miltiradiene as the substrate [32].

3. CYP76AH1 acts as a ferruginol synthase

The ability of these *S. miltiorrhiza* CYPs to react with miltiradiene was examined by recombinant expression in *S. cerevisiae*, followed by in vitro assays using microsomal preparations. Specifically, these six CYPs were expressed in the *S. cerevisiae* WAT11U strain, which expresses a CPR from *Arabidopsis thaliana* (AtCPR1), enabling a more efficient reduction of plant CYPs. Microsomal preparations from the resulting recombinant yeasts were then assayed with miltiradiene, in the presence of NADPH, followed by GC–MS analysis. Notably, only one of these CYPs (designated CYP76AH1 by the Cytochrome P450 Nomenclature Committee; GenBank accession no. JX422213) was able to convert miltiradiene to an oxidised derivative, which was determined to be ferruginol by comparison with an authentic standard (Fig. 8.3) [52].

4. Coincidence of CYP76AH1 transcription and ferruginol accumulation

To further strengthen the relevance of CYP76AH1 to ferruginol and tanshinone biosynthesis, the authors investigated ferruginol accumulation patterns. Tissue-specific analysis of whole plants demonstrated the expected accumulation of ferruginol in the rhizome, whereas none was detected in aboveground tissues, consistent with the CYP76AH1 transcript accumulation pattern noted above. Moreover, the authors found that the induction of hairy root cultures led to increased levels of ferruginol, tanshinones, and CYP76AH1 transcription [52], which is consistent with the hypothesised role for CYP76AH1 in ferruginol and, hence, tanshinone biosynthesis. When CYP76AH1 was RNAi-mediated silenced in the hairy roots of *S. miltiorrhiza*, the accumulation of tanshinones was indeed suppressed.

In addition, the genome of *Salvia miltiorrhiza* Bunge (Danshen) was reported in 2016 [61], which will facilitate our understanding of the biosynthetic pathway, as well as the relevant regulatory factors associated with tanshinone production. Moreover, access to the genome sequence is further expected to aid the molecular breeding possibilities of this important traditional medicinal herb [61].

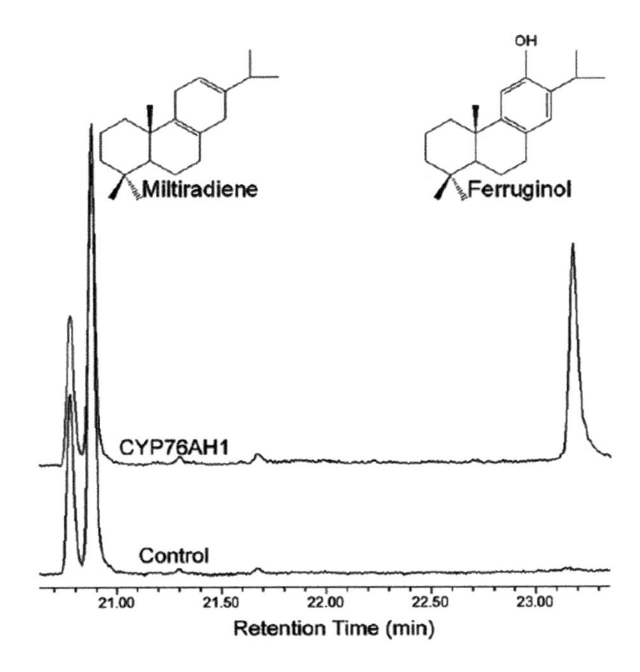

Fig. 8.3 Results of in vitro turnover of miltiradiene by CYP76AH1. GC-MS chromatogram of extracts from the reaction containing CYP76AH1 microsomes with NADPH (upper trace) and a control (lower trace)

References

1. Guo L, Winzer T, Yang X, et al. The opium poppy genome and morphinan production. Science. 2018;10:1126.
2. Kutchan TM. Alkaloid biosynthesis—the basic for metabolic engineering of medicinal plants. Plant Cell. 1995;7:1059–70.
3. Luca VD, Pierre BS. The cell and developmental biology of alkaloid biosynthesis. Trends Plant Sci. 2000;5(4):168–73.
4. World Health Organization. http://www.who.int/malaria/publications/world-malaria-report-2017/report/en/ (2017)
5. The Nobel Prize in Physiology or Medicine.. Nobel Foundation. Retrieved 2015-10-07 (2015).
6. Efferth T. Molecular pharmacology and pharmacogenomics of artemisinin and its derivatives in cancer cells. Curr Drug Targets. 2006;7:407–21.
7. Tin AS, Sundar SN, Tran KQ, et al. Antiproliferative effects of artemisinin on human breast cancer cells requires the downregulated expression of the E2F1 transcription factor and loss of E2F1-target cell cycle genes. Anti-Cancer Drugs. 2012;23:370–9.
8. Zheng H, Colvin CJ, Johnson BK, et al. Inhibitors of Mycobacterium tuberculosis DosRST signaling and persistence. Nat Chem Biol. 2017;13:218–25.
9. Li J, Casteels T, Frogne T, et al. Artemisinins target GABAA receptor signaling and impair a cell identity. Cell. 2017;168:86–100.e15.
10. A2S2.. http://www.a2s2.org/market-data/artemisinin-imports-into-india.html (2016)
11. Shen Q, Zhang L, Liao Z, et al. The genome of Artemisia annua provides insight into the evolution of Asteraceae family and artemisinin biosynthesis. Mol Plant. 2018;11:776–88.
12. Nafis T, Akmal M, Ram M, et al. Enhancement of artemisinin content by constitutive expression of the HMG-CoA reductase gene in high-yielding strain of Artemisia annua L. Plant Biotechnol Rep. 2011;5:53–60.

13. Ma D, Wang Z, Wang L, et al. A genome-wide scenario of terpene pathways in self-pollinated Artemisia annua. Mol Plant. 2015;8:1580–98.
14. Tan H, Xiao L, Gao S, et al. TRICHOME AND ARTEMISININ REGULATOR 1 is required for trichome development and artemisinin biosynthesis in Artemisia annua. L Mol Plant. 2015;8:1396–411.
15. Yu ZX, Li JX, Yang CQ, et al. The jasmonate-responsive AP2/ERF transcription factors AaERF1 and AaERF2 positively regulate artemisinin biosynthesis in Artemisia annua L. Mol Plant. 2012;5:353–65.
16. Yan T, Li L, Xie L, et al. A novel HD-ZIP IV/MIXTA complex promotes glandular trichome initiation and cuticle development in Artemisia annua. New Phytol. 2018;218(2):567–78. https://doi.org/10.1111/nph.15005.
17. Zhang F, Fu X, Lv Z, et al. A basic leucine zipper transcription factor, AabZIP1, connects abscisic acid signaling with artemisinin biosynthesis in Artemisia annua. Mol Plant. 2015;8:163–75.
18. Shen Q, Lu X, Yan T, et al. The jasmonate-responsive AaMYC2 transcription factor positively regulates artemisinin biosynthesis in Artemisia annua. New Phytol. 2016;210:1269–81.
19. Ma YN, Xu DB, Tang KX, et al. Jasmonate promotes artemisinin biosynthesis by activating the TCP14-ORA complex in Artemisia annua. Sci Adv. 2018;4(11):eaas9357.
20. Han J, Wang H, Kanagarajan S, et al. Promoting artemisinin biosynthesis in Artemisia annua plants by substrate channeling. Mol Plant. 2016;9:946–8.
21. Fuentes P, Zhou F, Erban A, et al. A new synthetic biology approach allows transfer of an entire metabolic pathway from a medicinal plant to a biomass crop. elife. 2016;5:e13664. https://doi.org/10.7554/eLife.13664.
22. Malhotra K, Subramaniyan M, Rawat K, et al. Compartmentalized metabolic engineering for artemisinin biosynthesis and effective malaria treatment by oral delivery of plant cells. Mol Plant. 2016;9:1464–77.
23. Cho K, Mahieu N, Ivanisevic J, et al. isoMETLIN: a database for isotope-based metabolomics. Anal Chem. 2014;86(19):9358–61. https://doi.org/10.1021/ac5029177.
24. Zou LQ, Kuang XJ, Sun C, et al. Strategies of elucidation of biosynthetic pathways of natural products. Zhongguo Zhong Yao Za Zhi. 2016;41(22):4119–23. https://doi.org/10.4268/cjcmm20162206.
25. Oliveira C, Aguiar TQ, Domingues L. Principles of genetic engineering. Oxford: Elsevier; 2017.
26. Corey EJ, Matsuda SPT, Bartel B. Isolation of an Arabidopsis thaliana gene encoding cycloartenol synthase by functional expression in a yeast mutant lacking lanosterol synthase by the use of a chromatographic screen. Proc Natl Acad Sci U S A. 1993;90:11628–32.
27. Vetter HP, Mangold U, Schroder G, et al. Molecular analysis and heterologous expression of an inducible cytochrome P-450 protein from periwinkle (Catharanthus roseus L.). Plant Physiol. 1992;100:998.
28. Wei YT. Principles and techniques of gene engineering. Beijing: Peking University Press; 2017.
29. Liu CX, Luo SZ. Experimental techniques of molecular biology. Beijing: Chemical Industry Press; 2018.
30. Lois AF, West CA. Regulation of expression of the casbene synthetase gene during elicitation of castor bean seedlings with pectic fragments. Arch Biochem Bilphys. 1990;276:270–7.
31. Dos Santos R, Schripdema J, Verpoorte R. Ajmalicine metabolism in Catharanthus roseus cell cultures. Phytochemistry. 1994;35:677–81.
32. Wang GY, Fang HJ. Plant genetic engineering. Beijing: Science Press; 2018.
33. Qiu DY, Zhu C, Zhu ZQ. Production of Trichosanthin from the hairy roots of Trichosanthes kirilowii Maxim. Acta Bot Sin. 1996;6:439–43.
34. Zhou LG, Wang JJ, Yang CR. Progress on plant hairy root culture and its chemistry. Nat Prod Res Dev. 1998;3:87–95.
35. Oksman-Caldentey KM, Inze´ D. Plant cell factories in the post-genomic era: new ways to produce designer secondary metabolites. Trends Plant Sci. 2004;9(9):433–40.

36. Oostdam A, Mol JNM, van der Plas LHW. Establishment of hairy root cultures of Linum flavum producing the lignan 5-methoxypodophyllotoxin. Plant Cell Rep. 1993;12:474–7.
37. Van der Krol AR, Lenting PE, Veenstra J, et al. An antisense chalcone synthase gene in transgenic plants inhibits flower pigmentation. Nature. 1988;333:866–9.
38. Waterhouse PM, Graham MW, Wang MB. Virus resistance and gene silencing in plants can be induced by simultaneous expression of sense and antisense RNA. Proc Natl Acad Sci U S A. 1998;95:13959–64.
39. Cogoni C, Macino G. Conservation of transgene-induced post-transcriptional gene silencing in plants and fungi. Trends Plant Sci. 1997;2:438–43.
40. Fire A, Xu S, Montagomery MK, et al. Potent and specific genetic interference by double-stranded in Caenorhabditis elegans. Nature. 1998;391:860–11.
41. Chuang CF, Meyerowitz EM. Specific and heritable genetic interference by double- stranded RNA in Arabidopsis thaliana. Proc Natl Acad Sci USA. 2000;97:4985–90.
42. Schweizer P, Pokorny J, Schulze P, et al. Double-stranded RNA interference with gene function at the single-cell level in cereals. Plant J. 2000;6:895–903.
43. Wang E, Wagner GJ. Elucidation of the functions of genes central to diterpene metabolism in tobacco trichomes using posttranscriptional gene silencing. Planta. 2003;216:686–91.
44. Ogita S, Uefuji H, Yamaguchi Y, et al. Producing decaffeinated coffee plants. Nature. 2003;423:823.
45. Davuluri GR, van Tuinen A, Fraser PD, et al. Fruit-specific RNAi-mediated suppression of DET1 enhances carotenoid and flavonoid content in tomatoes. Nat Biotechnol. 2005;23 (7):890–5.
46. Mangold U, Eichel J, Batschauer A, et al. Gene and cDNA for plant cytochrome P450 proteins (CYP72 family) from Catharanthus roseus, and transgenic expression of the gene and a cDNA in tobacco and Arabidopsis thaliana. Plant Sci. 1994;96:129–36.
47. Verpoorte R, van der Heijden R, Ten Hoopen HJG, et al. Metabolic engineering of plant secondary metabolite pathways for the production of fine chemicals. Biotechnol Lett. 1999;6:467–79.
48. Zhang W, Curtin C, Franco C. Towards manipulation of post-biosynthetic events in secondary metabolism of plant cell cultures. Enzym Microb Technol. 2002;30:688–96.
49. Blom TJM, Sierra M, van Vliet TB, et al. Uptake and accumulation of ajmalicine into isolated vacuoles of cultured cells of Catharanthus roseus (L.) G. Don. and its conversion into serpentine. Planta. 1991;183:170–7.
50. Zhou L, Zuo Z, Chow MS. Danshen: an overview of its chemistry, pharmacology, pharmacokinetics, and clinical use. J Clin Pharmacol. 2005;45:1345–59.
51. Yuan Y, Liu Y, Lu D, et al. Genetic stability, active constituent, and pharmacoactivity of Salvia miltiorrhiza hairy roots and wild plant. Z Naturforsch C. 2009;64:557–63.
52. Guo J, Zhou YJ, Hillwig ML, et al. CYP76AH1 catalyzes turnover of miltiradiene in tanshinones biosynthesis and enables heterologous production of ferruginol in yeasts. Proc Natl Acad Sci U S A. 2013;110(29):12108–13.
53. Cyr A, Wilderman PR, Determan M, et al. A modular approach for facile biosynthesis of labdane-related diterpenes. J Am Chem Soc. 2007;129(21):6684–5.
54. Cui G, Huang L, Tang X, et al. Candidate genes involved in tanshinone biosynthesis in hairy roots of Salvia miltiorrhiza revealed by cDNA microarray. Mol Biol Rep. 2011;38(4):2471–8.
55. Wang JW, Wu JY. Tanshinone biosynthesis in Salvia miltiorrhiza and production in plant tissue cultures. Appl Microbiol Biotechnol. 2010;88(2):437–49.
56. Hamberger B, Ohnishi T, Hamberger B, et al. Evolution of diterpene metabolism: Sitka spruce CYP720B4 catalyzes multiple oxidations in resin acid biosynthesis of conifer defense against insects. Plant Physiol. 2011;157(4):1677–95.
57. Ro DK, Arimura G, Lau SY, et al. Loblolly pine abietadienol/abietadienal oxidase PtAO (CYP720B1) is a multifunctional, multisubstrate cytochrome P450 monooxygenase. Proc Natl Acad Sci U S A. 2005;102(22):8060–5.

58. Wilderman PR, Xu M, Jin Y, et al. Identification of syn-pimara- 7,15-diene synthase reveals functional clustering of terpene synthases involved in rice phytoalexin/allelochemical biosynthesis. Plant Physiol. 2004;135(4):2098–105.
59. Xu M, Hillwig ML, Prisic S, et al. Functional identification of rice syn-copalyl diphosphate synthase and its role in initiating biosynthesis of diterpenoid phytoalexin/allelopathic natural products. Plant J. 2004;39(3):309–18.
60. Ma Y, Yuan L, Wu B, et al. Genome-wide identification and characterization of novel genes involved in terpenoid biosynthesis in Salvia miltiorrhiza. J Exp Bot. 2012;63(7):2809–23.
61. Xu H, Song J, Luo H, et al. Analysis of the Genome Sequence of the Medicinal Plant Salvia miltiorrhiza. Mol Plant. 2016;9(6):949–52. https://doi.org/10.1016/j.molp.2016.03.010.

Chapter 9
Synthetic Biology of Active Compounds

Yifeng Zhang, Meirong Jia, and Wei Gao

9.1 Overview

9.1.1 Concept and Development of Synthetic Biology

Synthetic biology is a rising discipline that combines biology, chemistry, computer science, engineering, and physics. In the early twentieth century, French physical chemist Stephane Leduc [1] put forward the idea that life can be simplified into a chemical reaction in his book *The Mechanism of Life*; however, because people's understanding has stayed at the early biological research stage, the level of understanding of molecular biology is insufficient, and synthetic biology has not been developed. Until 1962, Francois Jacob and Jacques Monod [2] proposed an operon model for *E. coli* gene expression, which was favoured by researchers for its precise regulation. With the rapid development of recombinant DNA technology in the 1970s and high-throughput sequencing in the 1980s, the construction of artificial biological systems has gradually changed from idea to reality, and people's understanding of synthetic biology has gradually deepened. In 1980, German scientist Barbara H-bomb [3] defined synthetic biology as a gene for bacteria using recombinant DNA technology in his long-form paper "Gene Surgery: On the Threshold of Synthetic Biology". In January 2000, *Nature* published two studies on the

Y. Zhang
School of Traditional Chinese Medicine, Capital Medical University, Beijing, China
e-mail: yifengzhang06@126.com

M. Jia
Department of Plant Biology, University of California, Davis, California, USA
e-mail: meirongj@iastate.edu

W. Gao (✉)
School of Pharmaceutical Sciences, Capital Medical University, Beijing, China
e-mail: weigao@ccmu.edu.cn

L.-q. Huang (ed.), *Molecular Pharmacognosy*,
https://doi.org/10.1007/978-981-32-9034-1_9

construction of the first artificial bistable gene regulatory network and synthetic gene oscillator in *E. coli* [4, 5]. So far, synthetic biology remains a new field. In the same year, Eric Kool and other spokespersons reintroduced the concept of synthetic biology at the American Chemical Society, defining synthetic biology as genetic engineering based on systems biology, from artificial base DNA molecules, gene fragments, gene regulatory networks with signal transduction pathways, to artificial design and synthesis in cells. There are many different opinions on the definition of synthetic biology; nowadays, scholars generally recognize that the use of engineering concepts rationally synthesizes complex, biologically meaningful systems of different levels, from individual biomolecules, to whole cells, tissues, and organs. Importantly, these biological systems can perform functions not found in nature.

Since the twenty-first century, people have begun to thoroughly study the cellular system by means of engineering methods, and synthetic biology has developed rapidly. D. Ewen Cameron [6] divided the development of synthetic biology into three phases in the book *A Brief History of Synthetic Biology*: The first phase, the basic development period (2000–2003), established classical experiments and disciplinary features in the field; in the second phase, the rapid development period (2004–2007), research in the field expanded rapidly, but the development of genetic engineering failed to keep up with scientific progress; in the third phase, the new era of accelerated innovation and transformation of practice (2008–2013), the emergence of new technologies and genetic engineering methods enabled synthetic biology to move in the direction of practical applications for biomedicine.

In recent years, the field of synthetic biology has gradually matured, and the speed and quality of development have made qualitative leaps. High-throughput DNA-assembly methods and the reduced costs of gene synthesis have driven its development. The emergence of post-translational control systems, the artificial protein skeleton being implanted into the artificial feedback loop, and the dynamic behaviour of the natural mitogen-activated protein kinase (MAPK) signalling pathway in yeast cells were selected as focal points for research. In a number of independently performed *E. coli* experiments, synthetic protein backbones were also used to reconstruct the signalling pathways composed of two different components. They can also be used to colocalize enzymes related to the mevalonate pathway, increase the production of glucaric acid, and reduce the production of toxic intermediate metabolites. At the same time, synthetic biologists have begun to use network engineering techniques to address fundamental issues, such as the formation, function, or evolutionary plasticity of natural networks. With the continuous development of synthetic biology, explosive growth of genomic data, decreasing cost of DNA synthesis, and rapid development of metabolic engineering, both the host's own metabolic system and known or predicted protein functions can be used to develop synthetic pathway prediction models to find suitable metabolic pathways. In early 2013, synthetic biology began to be applied to the industrial large-scale production of the antimalarial artemisinin and made milestones in its application [7]. Professor Jay Keasling of the University of California, in collaboration with Amyris Biotech, artificially designed an artemisinic acid (a synthetic precursor of artemisinin) synthesis pathway, through which artemisinic acid can be biosynthesized in yeast cells. It is worth noting that the birth of new technologies, such as the CRISPR-Cas-mediated genome-editing system, enables synthetic biologists to transform labour from the genome-wide to the system level.

9.1.2 Proposal of Synthetic Biology for Traditional Chinese Medicine

The ingredients of traditional Chinese medicine (TCM) are mainly derived from the secondary metabolites of medicinal plants, such as ephedrine in *Ephedra sinica* Stapf, the antimalarial component artemisinin of *Artemisia annua* L., and the anticancer ingredient taxol in *Taxus chinensis* (Pilger) Rehd. The current acquisition of TCM active ingredients relies mainly on chemical extraction methods, in which the compounds are directly extracted and separated from the plants (wild or cultivated products). For example, vincristine is extracted from cultivated *Catharanthus roseus* (vincristine content is about 0.0003% of the dry weight), taxol is extracted and isolated from wild or cultivated bark species (taxol content is about 0.02% of the dry weight), and the acquisition of glycyrrhizin is basically completely dependent on the wild liquorice *Glycyrrhiza uralensis* Fisch (glycyrrhizin content is about 2–8% of the dry weight). Traditional methods for obtaining active ingredients have many drawbacks, such as long plant growth cycles, low levels of secondary metabolites in plants, and the difficulty in compound purification. The acquisition of large quantities of compounds can also cause serious damage to biological resources, especially those of wild plants. At present, a few simple active ingredients, such as quinine and cinnamic acid, can be directly synthesized by chemical synthesis, which not only significantly reduces production costs, but also protects Chinese medicine resources and promotes its sustainable use. However, most active ingredients have complex structures due to their higher number of active sites, and chemical synthesis is difficult to achieve. In order to improve the synthesis efficiency, researchers have also begun to try semi-synthetic methods using intermediates as reaction substrates, such as Wuts [8], who synthesized taxol from 10-deacetylbaccatin III. Although the shortages of TCM resources have been solved to some extent, the fundamental problems of the chemical synthesis or semi-synthesis method are difficult to solve because of the high cost, low yield, high toxicity, complicated process, and serious environmental pollution. The plant tissue culture method that replaces the original plant greatly alleviates the demand for plants and avoids the disadvantages of the chemical synthesis method, but also has problems regarding its complicated operation, long growth cycles, high production costs, difficulty in industrialization, and can only currently be used on a small scale. High-value-added compounds are in production, for example Phyton Biotech, Inc. uses a *Taxus* plant cell line for the production of taxol.

In 1995, Liqi Huang published the article "Prospect of the Application of Molecular Biotechnology in Pharmacology", which first proposed the concept of molecular medicine in China, and introduced molecular biology technology into the field of TCM, bringing new ideas for the acquisition of active ingredients. In recent years, by digging out key genes in the biosynthesis pathway of active ingredients and designing and transforming microbial strains to produce natural active ingredients, synthetic biology has been recognized as one of the most promising sources of resources in the world. Scholars have achieved many results in their research. For example, in 2006, *Nature* published a strategy by Professor Keasling and his team, who, based on the codon usage of *E. coli*, used synthetic biology to synthesize the gene-encoding amorphadiene synthase (ADS) and clone *Artemisia annua* L. [9]. The cytochrome P450 (CYP450) oxidoreductase CYP71AV1/CPR introduces a heterologous yeast mevalonate pathway to obtain a more economical, environmentally friendly, high-yield, and reliable source of artemisinic acid with a yield of 115 mg/L. After fermentation optimization, the yield of artemisinic acid in yeast can be increased to 2.5 g/L [9], and artemisinin can be synthesized in five steps: hydrogenation reduction, esterification, oxidation, Hock cracking, and rearrangement. Professor Gregory Stephanopoulos of the Massachusetts Institute of Technology produced a precursor of taxol precipitate in *E. coli* with a yield of 1020 mg/L [10]. Yongjin Zhou et al. [11] obtained a high-yield, tanshinone-active, ingredient precursor, sub-danshenone diene, in *Saccharomyces cerevisiae* through the "modular pathway engineering" strategy, with a yield of 365 mg/L under optimized conditions.

9.1.3 Research Objects and Main Tasks of TCM Synthetic Biology

Starting from the sustainable use of TCM resources, synthetic biology is a new biological (plant or microbial) system with specific physiological functions. This system is artificially designed and constructed by elucidating and simulating the basic laws of the biosynthesis of active medicinal ingredients in living organisms to achieve the orientation, efficient cultivation, and production of active ingredients of TCM. Synthetic biology can be used to produce active ingredients of TCM and has the following advantages: short production cycles, easy regulation, single fermentation products, no restrictions on raw materials, easy separation and purification, and easy to realize large-scale industrial production, especially for rare and endangered source plants. The sustainable use of medicinal ingredients in plants is of great significance. At present, TCM synthetic biology has become a key development field of molecular medicine and pharmacology, and obtaining active medicinal ingredients through synthetic biology strategies will be important for the sustainable utilization of TCM resources.

The synthetic biosynthesis of TCM is based on original Chinese herbal medicine with clear active ingredients. In order to improve the yield of active ingredients, the genetic elements related to the biosynthesis of active ingredients are discovered, and the biosynthetic pathways and regulatory mechanisms in the original plants are studied. Artificial transformation or reconstruction of synthetic pathways can provide raw materials for the research and industrial production of TCMs, thereby ensuring the clinical supply of Chinese herbal medicines and proprietary Chinese medicines, as well as the sustainable use of Chinese medicine resources.

Synthetic biosynthesis is a sustainable utilization service for TCM resources. The main research tasks include the following five points: (1) the cloning and identification of biosynthesis gene components of active ingredients, (2) the analysis and regulation of the biosynthetic pathways of active ingredients, (3) the artificial design of active ingredient biosynthesis pathway strategies and key technologies, (4) the application and development of active ingredient cell chassis, and (5) the synthetic biological production of active ingredients.

9.2 Biological Basis of the Synthesis of Active Ingredients of TCM

The active ingredients of TCM are the material bases for its efficacy. They mainly come from the secondary metabolites of medicinal plants. With the continuous development of technologies, such as molecular biology, biochemistry, sequencing analysis, and bioinformatics, in recent years, the synthetic approach to the active constituents of TCMs has been gradually unveiling the mystery. It has become the main research direction of scholars both in China and abroad to excavate the gene elements of the biosynthesis of active ingredients of TCM, reconstruct the natural and existing biological systems of plants by using synthetic biology methods, and cultivate medicinal plants or engineered bacteria with high yields of the target active ingredients. This section will systematically introduce the research progress of synthetic biology in the production of active ingredients of TCM, such as terpenoids, flavonoids, and alkaloids.

9.2.1 Terpenoid Active Ingredients

The semi-synthesis of artemisinin is the most successful case of synthetic biology technology in the field of TCM. Artemisinin is a sesquiterpene lactone drug with a peroxide bridge, extracted from the stems and leaves of the medicinal plant *Artemisia annua* L. It is the most effective drug for the treatment of malaria. The discovery of artemisinin has saved millions of lives worldwide, especially in developing countries. In 1972, the Chinese scientist Tu Youyou successfully extracted and isolated artemisinin, and was awarded the 2015 Nobel Prize in Physiology or Medicine for his discovery that artemisinin can effectively reduce the mortality of malaria patients. The University of California scholars introduced amorphadiene synthase and the *S. cerevisiae* mevalonate (MVA) pathway into *E. coli* to overcome the technical obstacle of terpenoid synthesis in *E. coli* by heterologous expression of the *S. cerevisiae* MVA pathway, thus obtaining 24 mg/L of amorphadiene, a precursor of artemisinin [12]; with this artemisinin precursor, microorganism synthesis was first realized. In 2004, the Bill & Melinda Gates Foundation donated US

$42.5 million towards the study of artemisinin semi-synthesis at the University of California, Berkeley, Amyris Biotech, and the Universal Health Institute. With the joint efforts of many forces, the researchers reported the use of *Staphylococcus aureus*-derived 3-hydroxy-3-methyl glutaryl coenzyme A (HMG-CoA) synthase and HMG-CoA reductase to replace the yeast gene, which was optimized by the fermentation system to achieve an amorphadiene yield of 27.4 g/L [13]. Although *E. coli* is simple to genetically manipulate, it is suitable as a host for terpene biosynthesis. However, because of differences in the cell structure, it is impossible to express modified enzymes such as CYP450, and it is difficult to carry out the subsequent modification of complex terpenoid structures. *S. cerevisiae*, as a eukaryote, can withstand industrial fermentation conditions and is considered to be a more ideal host. On the basis of previous research, in 2006, Professor Keasling and other researchers optimized the MVA pathway metabolic gene in yeast, combined with the expression of the oxidase CYP71AV1, and introduced amorphadiene synthase, which showed a successful three-step oxidation reaction from amorphadiene to artemisinic acid and an engineered yeast to produce artemisinic acid at a yield reaching 115 mg/L [9]. Subsequently, a series of optimizations were carried out on the yeast synthesis system, including replacing the expression strain CEN.PK2, enhancing the MVA pathway gene expression, knocking out the galactose metabolism gene, and optimizing the fermentation process, to make the amorphadiene yield 40 g/L [14]. Through further gene mining, they found the cytochrome b5 gene, artemisinin dehydrogenase gene ADH1, and *Artemisia annua* aldehyde dehydrogenase gene and expressed these three genes in the yeast strain for the synthesis of amorphadiene. Meanwhile, when the CYP71AV1 reductase CPR1 was optimized, artemisinic acid production could be increased to 25 g/L. A four-step chemical conversion strategy of hydrogenation reduction, esterification, oxidation, Hock cracking, and rearrangement, was established to convert artemisinic acid into artemisinin with a total yield of 40–45%. In 2013, the French pharmaceutical giant *Sanofi* announced the start of the industrial production of artemisinin using an artemisinin production process developed by *Amyris*. In 2014, it produced 35 t of artemisinin, which could be used to treat 70 million people. It can be said that the application of synthetic biology in the production of artemisinin is of profound significance for the treatment of malaria and the reduction of drug prices.

Tanshinone is an active ingredient in the TCM *Salvia miltiorrhiza*, which aids in the prevention and treatment of cardiovascular and cerebrovascular diseases. A common precursor of the tanshinone compound is miltiradiene. By designing the combination method of modules and taking into account the problems of the precursor supply, rate-limiting step, substrate transport, and metabolic flow distribution, Zhou et al. expressed the mevalonate reductase (tHMG1) and FPPS–GGPPS fusion proteins and merged the copalyl diphosphate synthase (SmCPS) and kaurene synthase-like (SmKSL) genes to catalyse the synthesis of miltiradiene and obtain the high-yielding *S. cerevisiae* cell factory with a 365 mg/L yield of miltiradiene [11]. In 2013, Guo et al. co-transformed CYP76AH1 and SmCPR1 into a yeast strain producing miltiradiene to construct a yeast engineering strain capable of synthesizing ferruginol. The fermentation yield of ferruginol reached 10.5 mg/L after shaking the bottle [15]. In addition, Dai et al. first integrated the miltiradiene synthases, SmCPS and SmKSL, into the yeast genome and improved the supply of the precursors FPP and GGPP through the combined regulation of the functional modules of the terpenoid synthesis pathway (tHMGR-upc2.1 and ERG20-BTS1-SaGGPS), thereby increasing the synthesis of miltiradiene up to 488 mg/L [16]. In 2015, Guo et al. found that two new CYP450s, CYP76AH3 and CYP76AK1, which can catalyse ferruginol to produce 11,20-dihydroxy ferruginol and 11,20-dihydroxy cryptojaponol, provide valuable gene components for the construction of tanshinone synthesis biology modules [17].

Moreover, saponins are an important class of natural plant products with a wide range of biological activities. *Glycyrrhizin*, a triterpenoid saponin, is an important bioactive compound that is used clinically to treat chronic hepatitis. Xu et al. found a GuUGAT that can catalyse the continuous two-step glucuronosylation of glycyrrhetinic acid to directly yield glycyrrhizin. Li et al. found a diterpene glycosyltransferase from *Andrographis paniculate* (ApUGT), which could transfer a glucose to the C-19 hydroxyl moiety of andrograpanin to form the diterpenoid saponin neoandrographolide.

9.2.2 Flavonoid Active Ingredients

Yan et al. introduced a hydroxylase of p-hydroxy-cinnamic acid (C4H) from *Arabidopsis thaliana*, parsley-derived coenzyme ligase (4CL), *Petunia*-derived chalcone isomerase (CHI), and chalcone synthase (CHS) to construct the naringin biosynthesis pathway in yeast and realized the production of naringin at a yield of 28.3 mg/L by the addition of a 4-coumaric acid substrate. Genistein, kaempferol, and quercetin have also been biosynthesized in yeast using similar strategies. By overexpressing the red clover-derived isoflavone synthase (IFS) and the Asian rice-derived CPR, Kim et al. constructed a synthetic pathway from naringenin to genistein in yeast to obtain a genistein yield of 20 mg/L [18]. Eight plant-derived genes were introduced into yeast to synthesize flavonol compounds (kaempferol and quercetin); Trantas et al. achieved the production of kaempferol and quercetin in yeast by adding 0.5 mmol/L of naringin, and the yields of kaempferol and quercetin production were increased to 4.6 mg/L and 0.38 mg/L, respectively [19]. In order to save the cost of the fermentation system, Santos et al. constructed a naringenin biosynthesis pathway in L-tyrosine high-yielding *Escherichia coli*, and then increased the supply of another precursor, malonyl-CoA, and balanced the related metabolism; achieving the de novo synthesis of naringenin from glucose, the yield reached 29 mg/L.

9.2.3 Alkaloid Active Ingredients

Since the alkaloid biosynthesis pathway is extremely complex and most biosynthetic pathways have not been resolved, there are few reports on target alkaloids for heterologous biosynthesis, and the current focus is on the heterologous microbial synthesis of benzylisoquinoline alkaloids. Hawkins et al. first constructed a multi-step catalytic pathway in *S. cerevisiae* to synthesize a series of benzylisoquinoline alkaloids or their intermediates, (*R,S*)-reticuline, (*S*)-scoulerine, (*S*)-tetrahydrocolumbamine, and (*S*)-canadine with (*R,S*)-norlaudanosoline [20]. Subsequently, Fossati et al. constructed a metabolic pathway of more than 10 genes in *S. cerevisiaes* and realized the biosynthesis of the antitumor drug sanguinarine and its precursor dihydrosanguinarine with (R,S)-norlaudanosoline as a substrate. This study completed the longest alkaloid heterologous biosynthetic pathway in *S. cerevisiae*, and they are all complex reaction processes, providing a reference for the microbial synthesis of alkaloids [21]. Due to the low biosynthesis efficiency of the benzylisoquinoline alkaloid precursor tyrosine in *S. cerevisiae*, it is difficult to achieve the de novo synthesis of alkaloids. *E. coli* can solve the problem of low tyrosine synthesis efficiency; the tyrosine production of E. coli after metabolic engineering reached 14 g/L. In the tyrosine-high-yield strain of *E. coli*, the (S)-reticuline biosynthesis pathway was constructed. After optimizing the culture conditions, the yield of (S)-reticuline could reach 46 mg/L, achieving the de novo synthesis of alkaloids. For the heterologous synthesis of alkaloids involving complex post-modification, *E. coli* is not an ideal host, because it is difficult to express modified enzymes such as CYP450. If *S. cerevisiae* can achieve the efficient biosynthesis of tyrosine in the future, it will be beneficial to the economic heterogeneous synthesis of functional alkaloids.

9.3 Analysis of Key Links in the Study of Synthetic Biology in TCM

The key links in the study of synthetic biology in TCM include the following: the cloning of gene elements, selection and transformation of chassis cells, strategy of metabolic pathway construction, optimization of synthetic process systems, and strategic core and technical means of synthetic biology research.

9.3.1 Cloning of Genetic Elements

The screening, identification, and standardization of gene components are the primary tasks in the study of synthetic biology in TCM. A genetic element refers to an amino acid or nucleotide sequence having a specific function, including a basic biological component and a core biological element. The basic biological component is the simplest and most basic bioblock in the genetic system (BioBrick), which mainly includes promoters, terminators, transposons, enzyme-encoding genes, ribosome-binding sites, and transcriptional regulatory protein factors. The core biological element refers to a sequence which encodes a specific enzyme or protein function, often a functional gene that constitutes a metabolic pathway or gene loop. The basic biological component can be further combined with the core biological component into a biological device with a particular biological function in the design of a bioreactor or metabolic pathway for reconstruction or fabrication; for example, the ADS gene in the artemisinin precursor synthesis pathway [22], and the diterpene synthase genes, *SmCPS* and *SmKSL,* in the heterologous synthesis pathway of tanshinone [11].

The basic biological element can be obtained by DNA cloning or direct chemical synthesis, and artificial mutation elements can be obtained based on the modification, modification, and recombination of natural elements. The basic biological element promoter is where the gene initiates transcription. The strength and type of the promoter can control the expression level, expression time, and space of the gene. Engineered transformation of the promoter can construct a series of artificial promoters for different purposes to achieve an accurate regulation of gene expression. For example, by modulating the transcription factor-binding site of the *Pichia pastoris*-inducible promoter AOX1 [23, 24], an AOX1 artificial mutation promoter is obtained, which improves the quality and yield of the heterologous protein. Artificial terminator elements also have sequence diversity, and thus have different termination intensities. In addition, the ribosome-binding site is the initiation site of translation, and the use of different sequences of ribosome-binding sites enables the regulation of genes at the translational level. Due to its own limitations, the basic biological components currently collected and collated are still limited to typical elements of model organisms, such as *E. coli*-derived promoters, reporter genes, ribosome-binding site sequences, and the like.

The acquisition of core biological components related to the synthetic biology of TCM is in its infancy. At present, the functional genes in the biochemical metabolic pathways of the active ingredients of TCM, such as artemisinin, paclitaxel, tanshinone, and ginsenoside, have been analysed, and steroids, flavonoids, and organisms have been obtained. The reconstruction of heterologous synthetic pathways of base families has been carried out. The development of TCM genomics, transcriptomics, and related research provides a large number of candidate components and modules for the acquisition of active ingredients of TCM. For example, through genome-wide sequencing and assembly of the medicinal model species *Ganoderma lucidum,* all the genes for the skeleton synthesis of ganoderic acid were found [25].

9.3.2 Chassis Cell Selection and Transformation

Heterologous synthesis refers to the synthesis of genes, proteins, or natural products in non-source species or cells. Heterologous synthetic chassis cells have the advantages of rapid growth, a clear genetic background, simple nutritional requirements, simple and reliable operation, and easy large-scale culture. Therefore, the acquisition of active ingredients of TCM by heterologous synthesis has obvious advantages in both quality control and yield improvement. Taking the antimalarial drug artemisinin as an example, the source species is *Artemisia annua* L., which can be industrialized to produce artemisinin by the fermentation of an engineered *E. coli* strain. *E. coli* is the chassis cell of artemisinin heterologous synthesis. Common chassis cells are discussed below.

9.3.2.1 Escherichia coli

E. coli, a gram-negative bacterium, is the most widely used host strain for heterologous synthesis of natural products because of its good operability, clear genetic background, simple technical operation, and large-scale fermentation economy. Natural products of all kinds have been successfully reported for the heterologous synthesis of *E. coli*. In fact, *E. coli* itself can provide a small number of endogenous precursors. If the chassis cells synthesize complex natural products, some metabolic pathways need to be reconstructed.

Compared with eukaryotic hosts, *E. coli* lacks post-transcriptional modification and endoplasmic reticulum-assisted P450 reductase, making it difficult to functionally express some proteins of plant origin (e.g. CYP450), and the preference for codon usage of *E. coli* is very different from that of eukaryotes.

9.3.2.2 Saccharomyces

Saccharomyces sp. are single-cell eukaryotes with easy cultivation, rapid propagation, and easy genetic engineering operation. They also have the functions of protein processing and the post-translational modification of eukaryotes. It is more suitable for the functional expression of CYP450 proteins in plants than *Escherichia coli* [26]. The yeast species used as the heterologous synthetic host bacteria mainly include *Pichia pastoris*, *S. cerevisiae*, and the like. *S. cerevisiae* is the earliest eukaryotic expression system. At present, its whole genome has been sequenced, its genetic background is clear, and its genetic operation has been systematized and standardized; thus, it is easy to carry out heterologous synthesis.

Recently, Evolva Biotech of Switzerland reported that the pathway gene of the flavonoid synthesis pathway was cloned into the yeast gene expression cassette, and these expression cassettes were randomly integrated into the yeast artificial chromosome; the recombinant cloning of these new combinations of chromosomes can create different flavonoid synthesis pathways, and different flavonoids can be produced by feeding with different precursors.

9.3.2.3 Bacillus Subtilis

Bacillus subtilis is a heterogeneous synthetic chassis of natural products with three major advantages: (1) it has a metabolic network that synthesizes some key precursors of natural products, which effectively reduces the workload of reconstructing synthetic pathways; (2) it has the cell wall characteristics of gram-positive bacteria, in that it can secrete exogenous natural product molecules to the outside of cells, which is conducive to the recovery of target products; (3) it has no codon preference and can better express genes from eukaryotic organisms. It has been found that *Bacillus subtilis* can produce natural products, including terpenoids, polyketones, and non-ribosomal polypeptides.

However, as a heterologous host, *Bacillus subtilis* can only integrate the natural product synthesis-related genes directly into the chromosome due to the lack of a stable plasmid vector; this results in a small number of gene copies, and thus, affects the level of product synthesis. In addition, *Bacillus subtilis* has a significant protease background which may cause rapid degradation of the natural product synthetase.

9.3.2.4 Cyanobacteria and *Streptomyces*

In addition to the three microbial chassis cells mentioned above, Cyanobacteria (the oldest prokaryotes) and *Streptomyces* (the most developed actinomycetes) can also be used as chassis cells for research because of their own advantages. With the exception of microbial chassis cell synthesis systems, plant cells, such as those of *Camelina sativa* var. *suneson*, have also been developed for the heterologous synthesis of pharmaceutically active ingredients. *Camelina sativa* var. *suneson* is an ancient oilseed crop with a short growing season, winter and spring varieties, low water and fertilizer requirements, and high resistance to pests and diseases. In addition, it is genetically similar to the model plant *Arabidopsis thaliana*, with rich genomic and transcriptome data, and can be transformed by *Agrobacterium* infection using a simple dip method. Therefore, *Camelina sativa* var. *suneson* is an ideal metabolic engineering platform and has been successfully used as a heterologous host to synthesize active constituents of TCM such as monoterpenes, sesquiterpenes, and alkaloids in seeds. In the process of recombining the synthetic pathway, the *DsRed* gene (encoding red fluorescent protein) was simultaneously constructed in a vector and introduced into the plant seed, and the transgenic seed can be directly detected by fluorescence.

At present, the hosts used for the heterologous synthesis of active ingredients of TCM are mainly bacterial and fungal microorganisms. Although the use of codon-optimized synthetic genes and various protein-engineering methods can partially promote the effectiveness of heterologous biosynthetic pathways, the functional expression of medicinal plant genes in microbial hosts remains a challenging task; certain problems remain, such as the post-translational modification of proteins, subcellular organ localization, and optimization of plant gene expressions. Moreover, the growth of microorganisms requires nutrients, energy, and fermentation equipment to maintain. In order to meet the future needs of active ingredients for TCM, a more efficient and suitable heterogeneous synthetic plant cell platform needs to be established. Once a stable homozygous line is obtained, the seeds can be propagated from generation to generation, greatly reducing the cost of synthesis. Of course, compared to microbes as a heterologous host, plant cells require longer reproduction time, and the detection of active ingredients needs to be carried out after the crops are fully mature and harvested; there is also the problem of the intracellular localization of the metabolic pathways.

9.3.3 Metabolic Pathway Construction Strategy

At present, the strategies for constructing the metabolic pathways of active ingredients of TCM in different chassis cells are mainly divided into two types. The first one is the main construction strategy of synthetic biology research of TCM. It is based on the profound analysis of metabolic pathways and transfers, reconstructs, and engineers the inherent metabolic pathways of traditional Chinese medicine active ingredients. According to the construction idea, it can be divided into three levels: (1) Based on the principle of genetic engineering, only the biosynthetic specific pathway gene of the active ingredient is introduced, and the medicinal active ingredient is prepared by using the precursor supply inherent in the chassis cell. (2) Based on the principle of genetic engineering and modular design concept, introducing the substrate regulation module and bioactive synthesis module of TCM active ingredients and improving target product titers by regulating the synthesis of the substrate. (3) Based on the principles of genetic engineering, modular design concepts, and engineering ideas, simultaneously introducing substrate synthetic module, substrate control module, and active ingredient biosynthesis module, all of which comprise some detachable "plug and play" gene elements. Therefore, it is convenient to optimize the elements in the module, to easily use these modules for the construction of other metabolic pathways, and to transform many complex biosynthetic pathways into engineered biological systems that can be disassembled at any time. There is also a synthetic route, designed according to the chemical structure, design, screening, assembly, and programming of the new TCM active ingredient synthesis pathway construction strategy. According to this synthetic route, the enzyme genes that can participate in the designed synthetic route are screened from the gene database; these enzyme genes are introduced into the chassis cells, and a biosynthetic route is used to newly assemble the TCM active ingredient in the chassis cells.

The construction of metabolic pathways is a key step in the synthetic biology of TCM. No matter which construction strategy, different genetic elements (even sequences of completely different species) need to be assembled in an orderly manner according to the biosynthetic pathway of the target active ingredient to achieve the desired biological function. In the construction of metabolic pathways, on the one hand, it is necessary to exert a subjective initiative, and it is possible to randomly splice gene elements of different sources like a splicing circuit; on the other hand, to achieve repeatability of gene elements, standardization can enable all gene elements to be used normally under any circumstances. Here are some common methods for assembling gene elements:

9.3.3.1 Traditional Assembly

The traditional assembly method relies on a multiple cloning site (MCS) on a vector to enable the insertion of a single-gene element, and the same sticky end formed by the gene element and the vector is ligated to form a new plasmid. This method is suitable for assembling single or few genetic elements, and it is difficult to select a suitable restriction enzyme when the number of gene elements to be transferred is large.

Gene splicing by overlap extension PCR (SOEPCR) is an improvement on the traditional assembly method. Using homologous sequences, two or more genetic elements to be transferred are connected using SOEPCR (Fig. 9.1) and are then connected to the carrier by conventional methods. SOEPCR requires two rounds of PCR. The first round of amplification uses primers with complementary ends (primers b, c). After the first round of amplification, the two PCR products create overlapping regions, and then in the second round of amplification reactions, the overlapping regions complement each other and pass through the terminal primers (primers a, d) to amplify the deleted fragment (indicated by the dotted line in the figure), and the amplified fragments are spliced from different sources to obtain chimera AD [27]. The advantage of SOEPCR is that the complementary pairing of homologous sequences is used instead of the restriction endonuclease, not only for the assembly of individual gene elements, but also for the ability to assemble the

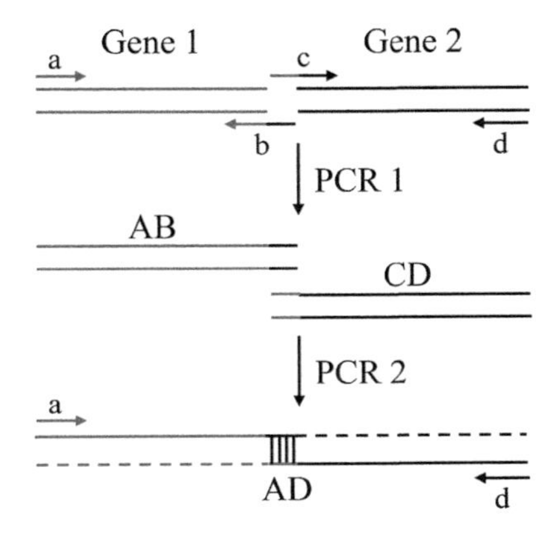

Fig. 9.1 Principle of SOEPCR. (a, b, c, and d are primers; AB, CD, and AD are DNA fragments; different sequences are represented by different colours)

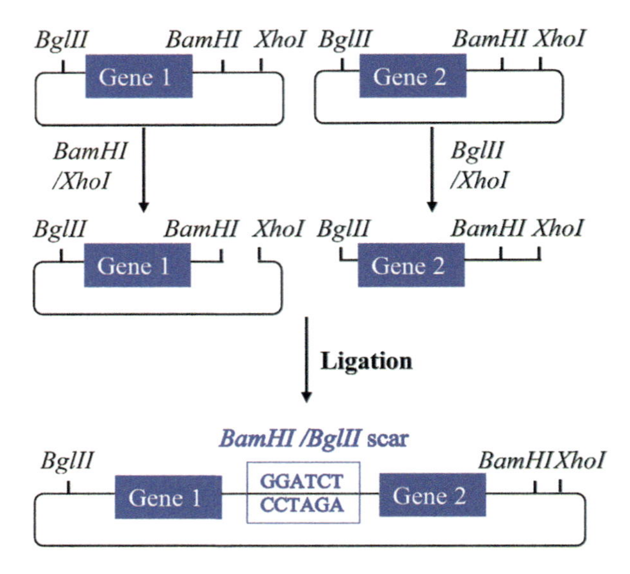

Fig. 9.2 Principle of BioBrick assembly

entire plasmid. The disadvantage is that it is not flexible enough to redesign the components for each assembly. Therefore, the method is suitable for the fine-tuning process of temporarily adding a small number of gene regulatory elements to a complete metabolic pathway.

9.3.3.2 BioBrick Assembly

BioBrick assembly adds standard flanking sequences (BamHI and BglII restriction endonuclease sites) to the ends of each individual element, then gene elements can be assembled in an orderly manner by a unified standardized digestion–ligation method (Fig. 9.2). Using the specificity of the two endonuclease recognition and cleavage sites, the digested products were ligated together through a 6-bp "scar" sequence to form a new sequence, "gene element A + 6-bp scar + gene element B", while the new scar sequence is no longer recognized by BamHI or BglII [28].

Compared to traditional assembly methods, BioBrick Assembly unifies the structure of gene elements and enables rapid reuse. At the same time, the types of restriction enzymes used are reduced. Almost all gene elements can be "BioBricks". However, BioBrick assembly also has certain disadvantages: First, the presence of scars between components constitutes a rigid structure, and it is not possible to arbitrarily change between individual gene elements; second, the assembly is based on the existing BioBrick. If there is no synthetic path module in the library, it can only be reconstructed; third, the influence of the scar sequence on the function of the genetic element is not clear.

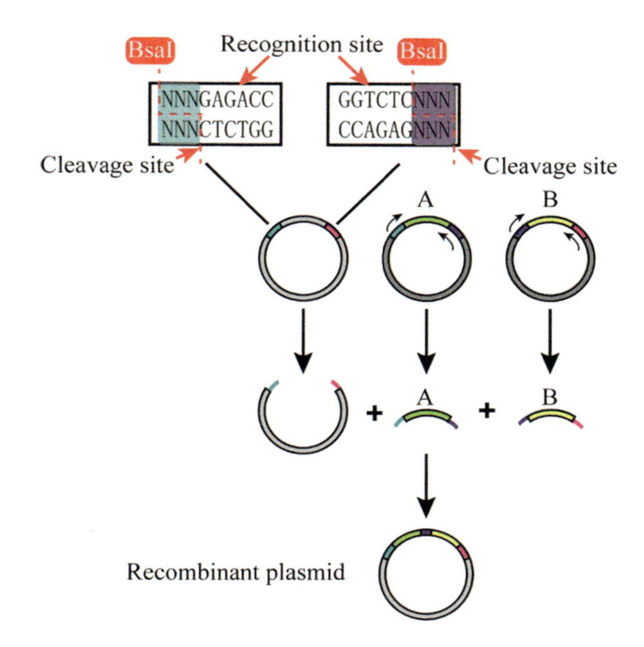

Fig. 9.3 Principle of Golden Gate assembly

9.3.3.3 Golden Gate Assembly

Golden Gate assembly is capable of cleavage adjacent to its DNA recognition site based on the action of a specific type II restriction enzyme; the recognition site and the cleavage site of the endonuclease are separated.

After artificially designing adjacent sequences with different recognition sites, different restriction enzymes can be used to generate different sticky ends, thereby assembling multiple fragments at once (Fig. 9.3). This method can overcome the limitation of the restriction enzyme species in conventional multi-segment assembly, and can be seamlessly connected [29]. At the same time, the method can also be used for the assembly of repeated sequences, such as transcription activator-like effector nucleases. The DNA-binding domain of the protein is composed of a plurality of modules with little difference in sequence and can be used to perform manual design assembly. Although this method also relies on the sticky ends produced by the endonuclease, it can also be used for the construction of multiple gene loops because of the longer overlap regions that can be produced.

9.3.4 Synthetic Process System Optimization

In order to improve target products titers, it is usually necessary to optimize the biosynthesis pathway or biosynthetic metabolic flux of the target products, which can be divided into three groups according to the level: local level, synthetic pathway level, and genomic/population level.

9.3.4.1 Local-Level Optimization

Local-level optimization usually refers to the optimization of a certain limiting factor in the entire synthetic system. Promoter modification, gene replacement, codon optimization, and gene copy number increase can be performed on the encoding rate-limiting enzyme gene, thereby increasing the absolute total amount of the substrate into the biosynthetic pathway of the active ingredient. For example, in the flavonoid biosynthesis pathway, acetyl-CoA carboxylase (ACC) is overexpressed from *Photobacterium*; this increases the intracellular malonyl-CoA content, as well as the naringenin and choline titers in the host *E. coli* from 0.45 mg/L and 0.75 mg/L to 119 mg/L and 429 mg/L, respectively [30]. In addition, protein fusion, enzyme organelle localization, and artificial protein scaffolding can be used to optimize the protein interaction distance and improve the synthesis efficiency of the target product. The fusion of the hydrazine synthase and 4-coumaroyl-CoA ligase in the yeast host increased resveratrol titers by 15-fold [31]. For sequential reactions catalysed by multiple enzymes, heterologous proteins can be assembled by constructing DNA scaffolds, RNA scaffolds, or protein scaffolds, and the efficiency of heterologous metabolism can be improved by improving the enzyme concentration or utilizing environmental advantages.

9.3.4.2 Synthetic Pathway Level

Metabolism in living organisms is usually an interconnected, intricate, and coordinated network structure, and biochemical synthetic pathways often have multiple branches. The target metabolite is usually referred to as the main product, while the product of the branched anabolic pathway is referred to as a by-product, and the path of the synthetic by-product is referred to as the competitive metabolic pathway. At a certain conversion rate, the sum of the main by-products is a fixed number, and an increase in the synthesis of by-products necessarily leads to a decrease in the main product. On the one hand, synthetic pathway level optimization inhibits the competition of by-products by optimizing or balancing the metabolic pathways regulated by multiple genes and promotes the synthesis of main products. The gene in the competitive metabolic pathway is used as a target gene, which is knocked out by either gene knockout or RNA interference. Alternatively, the expression level of the target gene is lowered, the branch metabolic pathway is cut off, or the production of the branch product is reduced, thereby changing the metabolic pathway. Thus, the metabolic flow proceeds in the direction of main product accumulation. By inhibiting the FabB and FabF genes, the synthesis pathway of malonyl-CoA to fatty acids is reduced, thereby maximizing the entry into the flavonoid synthesis pathway [32], and the production of naringenin and choline can be increased to 186 mg/L and 710 mg/L, respectively. On the other hand, in cases where the precursor is too expensive or unavailable, a new route that does not require the addition of a precursor is needed to bypass the synthetic route of the original plant. For example, when heterologously synthesizing flavonoids, expensive phenylpropionic acid or phenylalanine precursors are added according to their natural secondary metabolic pathways, while the newly developed *E. coli* host system utilizes 3-deoxy-D-arabinone glycol-7-phosphate synthase (DAHPS); pre-benzoate dehydratase and chorismate mutase, with glucose as a precursor additive, convert it to phenylpropionic acid, avoiding expensive precursor problems and greatly reducing costs.

9.3.4.3 Genomic or Population Level

The genomic or population level refers to balancing and regulating the entire metabolic system, increasing the synthesis of target products, and reducing the burden of heterologous cellular enzymes and product metabolism.

Scholars from the University of California introduced a promoter-based artificial regulator, glnAp2, to control the expression of two rate-limiting enzymes, IPP isomerase (IDI) and phosphate synthase (PPS); this alleviated the systemic metabolic imbalance and growth retardation, while greatly improving lycopene production in the heterologous host. With the ever-expanding availability of genome-wide datasets and large-metabolism models, the focus of system optimization has shifted from single-gene manipulation to genome-wide changes, and the regulation of cellular global transcriptional mechanisms has increased the cell-to-product tolerance (the active ingredients of TCM are mostly toxic to cells), which is also conducive to the coordination of the entire cell network and increased synthesis of target products. For example, in the paclitaxel biosynthesis pathway, the farnesene pyrophosphate (FPP)-inducible promoter was screened in *E. coli* by whole-genome transcription and corresponding array technology to reduce the toxicity of FPP. A balance between the FPP precursor supply and improved biosynthesis was also realized, which significantly increased amorphadiene titres [33].

Since the biosynthesis of active ingredients of TCM is affected by many physical and chemical factors, such as temperature, pH, oxygen concentration, and nutrient supply, the purpose of improving the titres of active ingredients is achieved by controlling and optimizing the physical and chemical factors in industrial fermentation. In addition, the system optimization of the synthesis process can use bioinformatics for metabolic flow control analysis, which can improve the secondary metabolic synthesis process of the chassis cells.

9.4 Applications of Synthetic Biology for the Sustainable Utilization of TCM Resources

Abstract The sustainable utilization of resources refers to a new concept for human development and aims to fully, rationally, economically, and efficiently use existing resources and continuously develop new alternative resources to meet the needs of current and future generations. As an important biological resource, TCMs are not only an important material for the prevention and treatment of diseases, but are also an important part of China's ecological environment. High-quality and abundant TCM resources are important for Chinese medicine research, development, and production. In recent years, researchers have carried out a series of studies on how to protect TCM resources and ensure their sustainable use.

9.4.1 Research Methods for the Sustainable Utilization of TCM Resources

The rapid development of the Chinese medicine industry has led to tremendous changes in the demand, reserves, and distribution of Chinese medicine resources. The overexploitation and utilization of TCM resources will lead to serious problems, such as the depletion of wild resources, deterioration of the ecological environment, and rapid decreases in rare material availability. There has been a sharp increase in the number of rare and endangered medicinal herb species, and there is difficulty in ensuring the quality of medicinal materials. Researchers in related fields have carried out a series of studies using a variety of methods, aiming to provide a theoretical and practical basis for the sustainable development and utilization of Chinese medicine resources.

9.4.1.1 Resource Survey

An investigation into traditional Chinese medicine resources has shown that there are more than 1000 kinds of commercial medicine in China, accounting for only 10% of all Chinese medicine resources.

In order to gain a deeper understanding of the distribution and storage of existing Chinese medicine resources in China, the fourth national survey of Chinese medicine resources has been carried out. This survey aims to strengthen the monitoring of Chinese medicine resources through the construction of information networks, to elucidate the basic condition of Chinese herbal medicine resources. It is not only conducive to the protection and sustainable development of traditional Chinese medicine resources, but also provides a basis to develop major strategic decisions for the protection, research, development, and reasonable use of Chinese medicine resources.

9.4.1.2 Factors Influencing the Quality of TCM Resources

Factors such as climate, soil, atmosphere, and irrigation are closely related to the variety, efficacy, and quality of TCM. Researchers have found that the medicinal plant origin, growth period, harvesting season, light and temperature during growth, and water and fertilizer management of cultivars have great impacts on the production of saponins in medicinal plants. Using research ideas and methods of TCM resource chemistry, and through the systematic evaluation of different production areas and harvesting periods, we have established a comprehensive multi-index evaluation model which objectively characterizes the phenological relationship between plant growth and development and environmental conditions and the correlation between the formation of medicinal materials and the biological yield of medicinal parts. We have also discussed methods of determining suitable harvesting periods for medicinal materials and establishing an objective evaluation from the aspect of temporospatial relationships and the dynamic accumulation of substances. These have provided theoretical guidance and methodological support for the guarantee and standardized production of Chinese medicinal materials.

9.4.1.3 Study on Genetic Diversity and Germplasm Resources of Medicinal Plants

Medicinal plant cultivation is the most direct and effective means to protect, expand, and recycle Chinese medicine resources. As the main source of Chinese medicine production, the excellent variety of medicinal plants has an important impact on the production of medicinal materials.

In the past 10 years, researchers have made remarkable achievements in the selection and evaluation of new varieties of Chinese medicinal materials and the study of genetic diversity in medicinal plants. For example, the inter simple sequence repeat (ISSR) analysis and internal transcribed spacer 2 (ITS2) DNA barcode segment analysis of the genetic diversity of different species and different varieties of the *Evodia* genus showed that among the three original varieties of *Evodia rutaecarpa* B., the phylogenetic relationship between *E. rutaecarpa* var. *officinalis* and *E. rutaecarpa* var. *bodinieri* is relatively close, but it is still affected by the environment, place of origin, and reproductive mode. The highest genetic diversity in the three original cultivars of *E. rutaecarpa* medicinal materials is in *E. rutaecarpa* B., followed by *E. rutaecarpa* var. *officinalis* and *E. rutaecarpa* var. *bodinieri* [34]. For example, the tracking, breeding, and evaluation of new medicinal plants of the genus *Curcuma*, such as *Curcuma wenyujin*, *C. longa* L., and *C. zedoariae*, provide a scientific reference for the development and utilization of high-quality medicinal plant breeding materials and guarantee the quality of TCM from *E. rutaecarpa* B. and *C. zedoariae* medicinal herbs.

9.4.1.4 Reconstruction or Production of Active Ingredients of TCM

The sustainable utilization of TCM resources faces two constraints regarding the socio-economic need and the supply of renewable resources, and with the application of synthetic biology technology in the biosynthesis of active ingredients, such as artemisinin, paclitaxel [35], and tanshinone [36], constraints on the supply of natural medicine resources to ensure their sustainable use have given cause for concern.

Biosynthesis technology, as used for the active preparation of TCM in a green and sustainable way, represents one of the future development directions of natural medicines. The application of synthetic biology can transform natural or heterologous hosts and achieve the large-scale production of active ingredients of TCM, providing new ideas and directions for the sustainable use of TCM resources.

9.4.2 Problems in the Sustainable Use of TCM Resources

With the development of society and the increasing development and utilization of medicines, the demand for Chinese medicine resources is increasing daily, and the lack of information on their rational development and utilization makes their sustainable utilization an enormous challenge.

9.4.2.1 Genetic Diversity and Variation

The genetic diversity of medicinal plant species resources affects germplasm resources of Chinese medicinal materials. Various different genes have formed during the long-term evolution of medicinal species. These are not only the basic materials for studying the origin, evolution, inheritance, and classification of TCM resources, but also for the breeding of medicinal plant species. Germplasm resources of Chinese medicinal materials are the carriers of genetic genes. Research on genetic diversity not only provides important information on the evaluation of the development of genetic resources, but also a basis for resource conservation, introduction, and cultivation, as well as original materials for the breeding of medicinal plants.

Rare and endangered medicinal resources, due to their own biological characteristics and genetic variability, have insufficient species abundance, and self-renewal is difficult. Therefore, for the effective ex situ conservation of these species, it is necessary to retain as much genetic diversity as possible to avoid poor environmental adaptability caused by low genetic variation, which could lead to a decrease or even complete loss of evolutionary potential.

9.4.2.2 Destruction of the Ecological Environment and Excessive and Disorderly Development and Utilization

With the continuous development of society, increased deforestation, mountain construction, vegetation destruction, and other activities have reduced the content of ecological resources and severely damaged the environment; the large-scale discharge of the "three wastes" from industrial production and the excessive and unreasonable use of pesticides and fertilizers have polluted the natural ecological environment. The destruction of the ecological environment and the degradation of resources have led to continuous reductions in the amount of biological resources of Chinese medicinal materials, decreased quality, and increased numbers of rare and endangered species. For example, medicinal materials with long-term cultivation histories, such as *Angelica sinensis* and *Panax pseudoginseng* var. *notoginseng,* are now difficult to find in the wild. The rapid development of the social economy has led to an increasing demand for natural resources, and the output of Chinese herbal medicines is far from meeting the huge market demand. At the same time, due to the opening of Chinese medicine resource markets and the huge price fluctuations of Chinese herbal medicine resources, some valuable Chinese herbal medicine resources are driven by disorderly predatory exploitation, and their resources are seriously threatened. Their wild resources are on the verge of exhaustion. In particular, some wild medicinal materials commonly used in clinical practice are almost in danger of extinction.

Nowadays, wild *Glycyrrhiza uralensis* resources have been difficult to find in their traditional places of origin, such as Xinjiang and Inner Mongolia. Wild reserves of rare resources, such as *Dendrobium nobile*, have also been decreasing.

9.4.3 Synthetic Biology Could Ensure the Sustainable Utilization of TCM Resources

The active ingredients of TCM are directly extracted and separated from the medicinal materials or decoctions. The traditional method is not only restricted by the germplasm source, growth environment, inheritance and variation, harvesting, and processing of the medicinal plants, but also consumes a lot of resources and costs a lot of manpower. Due to the extremely low content of some active ingredients in medicinal plants, the traditional extraction and separation method causes great damage and waste to rare and endangered medicinal plant resources. Therefore, the development of new methods is essential for the protection of Chinese medicine resources and quality. With the rapid development of molecular biology, plant physiology, biochemistry, and other disciplines, medicinal plant genetic engineering has gradually become one of the new research directions of TCM resources. Modern research techniques and methods using molecular biology and microbial genetics can not only change the genetic traits of TCM, cultivate excellent varieties, increase the content of active ingredients, and protect and reproduce rare and endangered TCM resources, but also use genetic engineering to cultivate new varieties of medicinal plants with insect resistance, disease resistance, and other beneficial properties. Among them, the use of transgenic medicinal plants or plant tissues as "bioreactors" can obtain a large number of secondary metabolites per plant, which lays a good foundation for the industrialized and large-scale production of TCM active ingredients.

As relevant components for the biosynthesis of the active constituents of TCM are continuously identified, the biosynthetic pathways of active constituents of medicinal plants are being continuously analysed. The use of synthetic biology methods to redesign natural, existing biological systems in plants enables the targeted genetic breeding of medicinal plants. By cultivating medicinal plants with high-yield target active ingredients, the pressure on medicinal plant resources can be effectively alleviated, and the extraction and production costs can be reduced. At the same time, the use of biological system integration in the microbial optimization and reconstruction of the active ingredient biosynthesis modules of TCM can achieve the heterogeneous and efficient synthesis of rare active ingredients and alleviate pressure on the demand for TCM resources for clinical use.

The advent of synthetic biology has given new opportunities to research natural products, especially monomeric compounds with well-defined pharmacological activities. Professor Keasling of the University of California, Berkeley, used synthetic biology to produce an industrial method for low-cost artemisinic acid production [37]. In addition, the use of synthetic biology to construct microbial cell factories and the successful synthesis of high value-added compounds, such as paclitaxel, ginsenosides, and vinblastine, has also seen breakthroughs. Increasing numbers of researchers are engaged in research on the synthetic biology of TCM, which could support new methods of sustainable use.

9.4.4 Strategy for the Application of Synthetic Biology in TCM for the Sustainable Utilization of Resources

Developing the sustainability of TCM resources is based on the coordinated development of cultivation techniques for Chinese herbal medicines based on "land" and the directional cultivation of raw materials for Chinese patented medicines with the aim of increasing the yield of "active ingredients" using synthetic biology. The use of synthetic biology can be efficient in the heterogeneous synthesis of complex and diverse TCM active ingredients, not only providing new strategies and technologies for the sustainable use of TCM resources, but also injecting new vitality into the development of the field of TCM. To maintain our leading position in international Chinese medicine, research has extremely important scientific significance and value.

The basic strategies for the application of synthetic biology in the sustainable use of traditional Chinese medicine resources can be divided into three points: (1) Cloning the key genes of biosynthesis pathways of active constituents of traditional Chinese medicines from medicinal plants and analysing the biosynthesis pathways of the active ingredients of TCM through gene function research (2). The plant-derived pathway designs and integrates the heterologous biosynthetic pathway, and the artificially designed pathway is loaded onto the genome of the cell to construct a microbial cell factory (3). Fermentation conditions are optimized to achieve the efficient fermentation production of the active constituents of Chinese medicine and its intermediates. The key to its successful application is to analyse the biosynthesis mechanism of complex natural products at the level of gene and protein function. Guided by biological and chemical knowledge, the biosynthetic pathway can be reconstructed in a fermentation-friendly microorganism through rational design. System optimization enables the efficient biosynthesis of target compounds.

In order to study the secondary metabolic pathways and their regulation of medicinal plants, the problems that need to be solved in synthetic biology include gene expression, the adaptability of the product to the host, and the separation and purification of the product. This requires multidisciplinary cooperation and the application needs, as a starting point, to not only be limited to the current scarce medicinal plant resources, but should also focus on resources that may be scarce or important in the future; focus should also be given to new drug discovery and creating new, original drugs with independent intellectual property rights.

To summarize, the correct use of modern synthetic biology techniques to rationally solve the shortage of TCM resources will certainly promote the modernization and internationalization of TCM in China and inject new vitality into international competition for China's TCM industry after joining the WTO.

References

1. Leduc S, Butcher WD. The mechanism of life. Br Med J. 1923;58:141.
2. Jacob F, Monod J. Genetic regulatory mechanisms in the synthesis of proteins. J Mol Biol. 1961;3:318–56.
3. Hobom B. Gene surgery: on the threshold of synthetic biology. Med Klin. 1980;75:834.
4. Gardner TS, Cantor CR, Collins JJ. Construction of a genetic toggle switch in Escherichia coli. Nature. 2000;403:339.
5. Elowitz MB, Leibler S. A synthetic oscillatory network of transcriptional regulators. Nature. 2000;403:335–8.
6. Ewen CD, Caleb JB, James JC. A brief history of synthetic biology. Nat Rev Microbiol. 2014;12:381–90.
7. Paddon CJ, Westfall PJ, Pitera DJ, Benjamin K, Fisher K, Mcphee D, Leavell MD, Tai A, Main A, Eng D. High-level semi-synthetic production of the potent antimalarial artemisinin. Nature. 2013;496:528.
8. Patel RN, Banerjee A, Howell JM, Mcnamee CG, Brozozowski D, Mirfakhrae D, Nanduri V, Thottathil JK, Szarka LJ. Microbial synthesis of (2R,3S)-(−)-N-benzoyl-3-phenyl isoserine ethyl ester-a taxol side-chain synthon. ChemInform. 2010;25. no–no

9. Ro DK, Paradise EM, Ouellet M, Fisher KJ, Newman KL, Ndungu JM, Ho KA, Eachus RA, Ham TS, Kirby J. Production of the antimalarial drug precursor artemisinic acid in engineered yeast. Nature. 2006;440:940–3.
10. Parayil Kumaran A, Wen-Hai X, Tyo KEJ, Yong W, Fritz S, Effendi L, Oliver M, Too Heng P, Blaine P, Gregory S. Isoprenoid pathway optimization for Taxol precursor overproduction in Escherichia coli. Science. 2010;330:70–4.
11. Zhou YJ, Wei G, Qixian R, Guojie J, Huiying C, Wujun L, Wei Y, Zhiwei Z, Guohui L, Guofeng Z. Modular pathway engineering of diterpenoid synthases and the mevalonic acid pathway for miltiradiene production. J Am Chem Soc. 2012;134:3234–41.
12. Vjj M, Pitera DJWithers ST, Newman JD, Keasling JD. Engineering a mevalonate pathway in Escherichia coli for production of terpenoids. Nat Biotechnol. 2003;21:796–802.
13. Tsuruta H, Paddon CJ, Eng D, Lenihan JR, Horning T, Anthony LC, Regentin R, Keasling JD, Renninger NS, Newman JD. High-level production of Amorpha-4,11-diene, a precursor of the antimalarial agent artemisinin, in Escherichia coli. PLoS One. 2009;4:e4489.
14. Westfall PJ, Pitera DJ, Lenihan JR, Eng D, Woolard FX, Regentin R, Horning T, Tsuruta H, Melis DJ, Owens A. Production of amorphadiene in yeast, and its conversion to dihydroartemisinic acid, precursor to the antimalarial agent artemisinin. Proc Natl Acad Sci U S A. 2012;109:655–6.
15. Juan G, Zhou YJ, Hillwig ML, Ye S, Lei Y, Yajun W, Xianan Z, Wujun L, Peters RJ, Xiaoya C. CYP76AH1 catalyzes turnover of miltiradiene in tanshinones biosynthesis and enables heterologous production of ferruginol in yeasts. PNAS. 2013;110:12108–13.
16. Dai Z, Liu Y, Huang L, Zhang X. Production of miltiradiene by metabolically engineered Saccharomyces cerevisiae. Biotechnol Bioeng. 2012;109:2845–53.
17. Guo J, Ma X, Cai Y, Ma Y, Zhan Z, Zhou YJ, Liu W, Guan M, Yang J, Cui G. Cytochrome P450 promiscuity leads to a bifurcating biosynthetic pathway for tanshinones. New Phytol. 2016;210:525–34.
18. Kim DH, Kim BG, Jung NR, Ahn JH. Production of genistein from naringenin using Escherichia coli containing isoflavone synthase-cytochrome P450 reductase fusion protein. J Microbiol Biotechnol. 2009;19:1612–6.
19. Trantas E, Panopoulos N, Ververidis F. Metabolic engineering of the complete pathway leading to heterologous biosynthesis of various flavonoids and stilbenoids in Saccharomyces cerevisiae. Metab Eng. 2009;11:355–66.
20. Hawkins K, Smolke C. Production of benzylisoquinoline alkaloids in Saccharomyces cerevisiae. Nat Chem Biol. 2008;4:564–73.
21. Fossati E, Ekins A, Narcross L, Zhu Y, Falgueyret JP, Beaudoin GAW, Facchini PJ, Martin VJJ. Reconstitution of a 10-gene pathway for synthesis of the plant alkaloid dihydrosanguinarine in Saccharomyces cerevisiae. Nat Commun. 2014;5:3283.
22. Mercke P, Bengtsson M, Bouwmeester HJ, Posthumus MA, Brodelius PE. Molecular cloning, expression, and characterization of amorpha-4,11-diene synthase, a key enzyme of artemisinin biosynthesis in Artemisia annua L. Arch Biochem Biophys. 2000;381:173–80.
23. Stephen GA, Jodie Y, Andrew W, Yue W, Srinivas C, Rupeng Z, Patina MH, Yenphuong TT, Qinghai Z, Ina LU. Structure of P-glycoprotein reveals a molecular basis for poly-specific drug binding. Science. 2009;323:1718–22.
24. Bill RM, Henderson PJF, So I, Kunji ERS, Hartmut M, Richard N, Simon N, Bert P, Tate CG, Horst V. Overcoming barriers to membrane protein structure determination. Nat Biotechnol. 2011;29:335–40.
25. Chen S, Xu J, Liu C, Zhu Y, Nelson DR, Zhou S, Li C, Wang L, Guo X, Sun Y, Luo H, Li Y, Song J, Henrissat B, Levasseur A, Qian J, Li J, Luo X, Shi L, He L, Xiang L, Xu X, Niu Y, Li Q, Han MV, Yan H, Zhang J, Chen H, Lv A, Wang Z, Liu M, Schwartz DC, Sun C. Genome sequence of the model medicinal mushroom Ganoderma lucidum. Nat Commun. 2012;3:913.
26. Krivoruchko A, Nielsen J. Production of natural products through metabolic engineering of Saccharomyces cerevisiae. Curr Opin Biotechnol. 2015;35:7–15.

27. Heckman KL, Pease LR. Gene splicing and mutagenesis by PCR-driven overlap extension. Nat Protoc. 2007;2:924–32.
28. Røkke G, Korvald E, Pahr J, Øyås O, Lale R. BioBrick assembly standards and techniques and associated software tools. Methods Mol Biol. 2014;1116:1.
29. Walhout AJM, Temple GF, Brasch MA, Hartley JL, Lorson MA, Heuvel SVD, Vidal M. [34] GATEWAY recombinational cloning: application to the cloning of large numbers of open reading frames or ORFeomes. Methods Enzymol. 2000;328:575–92.
30. Effendi L, Kok-Hong L, Phan-Nee S, Koffas MAG. Engineering central metabolic pathways for high-level flavonoid production in Escherichia coli. Appl Environ Microbiol. 2007;73:3877–86.
31. Yechun W, Hankuil Y, Melissa W, Oliver Y, Jez JM. Structural and kinetic analysis of the unnatural fusion protein 4-coumaroyl-CoA ligase::stilbene synthase. J Am Chem Soc. 2011;133:20684–7.
32. Yang Y, Lin Y, Li L, Linhardt RJ, Yan Y. Regulating malonyl-CoA metabolism via synthetic antisense RNAs for enhanced biosynthesis of natural products. Metab Eng. 2015;29:217–26.
33. Baadhe RR, Mekala NK, Parcha SR, Prameela DY. Combination of ERG9 repression and enzyme fusion technology for improved production of amorphadiene in Saccharomyces cerevisiae. J Anal Methods Chem. 2013;2013:140469.
34. Jing-Yuan XU, Zhu Y, Ze YI, Gang WU, Xie GY, Qin MJ. Molecular diversity analysis of Tetradium ruticarpum (WuZhuYu) in China based on inter-primer binding site (iPBS) markers and inter-simple sequence repeat (ISSR) markers. Chin J Nat Med. 2018;16:1–9.
35. Ajikumar PK, Xiao WH, Tyo KEJ, Wang Y, Simeon F, Leonard E, Mucha O, Phon TH, Pfeifer B, Stephanopoulos G. Isoprenoid pathway optimization for Taxol precursor overproduction in Escherichia coli. Science. 2010;330:70–4.
36. Ma XH, Ma Y, Tang JF, He YL, Liu YC, Ma XJ, Shen Y, Cui GH, Lin HX, Rong QX. The biosynthetic pathways of tanshinones and phenolic acids in Salvia miltiorrhiza. Molecules. 2015;20:16235.
37. Paddon CJ, Keasling JD. Semi-synthetic artemisinin: a model for the use of synthetic biology in pharmaceutical development. Nat Rev Microbiol. 2014;12:355.

Chapter 10
The Mechanism of Formation of Daodi Herbs

Sheng Wang, Chuan-zhi Kang, Lan-ping Guo, and Thomas Avery Garran

Abstract "*Daodi*" is a term used to define traditional Chinese herbs selected for long-term clinical application in traditional Chinese medicine (TCM). These herbs are produced in certain areas with better quality and efficacy compared to when they are produced in other regions and have stable quality and high popularity. Meanwhile, Daodi herbs have long-standing reputations for their high quality and excellent treatment effects.

Distinguishing features of Daodi herbs include special geographic variations, specific quality standards, historical and cultural implications, and comparatively high economic value.

Viewing through the lens of biology, a Daodi herb is a particular species growing under specific environmental influences, thus the interaction of nature and genetics creates a biologically unique herb. The interaction of genotypes and the environment, over time, develop phenotypes, which create Daodi herbs.

The chemical profile of Daodi herbs is unique. The regional characteristics of Daodi herbs are positively correlated with their specialized genotypes. Suboptimal environments promote the formation of Daodi herbs.

The molecular mechanisms of Daodi herbs formation are specific genotypes; complex regulations in specific environments caused temporospatial differences in the expressions of genes encoding key enzymes in the secondary metabolism processes. The diversity of patterns and types of products of secondary metabolism are caused by hereditary variations.

10.1 Introduction

"Daodi herbs" is a term used to define traditional Chinese herbs selected by the long-term clinical application of traditional Chinese medicine (TCM), produced in certain areas with better quality and efficacy compared to the same species produced in other

S. Wang · C.-z. Kang · L.-p. Guo (✉) · T. A. Garran
National Resource Center for Chinese Materia Medica, China Academy of Chinese Medical Sciences, Beijing, China
e-mail: mmcniu@163.com; kangchuanzhi1103@163.com; glp01@126.com; tag.plantgeek@gmail.com

L.-q. Huang (ed.), *Molecular Pharmacognosy*,
https://doi.org/10.1007/978-981-32-9034-1_10

regions, and thus, with stable quality and high popularity (according to the Traditional Chinese Medicine Law of the People's Republic of China enacted on December 25, 2016). Daodi herbs were considered as distinctive high-quality Chinese herbs established by ancient doctors. And, Daodi herbalism is the generic term for various merits possessed by Daodi herbs. Among the 600 commonly used TCMs, there are 200 with obvious Daodi herbalism, the usage of which accounts for 80% of the total usage of TCMs, giving Daodi Herbs the highest market occupancy and economic benefit of all TCMs. "Geoherbs in China", edited by Hu SL, reports 160 common Daodi herbs consisting of 132 plant medicines, 20 animal medicines, and 8 mineral medicines [1].

10.2 The Connotation of Daodi Herbs

10.2.1 History Evolution of Daodi Herbs

In the pre-Qin period (before 221 B.C.), perception of the production location and growing environment of herbs was formed, while the importance of herb quality was still not clear. During the Qin and Han dynasties (221 B.C. to 220 A.D.), the origin of a herb was highly valued and connected to its quality, as recorded in "Ming Yi Bie Lu", where most herbs were marked to have the best quality in specific production locations. Moreover, in "Shen Nong Ben Cao Jing" of the western Han dynasty, the meaning of Daodi herbs was firstly reflected. Among the 365 kinds of herbs recorded, many can be distinguished as Daodi herbs by their names, such as Bajitian, Shujiao, Shuzao, Qinjiao, Qinpi, Wuzhuyu, Ejiao, Daizheshi, etc. (Ba, Shu, Qin, Wu, Dong'e, and Daizhou, are all names of old places around the western Zhou Dynasty), with growing conditions such as in the valley, marsh, pond, hill, field, and flat land being mentioned. "Huang Di Nei Jing", a contemporary medicinal monograph as "Shen Nong Ben Cao Jing", expounded the meaning of Daodi herbs in theory, and "Shang Han Lun" was a medicinal monograph that first recorded lots of Daodi herbs, such as Ejiao, Badou, Daizheshi, etc. Since the southern Song dynasty (1127–1279 A.D.), the word Daodi has been combined with herb names to describe high-quality and better clinical effects than herbs from other production places.

10.2.2 Concept of Daodi Herbs

The concept of Daodi herbs came from the clinical practice of TCM and had been proved by numerous clinical practices for thousands of years. It has a rich scientific connotation. As a conventional concept for the standardization of ancient drugs,

Daodi herbs, stemming from ancient reports, are not only of high-quality comprehensive standards to evaluate the quality of Chinese herbal medicines, but also comprehensive standards to control the quality of medicinal materials in Chinese *materia medica*. Generally speaking, Daodi herbs refer to medicinal materials growing in habitats with particular natural conditions and ecological environment. The production of Daodi herbs is usually concentrated, and their cultivation and harvest-processing techniques usually have certain requirements. Meanwhile, these drugs have better quality and curative effects in normal condition, than those drugs originating from the same species but from different habitats. The names of Daodi herbs are usually shown together with their region, for example, Xining Dahuang, Ningxia Gouji, Chuan Beimu, Chuan Shao, Qin Jiu, Liao Wuwei, Guan Fangfeng, Huai Dihuang, Mi Yinhua, Hao Juhua, Xuan Mugua, Hang Baizhi, Zhe Xuanshen, Jiang Zhiqiao, Su Bohe, Mao Cangzhu, Jian Zexie, Huang Chenpi, Taihe Wuji, A Jiao, Dai Zheshi, etc. There are some special cases. For example, Guang Muxiang is not produced in but imported via Guangdong, while Zang Honghua is imported via Tibet.

10.2.3 Attributes of Daodi Herbs

10.2.3.1 Specialized Genotype

Daodi herb refers to a special population affected by its habitat during its long-term evolution. Different groups belonging to the same species, called populations in biology, are formed in different regions. Daodi herbs are stable natural or cultivated populations grown in specific locations and at specific times. Traditionally, Daodi herbs are considered to have similar genetic backgrounds and better qualities than their non-Daodi herb counterparts. Through DNA analysis technology, an in-depth study of specialized genotypes of Daodi herbs could be carried out, and the nature of Daodi herbs could be clarified. For example, RAPD analysis of *Angelica sinensis* (danggui) grown in its "famous" region showed that the smaller its geographical distribution is, the smaller its genetic diversity would be [2]. Yuan QJ et al. [3] used molecular phylogeography to analyze the chloroplast haplotypes of 602 and 451 individuals of Chinese skullcap *Scutellaria baicalensis*, representing 28 wild and 22 cultivated populations across Daodi-herb and non-Daodi-herb areas, respectively. This study showed that the genetic differentiation between Daodi-herb populations and other populations was noticeable.

10.2.3.2 Obvious Regional Characteristics

The formation and development of the concept of Daodi herbs are closely related to specialized regions. The nature of Daodi herbs is the material with the best quality among products belonging to a same species, but with different origins. Thus, the names of Daodi herbs are always prefixed with a region name; for example, Yang Chunsha, Xuan Mugua, Mao Cangshu, Min Danggui, Guan Shanbo, Si Da Chuan Yao (four famous drugs from Sichuan), Si Da Huai Yao (four famous drugs from Jiaozuo), and Zhe Ba Wei (eight famous drugs from Zhejiang). Environmental factors, including soil, water, light, temperature, and terrain, play important roles in the formation of quality Daodi herbs. There would be no Daodi herbs without famous regions. The specialized environments in famous regions lead to the good quality and high yields of Daodi herbs. As a result, Daodi herbs usually have good reputations in the market.

10.2.3.3 Unique Quality Standards

Regional characteristics result in differences in the characteristics, chemical composition, and efficacy of Daodi herbs. For example, Chuanxinlian from Guangzhou has a better antibacterial effect than materials from Fujian and Anhui; Chuan Maidong from Mianyang and Santai in Sichuan Province are worse in appearance and taste than Hang Maidong from Yuyao and Hushan in Zhejiang Province; and there is a high content of anthraquinones in Dahuang from northwest China, while Dahuang from Heilongjiang contains more tannins and has a greater antidiarrheal effect. The content of active components in Qinpi from Shanxi is higher than that in material from Sichuan. The phenolic content in Houpu from Sichuan is six times that in materials from Jiangxi [4]. Shanxi is the famous region of Huangqi, and the characteristics and chemical composition of Huangqi materials from Hubei are much different from Daodi herbs from Huangqi. The appearance and taste of Gouji from Ningxia are much better than those of materials from other regions. Daodi herbs should have unique chemical compositions. Mao Cangzhu is much different from Nan Cangzhu in its volatile oil contents. The total volatile oil content in Mao Cangzhu is significantly lower, but it has extremely high atractylon and atractydin contents. In addition, the ratio of atractylon, atractydin, hinesol, and b-eucalyptol in Mao Cangzhu shows a unique composition [5].

10.2.3.4 Particular Technology or Production Process

The cultivation (or aquaculture), maintenance, harvesting, processing, and other processes of Chinese medicinal herbs affect the formation of quality Daodi herbs, and sometimes even become one of the decisive factors. Jiangyou in Sichuan and Hanzhong in Shannxi have always mass-produced Fuzi and Fupian, but their qualities are very different, and Jiangyou Fuzi is considered to be the Daodi herb. Famous regions are always aligned with particular processes and histories. The longer history it has, the more mature the technique will be, and the more obvious the authenticity will be.

10.2.3.5 Rich Cultural Connotations

The formation of Daodi herbs is related to the unique geography and national cultural background, as well as Chinese medicine theory. There are no concepts similar to Daodi herbs mentioned in any other nations or countries in the world. The use of Daodi herbs reflects the deep understanding of Chinese medicine; the concept of "Zheng" and the state of the patients' body; and theories on the properties, taste, effectiveness, clinical characteristics, and adverse effects of medicinal materials. The formation of Daodi herbs promotes the development of traditional culture in famous regions to some extent. Nowadays, the prevalence of various festival activities named after Daodi herbs has become one of the most important forms of expressing its cultural connotations; one example of such a festival is the Ningxia Gouqi festival. Therefore, the cultural connotation is an important feature of Daodi herbs, making them a special class of natural product with intellectual property rights. Daodi herbs are especially valuable because of their cultural importance.

10.3 The Hypothesis of the Formation Mechanisms of Daodi Herbs

Each Daodi herb has a typical quantitative trait in its phenotype, which results from the interaction between its heredity and environment. Thus, phenotype equals gene plus environment plus interaction between gene and environment. This shows that genes have the potential to develop some definite phenotypes depending on the habitat. According to the nature of Daodi herbs, this hypothesis of their formation was proposed.

10.3.1 Unique Adaptive Features Appear in the Chemical Compositions of Daodi Herbs

Secondary metabolites are the substantial bases of authenticity of Daodi herbs and result in them having differential chemotypes and better efficacies compared to those of other populations of the same species. The chemotype, a kind of ecotype, emphasizes the significant difference in chemical composition generated by phenotypic plasticity, which is the environmental modification of genotypic expression and an important means by which individual plants respond to environmental heterogeneity. Phenotypic plasticity helps explain the difference in quality and clinical effects of the same herb species from different production regions. As an open and complicated system, the chemotype of a Daodi herb (one of its phenotypes) is the result of its long-term adaptation to the surrounding environment. Therefore, the unique adaptive feature appears in the chemical compositions of Daodi herbs that grow in unique habitats with specific environments.

For example, the variation in chemical compositions of widespread species is always continuous. The variations in volatile oil in mint (Bohe), *Chrysanthemum* (Juhua), and *Atractylodes* (Cangzhu) might change from continuous to discontinuous when the climate of their distribution domain, especially dominant factors that could affect their chemical composition, changes significantly. This change in the chemical composition of Daodi herbs is the result of their adaptation to their habitat, which fully demonstrates their adaptiveness to complicated systems. The composition features of volatile oil in the *Atractylodes* sp. Daodi herbs and those from other regions are very different. An analysis of environmental factors showed that famous production regions featured drought and potassium deficiency compared to other production areas. The volatile oil content in Daodi *Atractylodes* is significantly low ($P < 0.01$). Components whose contents were higher than 1% when treated by unitary processing are significantly more ($P < 0.01$); the contents of atractylon and atractydin were extremely high, and the contents of hinesol and b-eucalyptol were extremely low ($P < 0.01$) in Daodi *Atractylodes*. Besides, the ratio of atractylon, atractydin, hinesol, and b-eucalyptol was unique (0.70–2.0):(0.04–0.35):(0.09–0.40):1 [5]. These results indicate that the basic substance of the high clinical efficacy of Daodi *Atractylodes* is its unique chemical composition accumulated under suboptimal environmental conditions.

10.3.2 The More Obvious the Authenticity of Daodi Herbs Is, the More Obvious Its Gene Specialization Will Be

In the long-term process of breeding and domesticating, Daodi herbs will form unique genotypes. Good-quality Daodi herbs (e.g., *Rehmannia* populations, such as Jin Zhuang Yuan, Beijing No. 1, and 85-5; and ginseng populations such as Da Ma Ya and Er Ma Ya) are very closely related to their unique genetic background. For wild Daodi herbs, the genotypes of populations in different areas are usually different when distributed throughout a large area, which is called a locally specialized genotype. These genotypes are the result of long-term selection under different ecological or geographical conditions and are essential for the formation of Daodi herbs. That is to say, the more widely the species is distributed, the more obvious its authenticity will be, and the more specialized its genotype will be. Studies on intraspecific genetic structure and genetic differentiation could provide some proof of the genotype specialization of Daodi herbs. For example, Yuan QJ et al. [3] found that the genealogical structures and frequencies of 25 chloroplast haplotypes of *S. baicalensis* presented a significant divergence: haplotypes G, B, and C connected to 10, 6, and 4 other haplotypes by one mutation, respectively. The haplotype distributions of wild populations were noticeably structured. The ancestral lineage formed by haplotype G with closely connected haplotypes was mainly restricted to the central range of this species, Chengde in Hebei Province and nearby areas, which was exactly the famous area of *S. baicalensis*. Thus, this lineage could be named a Daodi-herb lineage.

The features of Daodi herbalism, "good appearance" and "good quality", are developed through the microevolution of quantitative genetics controlled by multiple genes under environmental stress. Microevolution, the change in allele frequencies that occur over time within a population, is the foundation of the ability of Daodi herbs to adapt to their environment. Epigenetic inheritance, caused by interactions between genotypes and the environment, plays a vital role in the formation of Daodi herbs. Rates of epigenetic mutations are higher than those of gene mutations under environmental stress, revealing that environmental factors are the basic driving force for the formation of Daodi herbs and the continuity in time and location leads to continuity in the inheritance and phenotype.

10.3.3 Stress Will Promote the Formation of Daodi Herbs

The active components in Chinese *materia medica* are usually low molecular weight compounds produced by secondary metabolism. It is hypothesized that stress would promote the biosynthesis of secondary metabolites. For instance, drought stress can increase the quercetin content in ginkgo to some extent [6]; the polyphenol content in marigold under drought stress is higher than that under normal conditions [7]; the polyphenol content in sunflower increases with increased nutrition stress [8]; and the monoterpene content in rosemary increases with increased CO_2 concentrations [9]. The most obvious change in the secondary metabolism of plants fed upon by pests is increased polyphenol content [10].

Many studies on the gene expression in plants under stress have revealed the effect of stress on plant secondary metabolism at the genetic level. Dehydration-responsive element-binding (DREB) transcription factors, which are induced by environmental stress, might activate 12 other genes depending on the ability of the cis-acting elements for DREB transcription factors to withstand adversity; from this, the proline and sucrose contents increase and resistance to a variety of stress (drought, cold, and salinity) conditions is enhanced [11]. In one study, northern blot was carried out on the *Phaseolus vulgaris* stress-related (PvSR6) gene, which encodes a DnaJ-like protein in kidney beans. The results indicated that the expression of the PvSR6 gene is weak in nonprocessed kidney bean leaves. Environmental stresses, such as heavy-metal exposure (Hg^{2+} and Cd^{2+}), mechanical damage, UV exposure, high temperature, and salicylic acid exposure, could promote the transcription of this gene. It can be inferred that the DnaJ-like protein plays an important role in protecting the structure and function of membranes and enzymes, as well as increasing the resistance of plants. These results provide reference for discovering the effect of stress on the gene expression in Daodi herbs [12]. Lanping Guo et al. [13, 14] used canonical correlation and stepwise regression analysis to figure out the leading factors affecting the contents of essential oil components in *A. lancea* and formed six regressive models between climate factors and the six main essential oil components. They found that high temperature was the main limiting factor for *A. lancea*, and the interaction between temperature and precipitation was the key factor in forming the essential oil components of *A. lancea*. They used the GIS software IDIRIEIW to extract the climate characteristics on Mt. Maoshan, the famous habitat of *A. lancea*, and found that the habitat of Maocangzhu was characterized as having the highest temperature, shorter drought season, and greater precipitation compared to those of habitats of other populations of *A. lancea*, and the formation of *Mao-A. lancea* was related to high-temperature stress. They also found that Daodi *A. lancea* faced nutrient stress and proved that high-temperature and low-K stress conditions could produce plants with essential oil compounds more similar to those of the Daodi herbs than those of the other populations through experiments in a green house.

Effective constituents are the material bases that could determine the curative effects of TCM. Some effective constituents do not exist under normal conditions and are only produced under certain external stimuli (e.g., drought, freeze, injury); these constituents belong to a group of abnormal secondary components called phytoalexins.

In short, on one hand, environmental change selects genotypes of Daodi herbs through long-term action, and on the other hand, it affects the formation and accumulation of secondary metabolites by affecting their gene expression. Through long-term adoption, medicinal plants have chosen their preferred natural environments. When environmental changes or stresses appear, plants' competitiveness for the limited resources by physical means would drop, and chemical means would become more important. Secondary metabolites can protect the plants, and they will be released externally to limit the growth of other plants when stresses appear. In this situation, it seems that the reaction of Daodi herbs to environmental change is more likely an adverse effect.

10.4 Tips for Studying the Molecular Mechanisms of Daodi Herb Formation

The morphology of Daodi herbs and their multifarious secondary metabolites are caused by genetic variations. Studies on the molecular mechanisms of Daodi herb formation aim to reveal the genetic variation between populations of Daodi herbs at the molecular level, discover the effects of genotype and environment on gene expression, and reveal the contribution of genetic factors to the formation of Daodi herbs. Studying the relationship between genetic variation and secondary metabolites is the most direct way to reveal the authenticity of Daodi genes. Analyzing the genetic differences and genetic structure at the interspecies population level can help to determine the genetic variation in Daodi herbs at the molecular level. Sequence analyses of nDNA and cpDNA have been used in studies on the DNA barcoding of medicinal plants in *Angelica* based on molecular systematics and the exploration of new medicine sources in *Dendrobium* [15, 16].

Remaining unknowns should be paid attention to in further studies on the molecular mechanisms of Daodi herbs:

10.4.1 Secondary Metabolism Is a Typical Multigene Trait

The steps of plant secondary metabolism are numerous and complex. Each step in the metabolic process has at least one gene at work. Thus, secondary metabolism is typically controlled by many genes. As has been established, biosynthesis of the various types of preliminarily investigated secondary metabolites, for example, terpenes, alkenes, alkaloids, flavonoids, anthraquinones, and coumarins, involve a considerable number of metabolic steps and large number of key genes, regardless of the secondary metabolic pathway used. For example, more than 20 genes and 20 steps are involved in the biosynthesis of tanshinone.

10.4.2 Variation of Secondary Metabolites Is Continuous

According to the multigene theory, genes are passed on in accordance with Mendelian genetics. The effect of each gene is weak; thus, noncontinuous and quantitative gene mutation may produce a smooth and continuous phenotypic variation. Modification of the environment may cover the noncontinuous variation of genotype; thus, continuous phenotypic variation is shown and trait variations become smooth and difficult to detect. As a typical multigene trait, secondary metabolism has shown continuous variation; this not only brings about confusion in the quality evaluation of Daodi herbs, but also increases the difficulty of genotype detection.

10.4.3 Interactions Between Genes and the Environment Should Be Taken Seriously

Interactions between genes and the environment are important features of multigene genetics. There are many forms of these interactions, mainly including (1) the environment affecting the genetic structure of a group through the choices imposed on the group; (2) genotypes and the environment affect the decisions of individuals or populations we observed directly, because of the nongenetic effects caused by interactions between genotypes and the environment in the development of Daodi herbs. Both of them play important roles in the formation of Daodi herbs, and the second is an important reason why the production of Daodi Herbs is inseparable from its specialized habitats. Studies on the effects of interactions between genes and the environment on the formation of authenticity have just begun, but the rapid development of molecular geography, molecular ecology, and other related fields has provided good bases and methods for further studies.

10.4.4 Quantitative Genetics Theories Should Be Taken Seriously

According to the forms of expression, genetic variation can be divided into two types: qualitative variation and quantitative variation. The former, which shows discontinuous variation, is decided by a few major genes and is consistent with Mendel's law; the latter, which shows continuous variation, is decided by many genes and follows a normal distribution. Genetic variations determined by quantitative traits are very susceptible to environmental effects and will interact with the environment. That is to say quantitative traits comprise three parts: genetic variation, environmental variation, and the interactions between genes and the environment.

The phenotypic characteristics of famous Daodi herbs are closely related to the environment first, followed by the trait of continuity, which suggest that the genetic nature of Daodi herbs may be a quantitative trait. Therefore, the characteristics of continuous variations should be considered when looking for the specialized genes of Daodi herbs. In addition, quantitative genetics should be studied deeply as a main point of focus.

Quantitative genetics is a combined product of genetics and biostatistics. It mainly studies the continuous variation decided by multigene genetics and its relationship with quantitative traits. One of the most important research fields in quantitative genetics is studying the interactions between genetics and the environment. It can be expected that quantitative genetics technology will become an important means for studying the molecular mechanisms of Daodi herb formation.

References

1. Shi lin H. Geoherbs in China. Haerbin: Heilongjiang Science and Technology Press; 1989.
2. Gao W y, En q Q, Xiao X h, et al. Analysis on Genuineness of *Angelica sinensis* by RAPD. Chin Tradit Herb Drug. 2011;32(10):926–9.
3. Yuan Q j, Zhang Z y, Juan H, et al. Impacts of recent cultivation on genetic diversity pattern of a medicinal plant, Scutellaria baicalensis (Lamiaceae). BMC Genet. 2010;11(1):29.
4. Wen F, Zhuang J. The characteristics of geoherbs. J Beijing Univ Tradit Chin Med. 2001;24 (1):47.
5. Guo L p, Liu J y, Li J, et al. The naphtha composing characteristics of geoherbs of *Atractylodes lancea*. China J Chin Mater Med. 2002;27(11):814–9.

6. He B h, Zhong Z c. Study on variation dynamics of modular population of *Ginkgo biloba* under different conditions of environmental stress. J Southwest Agric Univ. 2003;25(1):7–10.
7. Tang C s, Cai W f, Kohl K, et al. Plant stress and allelopathy. ACS Symp Ser. 1995;582:142–8.
8. Hall AB, Blum U, Fites RC. Stress modification of allelopathy in Helianthus annuus L. debris on seed germination. Am J Bot. 1982;69:776–81.
9. Peñuelas J, Llusià J. Effects of carbon dioxide, water supply, and secondary on terpene content and emission by Rosmarinus officinails. J Chem Ecol. 1997;23:979–800.
10. Bryant JP, Reichardt PB, Clausen TP, et al. Effects of mineral nutrition on delayed inducible resistance in Alaska paper birch. Ecology. 1993;74:2072–84.
11. Wang S x, Wang Z y, Peng Y k. Dehydration responsive element binding (DREB) transcription activator and its function in plant tolerance to environmental stresses. Plant Physiol Commun. 2004;40(2):7–13.
12. xiu Zhang Y, Chai T y, Li J c, et al. Expression analysis of DnaJ-like protein gene from bean under various environment stresses. Acta Botan Boreali–Occiden Sin. 2000;20(2):171–4.
13. Guo L p, qi Huang L, Yan H, et al. Habitat characteristics for the growth of *Atractylodes lancea* based on GIS. China J Chin Mater Med. 2005;30(8):565–9.
14. Guo L p, qi Huang L, Jiang Y x, et al. Selection of dominant climate factors and climate suitability regionalization of components of rhizoma atractylodis oil. China J Chin Mater Med. 2007;32:888–93.
15. Yuan Q j, Zhang B, Jiang D, et al. Identification of species and materia medica within Angelica L. (Umbelliferae) based on phylogeny inferred from DNA barcodes. Mol Ecol Resour. 2015;15:358–71.
16. Xiang X g, Schuiteman A, De z L, et al. Molecular systematics of Dendrobium (*Orchidaceae, Dendrobieae*) from mainland Asia based on plastid and nuclear sequences. Mol Phylogenet Evol. 2013;69(3):950–60.

Suggested reading

Huang L q, Guo L p, Ma C y, et al. Top-geoherbs of traditional Chinese medicine: common traits, quality characteristics and formation. Front Med. 2009;5(2):185–94.

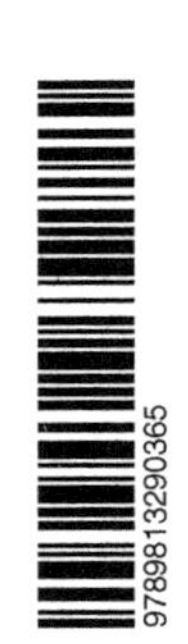